VIRUS PERSISTENCE

SYMPOSIA OF THE
SOCIETY FOR GENERAL MICROBIOLOGY*

* Published by the Cambridge University Press, except for the first Symposium, which was published by Blackwell's Scientific Publications Limited.

VIRUS PERSISTENCE

EDITED BY

B. W. J. MAHY, A. C. MINSON AND
G. K. DARBY

THIRTY-THIRD SYMPOSIUM OF
THE SOCIETY FOR GENERAL MICROBIOLOGY
HELD AT
THE UNIVERSITY OF CAMBRIDGE
MARCH 1982

Published for the Society for General Microbiology

CAMBRIDGE UNIVERSITY PRESS
CAMBRIDGE
LONDON   NEW YORK   NEW ROCHELLE
MELBOURNE   SYDNEY

Published by the Press Syndicate of the University of Cambridge
The Pitt Building, Trumpington Street, Cambridge CB2 1RP
32 East 57th Street, New York, NY 10022, USA
296 Beaconsfield Parade, Middle Park, Melbourne 3206, Australia

First published 1982

Printed in Great Britain at the Pitman Press, Bath

Library of Congress catalogue card number: 81–15487

British Library Cataloguing in Publication Data

Virus persistence. – (Symposia of the Society for
General Microbiology; 33)
1. Viral genetics – Congresses
I. Mahy, B. W. J.   II. Minson, A. C.
III. Darby, G. K.   IV. Society for General
Microbiology   V. Series
576'.64   QH434

ISBN 0 521 24454 4

CONTRIBUTORS

BERNS, K. I., Department of Immunology and Medical Microbiology, University of Florida College of Medicine, Gainesville, Florida 32610, USA

BLUMBERG, B. M., Department of Microbiology, Medical School, University of Geneva, 64 Avenue de la Roseraie, CH-1205 Geneva, Switzerland

CARTER, M. J., Institut für Virologie and Immunbiologie, Versbacher Straße 7, D-8700 Würzburg, West Germany

CHEUNG, A. K.-M., Marjorie B. Kovler Viral Oncology Laboratories, University of Chicago, Chicago, Illinois 60637, USA

DOUGLAS, A. R., Division of Virology, National Institute for Medical Research, Mill Hill, London NW7 1AA, UK

EPSTEIN M. A., Department of Pathology, University of Bristol Medical School, University Walk, Bristol BS8 1TD, UK

EVANS, H. F., Natural Environment Research Council, Institute of Virology, Mansfield Road, Oxford OX1 3SR, UK

FIELD, H. J., Department of Pathology, Tennis Court Road, Cambridge CB2 1QP, UK

FUJINAMI, R. S., Department of Immunopathology, Research Institute of Scripps Clinic, Scripps Clinic and Research Foundation, La Jolla, California 92037, USA

GRIFFIN B., Imperial Cancer Research Fund, Lincoln's Inn Fields, London WC2A 3PX, UK

HARRAP, K. A., Natural Environment Research Council, Institute of Virology, Mansfield Road, Oxford OX1 3SR, UK

KOLAKOFSKY, D., Department of Microbiology, Medical School, University of Geneva, 64 Avenue de la Roseraie, CH-1205 Geneva, Switzerland

LEWIS, M., Department of Immunology and Medical Microbiology, University of Florida College of Medicine, Gainesville, Florida 32610, USA

TER MEULEN, V., Institut für Virologie and Immunbiologie, Versbacher Straße 7, D-8700 Würzburg, West Germany

MIMS, C. A., Department of Microbiology, Guy's Hospital Medical School, London Bridge SE1, UK

NASH, A. A., Department of Pathology, Tennis Court Road, Cambridge CB2 1QP, UK

OLDSTONE, M. B. A., Department of Immunopathology, Research Institute of Scripps Clinic, Scripps Clinic and Research Foundation, La Jolla, California 92037, USA

OSTROVE, J. M., Department of Microbiology, Johns Hopkins University School of Medicine, Baltimore, Maryland, 21205, USA

PEREIRA, M. S., Virus Reference Laboratory, Central Public Health Laboratory, Colindale, London NW9 5HT, UK

SKEHEL, J. J., Division of Virology, National Institute for Medical Research, Mill Hill, London NW7, UK

WEISS, R., Chester Beatty Research Institute, Fulham Road, London, SE3 6JB, UK

WILDY, P., Department of Pathology, Tennis Court Road, Cambridge CB2 1QP, UK

WILEY, D. C., Gibbs Laboratory, Harvard University, Cambridge, Mass. 02138, USA

WILSON, I. A., Gibbs Laboratory, Harvard University, Cambridge, Mass. 02138, USA

ZUCKERMAN, A. J., Department of Medical Microbiology and WHO Collaborating Centre for Reference and Research on Viral Hepatitis, London School of Hygiene and Tropical Medicine (University of London), London WC1E 7HT, UK

CONTENTS

EDITORS' PREFACE

The last two SGM symposia devoted to viruses, held in 1968 and 1975, dealt primarily with the molecular biology of virus replication using model systems. The present symposium deals with the much less tractable problem of how certain virus infections persist and cause recurrent or chronic diseases despite the existence of established host defence mechanisms.

The interaction of viruses with permissive cells normally leads to productive infection and cell death, and this is reflected in the individual as an acute infection against which an immune response is mounted. For virus infection to persist, a degree of equilibrium must be established between virus and host, but this equilibrium may break down after a variable period of time, with resultant cellular injury and disease. The factors influencing the maintenance of a persistent infection are complex, and vary depending upon the genetic background and immune responsiveness of the host, and the strategy for persistence employed by different viruses. We have tried in this symposium to bring together examples of each of the strategies for which definitive evidence exists: antigenic variation, latent infection of nerve cells, and integration into host chromosomal DNA. In addition to these, studies of the interaction of viruses with cells in culture suggest the existence of other mechanisms, such as production of temperature-sensitive mutants or defective interfering viruses in the generation of persistent infection. The significance of such phenomena in virus persistence within the host has yet to be established.

We have grouped the contributions to the symposium so that, after an introductory chapter, the mechanisms underlying virus persistence are considered at three levels: in the population as a whole (chapters 2 to 4), in the individual host (chapters 4 to 8), and at the molecular level (chapters 9 to 13). These divisions are not rigid, but they may help to orientate the reader towards the emphasis which was intended.

We are extremely grateful to our thirteen contributors, each of whom survived the stringent deadlines associated with a book of this kind. Our thanks must also go to Fay Bendall and Peter Silver of

Cambridge University Press, and to Roger Berkeley of the SGM, whose persistence was the principal contributory factor in producing the volume on time.

Division of Virology,   B. W. J. MAHY
Department of Pathology,   A. C. MINSON
University of Cambridge,   G. K. DARBY
Cambridge CB2 2QQ

ROLE OF PERSISTENCE IN VIRAL PATHOGENESIS

C. A. MIMS

Department of Microbiology, Guy's Hospital Medical School, London Bridge SE1, UK

At first I was not too happy about the title of my contribution. My first impulse was to take the easy way by inverting it and talking about the role of pathogenesis in viral persistence. But would it then be a boring prelude to the definitive talks which followed? An introductory chapter rich in platitudes, to be missed out by the reader as he turns eagerly to the main contributions. I therefore resisted this first impulse and wondered whether by providing this particular title the organizers, accidentally or in a moment of inspiration, had posed a very interesting problem. This did indeed turn out to be the case, and I propose to take the novel approach suggested in the title. I shall itemize and discuss the ways in which viral persistence itself influences pathogenesis.

TYPES OF PERSISTENCE

Three types of persistence can be distinguished. In the first, the virus persists but at no time has the opportunity to be shed to the outside world. This means that the persistence, although possibly of immense interest for the infected individual and for the veterinarian or physician, is of no significance for the virus or for its maintenance in the host species. Examples include persistent infections where viral antigens or virions are produced, for example, the brain in subacute sclerosing panencephalitis (SSPE) in man, progressive multifocal leukoencephalopathy (PML) in man, Visna in sheep and others where no more than viral nucleic acid sequences are present. The infection is a dead end from the virus point of view.

In the second type of persistence, shedding to the outside world occurs, but only sporadically, for instance as a result of reactivation of herpesviruses in dorsal root ganglia, or as a result of reactivation of human papovaviruses in the kidney (Coleman *et al.*, 1980).

Thirdly there are viruses that offer the ultimate insult to the host immune defences by replicating and being continuously shed to the outside world, often throughout the life of the individual. Examples include lymphocytic choriomeningitis (LCM) virus in mice, that classic of virus persistence which at the same time provides examples of almost everything that is possible in immunopathology. In carrier mice, LCM virus persists in all organs and tissues (Mims, 1966) being continuously shed from respiratory, urinary and intestinal tracts.

It is the last two types of persistence that are of significance in evolutionary terms, giving a basis for the maintenance of the infection in small groups of individuals (Black, 1966). By persisting in the individual these viruses are able to persist in the species.

In each of the above types of persistence virus nucleic acid is present, often with virus antigens or infectious virus. There are reasons for supposing that following many virus infections there is a persistence of viral antigens in the body in the absence of viral nucleic acid or viral replication. This is not different from the persistence of any antigen in the body. Bovine serum albumin, for instance, has been detected in the local lymph node of a rabbit 18 months after injection (Tew, Phipps & Mandel, 1980). Many years ago, it was observed that individuals infected with yellow fever who recovered and subsequently went to live in countries where yellow fever did not exist nevertheless maintained high levels of specific yellow fever neutralizing antibodies in their serum after as long as 75 years (Sawyer, 1931). It seems very unlikely that yellow fever virus itself persists in the body, any more than smallpox virus persists in individuals who maintain antibodies to smallpox for many years. Memory cells too would not be expected to persist for such long periods, but the phenomenon can be explained if antigens are retained in the body. One theoretically attractive site would be in the follicular dendritic cells referred to by Mandel, Phipps, Abbot & Tew (1980). These large and fragile cells are present in lymphoid follicles. They are distinct from macrophages and they trap immune complexes. They have extensive dendritic processes, giving a huge surface area for contact with migrating lymphoid cells. This is the type of cell that could retain viral antigens, act as a continuous source of antigen for reactive lymphocytes, and provide a mechanism for life-long immune responses to acute, non-persistent viruses. More will be learnt of these fascinating cells when the technical difficulties in their *in vitro* cultivation and study are overcome.

PERSISTENT VIRUSES PROVIDE THE BASIS FOR IMMUNE COMPLEX DISEASE

When viruses persist and continue to produce antigens in the infected host there is generally some degree of immune tolerance. This means that immune responses, although not necessarily zero, are less powerful than they might otherwise have been. In mice infected as embryos or neonatally with LCM virus, immune tolerance is not complete and small amounts of antibody of low affinity are produced (Oldstone & Dixon, 1967). Complexes of these antibodies with excess LCM virus antigen are formed in the blood, deposited in kidney glomeruli and elsewhere, and by a fascinating series of immunopathological changes, this leads to the development of serious kidney damage in the form of glomerulonephritis (Oldstone & Dixon, 1969). The circulating immune complexes must be deposited in the kidney for a long period and at a high enough rate if the kidney lesions are to be produced. In mice persistently infected with lactic dehydrogenase (LDH) virus there is a long-term formation of immune complexes in the blood and deposition in the kidney, but apparently not fast enough to cause pathological changes (Oldstone & Dixon, 1971). Immune complexes are powerful pathogenic products, and presumably the host's ability to remove and dispose of them in the kidney must be overcome if long-term effects are to be produced.

Circulating immune complexes are also formed in mink persistently infected with Aleutian disease virus (ADV), and over long periods the complexes are deposited in glomeruli and in blood vessels, causing damage which is manifested by glomerulonephritis and sometimes haemorrhage from affected arteries (Porter, Larsen & Porter, 1980).

Persistent virus infection therefore, by providing a continuous source of antigen for the continued generation of circulating immune complexes, can be a cause of glomerulonephritis, a vasculitis, and other pathological changes. Most cases of chronic glomerulonephritis in man are thought to result from the long term deposition of immune complexes, and if the antigen was viral in nature, it would have to be a persistent virus that acted as a continuous source of circulating antigen.

PERSISTENT VIRUS INFECTIONS PROVIDE AN OPPORTUNITY FOR LONG-TERM PATHOLOGICAL RESULTS TO OCCUR

The pathological consequences of acute and limited virus infections are fairly easy to follow. Polioviruses reach the anterior horn cells of a susceptible individual during the first week of the infection. These cells are destroyed within hours and if enough are affected, muscles are paralysed and the characteristic disease is produced. The virus is rapidly eliminated in the body by host defences, but the damage has already been done. The missing neurones are not replaced and the patient shows residual weakness or paralysis. Sometimes however, a virus is capable of damaging cells or tissues only after infecting them for a long period. There would be no opportunity to express this potential during the course of an acute and limited infection, where the virus is eliminated within a week or two. Persistent viruses on the other hand, would be able to give rise to any long-term effects that they were capable of.

Strictly speaking a virus infection could give long-term effects either at the level of the infected cell, or because of slowly evolving immune and other changes in tissues. Immune complex glomerulonephritis comes into this last category. In most cases the distinction cannot yet be made. Long-term changes in the infected cell could include a shortening in life-span, obviously important in the non-renewing cell populations in the central nervous system. Also there could be alterations in the quality or rate of production of extracellular products such as collagen, myelin, immunoglobulins, hormones etc.

There are a number of fascinating human diseases of obscure origin which have stimulated hypotheses about virus aetiology. In some instances, such as Coxsackie viruses and juvenile diabetes, the infection is an acute one and has at times been associated unequivocally (Yoon, Austin, Onodera & Notkins, 1979) although so far not regularly with the disease. But in other cases, a long period would be expected between infection and the onset of disease, so that by definition these would be slow virus diseases. Diseases such as multiple sclerosis and rheumatoid arthritis have been considered from this point of view. The discovery that other neurological conditions such as SSPE and PML in man and Visna in sheep were caused by slow virus infections gave great power and promise to the search for a multiple sclerosis virus. So far, in spite of a great deal of investigation and many false alarms there is no acceptable evidence for a viral aetiology. Various other diseases, however, are fairly well associated with slow virus infections. These include the diabetes seen in those with congenital rubella infection who survive into adult life, although the pathogenesis is unknown (Menser, Forrest & Bransby, 1978). Another example, which awaits confirmation from other laboratories, is the report of measles virus infection in the osteoclasts of patients suffering from Paget's disease of bone. Viral antigens have been seen by immunofluorescence and nucleocapsids by electron microscopy (Rebel et al., 1980), although measles antibody levels (HA1 and CF) are not raised (Morgan-Capner et al., 1981). Evidently under certain circumstances and in certain individuals a number of common virus infections can persist and cause disease.

Even rabies virus would be incapable of causing its characteristic disease if it were eliminated from the body as readily as is poliovirus, influenza or yellow fever virus. By persisting in the body, sometimes for months, there is an opportunity for the unfolding of the poorly understood sequence of events (Murphy, 1977) that results in a devastating and fatal disease.

PERSISTENT VIRUSES THAT REACTIVATE AND ARE SHED MAY THEREBY CAUSE DISEASE

Certain persistent viruses reactivate at intervals, and then undergo a period of renewed replication in the host, as a result of which they are shed to the outside world. This is of immense importance for the maintenance of these viruses in nature (Mims, 1978), and is often associated with the development of pathological lesions. The herpesviruses reactivating in dorsal root ganglia are in this category. If the reactivating virus is to be shed it must replicate in mucous membranes or the skin, and this results in direct damage to cells and immunopathological changes, which ensure that lesions are produced. Herpes simplex virus is often detected in the saliva of healthy individuals, and it is possible that non-pathogenic reactivation occurs in sites giving access to the mouth, but little is known about this.

The immunosuppression of kidney transplant patients has given much information about the ability of persistent viruses such as cytomegalovirus, papillomavirus etc. (Mims, 1978) to reactivate when immune constraints are removed. Herpes simplex reactivation is often more frequent under these circumstances, and when it does occur the lesions are more severe (Korsager, Spencer & Mordhorst, 1975). But if the reactivation is to be important under natural circumstances there must be episodes of reactivation during the normal life-span. One of these circumstances, old age, is well charted as a time for reactivation and shedding of varicella-zoster virus. Cell-mediated immune responses, particularly to this virus (Miller, 1980), are weakened in old people. Perhaps the varicella-zoster response is for some reason especially vulnerable, because a similarly impaired cell-mediated immunity to varicella-zoster, but not to herpes simplex or to cytomegalovirus, is found in patients with lymphomas (Arvin, Pollard, Rasmussen & Merigan, 1978). The impaired immune responsiveness allows spontaneously reactivating virus from dorsal root ganglia to travel down peripheral nerves, reach the skin, multiply, and produce virus-rich lesions. The incidence of zoster (shingles) increases steadily after the age of 50 to 60 years (Hope-Simpson, 1965) and the virus shed from the sufferer is a source of infection of susceptible young people in the community. This sort of reactivation has enabled varicella-zoster virus to persist in the host species throughout much of human evolution, and on general principles it might be expected that other viruses would have developed a similar capacity to reactivate in old age. So far there are no known examples but it will be worth looking more closely at the saliva, urine etc. of old people for reactivated viruses.

Pregnancy is another time of life at which viruses are known to reactivate. The mysterious human papovaviruses BK and JC infect most people by the time they are adults, and may be recoverable from urine of pregnant women. In a recent study of the urine samples of 1235 pregnant women 40 (3.2%) contained characteristic inclusion-bearing cells, and from many of these JC or BK virus was isolated or seen by electron microscopy (Coleman *et al.*, 1980). Venereal warts are also reactivated in pregnancy. As with BK and JC reactivation, this does not strictly lead to a disease but to a renewed flowering of these ornamental papillomas on the genital tract. Herpes simplex virus is more commonly isolated from the cervix of pregnant women but not in all studies (Knox *et al.*, 1979). Cytomegalovirus, although isolated less frequently than in normal women during the first trimester (Stagno *et al.*, 1975) is commoner later in pregnancy. The persistent virologist might profitably make further observations on the secretions and excretions of pregnant and lactating women.

PERSISTENCE POSES REQUIREMENTS IN THE TOPOGRAPHY OF INFECTION

Persistence is often favoured if as few antigens as possible are produced by the persisting virus. Host immune defences that would threaten the infection are then less likely to be aroused. This operates first at the level of the individual infected cell. As discussed elsewhere in this volume (Oldstone) there are advantages when minimal amounts of viral antigen are liberated or expressed on the infected cell surface. Infected cells then become less vulnerable to immune lysis by cytotoxic T cells, by natural killer (NK) cells, by complement, by antibody plus complement or by antibody plus killer (K) cells. It also operates at the histological level. As has been pointed out elsewhere (Mims, 1978) there is a striking tendency for persistent viruses to infect epithelial sites in glands and other body surfaces facing the outside world. This is true of the skin with papilloma viruses, feather follicles with Marek's disease virus, kidney tubules with cytomegalovirus and polyoma viruses, bile ducts with reovirus 3 in mice and salivary glands in vampire bats with rabies. From these sites the virus is discharged directly to the exterior and can infect other individuals. At the same time it is protected from circulating antibodies, which fail to reach the luminal surface of infected cells where infectious particles are being released. Viruses often show a striking polarity in epithelial cells, viral antigens and budding being confined to the luminal surface of the cell (Boulan & Sabatini, 1978). The infected cell also escapes attack by sensitized lymphocytes which are unable to traverse the basement membrane and the epithelial cell sheet unless there is local inflammation or cell damage. The nervous system is another site that is to some extent protected from immune defences and is a well-established site of persistent infection as discussed above. Bone might be considered in the same category because there is evidence for persistent measles virus infection in Paget's disease of bone (see above).

Table 1. *Infection of lymphoreticular tissues by viruses exhibiting systemic infection or persistence*

Virus	Host	Target Cells
Adenovirus	Man	L[a]
African swine fever virus	Pig	M[b]
Epstein-Barr virus	Man	L
Varicella-zoster virus	Man	L
Cytomegalovirus	Man, mouse, pig	LM
Marek's disease virus	Chicken	L
Mouse thymic virus	Mouse	L
Pseudorabies virus	Pig	M
Rubella virus	Man	LM
Lactic dehydrogenase virus	Mouse	LM
Mouse hepatitis virus	Mouse	LM
Measles virus	Man	L
Canine distemper virus	Dog	L
Lymphocytic choriomeningitis virus	Mouse	LM
Visna virus	Sheep	LM
Leukaemogenic retrovirus	Mouse, etc.	LM
Reovirus type 3	Man, mouse	L
Aleutian disease virus	Mink	M
Equine infectious anaemia virus	Horse	M
Scrapie agent[c]	Mouse	?

[a] Lymphocytes known to be infected; [b] Macrophages known to be infected; [c] Scrapie is not a conventional virus. Its apparent failure to induce immune responses presents a challenge to the immunologist, and is associated with replication in lymphoreticular tissue.

PERSISTENCE REQUIRES REPLICATION IN LYMPHORETICULAR TISSUES

I have pointed out elsewhere (Mims, 1978) that persistent viruses generally invade or persist in lymphoreticular tissues (Table 1), and suggested that this was because the infecting virus could thereby influence immune responses to its own advantage. Of course, this is not the only reason why lymphoreticular tissues are invaded. The lymphatic system focuses infecting viruses at body surfaces into lymph nodes, and viruses in blood are often taken up into lymphoreticular tissues in liver and spleen. Indeed, some ability to avoid destruction in these sites is needed if viruses are to spread through the body. But I am suggesting that localization and replication in lymphoreticular tissues may also have a deeper significance.

Viruses have a unique ability to invade tissues and infect cells without causing inflammation and necrosis. In lymphoid tissues there are normally strict controls on the movement and handling of antigens, and viruses that invade these tissues disturb this well-ordered process. They can infect the very macrophages and lymphocytes in which the immune response is unfolding and thereby are in a position to weaken the response to the infecting virus. They *invade* lymphoreticular tissues in order to *evade* the immune defences that are being generated there. Immunology is now equipping us to study the various ways in which this can occur. We can investigate the generation of antigen-specific suppressor cells, the deletion of clones of cells reacting with viral antigen, or the impaired handling and processing of viral antigen by macrophages (Mims, 1981; Dunlop & Blanden, 1977; Liew & Russell, 1980; Nash & Gell, 1981).

The close interaction between persistent viruses and lymphoreticular tissues is reflected in two interesting ways. First a generalized immunosuppression to unrelated antigens is common, at least in the initial stages of the infection (Notkins, Mergenhagen & Howard, 1970). Second, autoimmune phenomena are common, as in Epstein-Barr virus and cytomegalovirus infections, indicating a degree of immunological chaos. Recently much interest has focused on NK cells, representing a possibly important antiviral defence that is expressed very early after infection and which is induced by interferon. NK cells are located in lymphoid tissues and can attack uninfected as well as infected cells. A major NK cell attack in lymphoid tissues which had been invaded by a persistent virus could possibly lead to serious tissue necrosis. Perhaps it is no accident that gross necrosis of lymphoid tissue has been observed in certain infections with persistent viruses, including LCM virus in mice (Mims & Tosolini, 1969), cytomegalovirus in mice (Mims & Gould, 1978), and Lassa fever virus in man (Edington & White, 1972). In cytomegalovirus infection the lymphoid necrosis appeared four days after infection and was not prevented by various forms of immunosuppression. But the connection between NK activity and lymphoid tissue necrosis is purely conjectural. Lymphoid tissue necrosis is severe in MHA strain hamsters infected with Pichinde virus (Murphy, Buchmeier & Rawls, 1977) and this has been attributed to a direct cytopathic effect in infected macrophages and reticular cells – indeed this virus selectively destroys or inactivates NK cells in MHA strain hamsters (Gee, Clark & Rawls, 1979). Necrosis could also be caused by antigen-specific T cell or K cell cytotoxicity, if there were enough non-cytopathically infected target cells present in lymphoid tissues. In the case of LCM virus it was shown that simultaneous replication of virus and generation of immune responses in lymphoid tissue was necessary for necrosis.

All that has been said about persistent viruses also holds, perhaps less inevitably, for non-persistent but systemic virus infections. Particularly when the incubation period is 2–3 weeks or more, some interference with host defences may be necessary if the virus is to achieve the full cycle of pathogenesis and be shed to the exterior. Measles, for instance, undergoes fascinating encounters with lymphoid cells (Huddlestone, Lampert & Oldstone, 1980) and can be persistent. In hepatitis B infections the incubation period is 2–5 months. Control of the infection presumably depends on antibody and especially cell-mediated immune responses to viral antigens (HBs) present on the surface of the virion and the infected liver cell. We can ask why it takes several months for these responses to develop. Is it possible that the infecting virus interferes with their development, organizing as it were, its own freedom to multiply for at least a month or two before being suppressed and eliminated? Rabies virus has a long incubation period, sometimes measured in months. For its continued survival in nature the virus depends on a lengthy period of growth and spread in the central nervous system as well as the salivary glands. Only after invading the appropriate parts of the brain can it achieve its fiendish goal of inducing the infected individual to bite another and thus spread the infection. Does rabies virus interfere with host immune defences? There have been reports of immunosuppression by virulent, but not by attenuated strains of rabies virus (Wiktor, Doherty & Koprowski, 1977) and studies of the interactions of this virus with lymphoid cells would certainly be worthwhile.

PERSISTENCE GENERALLY INVOLVES MINIMAL PATHOGENICITY

Persistent viruses, if they are to remain for long periods in the host, cannot afford to be very pathogenic. This is true at the level of the infected cell (Mims, 1974), and also applies to the infected host. Sometimes pathological changes are an inevitable result of reactivation and shedding to the exterior, as when certain herpesviruses, after travelling down peripheral nerves, must replicate extensively in dermal and epidermal tissues. On the other hand, viruses that persist in the acinar cells of salivary gland or in tubular epithelial cells in kidneys achieve the same result without significant pathological damage. The most successful persistent viruses are ancient associates of the host species that have had time to settle down to a state of balanced pathogenicity, inducing the minimal damage that is compatible with their survival and continued transfer. Viruses such as cytomegalovirus in man, polyoma-like viruses (JC and BK) in man, papillomaviruses in man, Lassa fever virus in the African rat *Mastomys natalensis* (Walker, Wulff, Lange & Murphy, 1975) or LDH and LCM viruses in mice, are strikingly successful persistent parasites, and give few if any signs of their presence in these hosts.

CONCLUSIONS

I have discussed some of the ways in which the phenomenon of persistence influences pathogenicity, and suggested some possibly important areas for investigation. It is both useful and relevant to think of persistent viruses as successful parasites and look at them just as we would, for instance at, *Sacculinus*, that extraordinarily successful copepod parasite of crabs which eventually constitutes a major part of the host's body. With caution we can indulge in unashamedly teleological thoughts because these give added interest and sometimes profitable insight to virological studies.

I have referred to the persistent virologist, who is prepared to spend much time and effort studying a single phenomenon. I will conclude with an example from another sphere of biological research. Dr W. B. Bean, a distinguished American physician, charted in meticulous fashion the growth of his left thumbnail over the course of 36 years (Bean, 1980). Over this time the nail, if not cut, would have grown to be several feet long. This persistent investigator made many interesting observations, such as that the nail on the middle finger grew faster than that on the thumb and other fingers, that nails grow faster in a warm environment, faster when he was younger, and (to justify this diversion from virology) that nail growth ceased completely when he suffered from mumps.

REFERENCES

ARVIN, A. M., POLLARD, R. B., RASMUSSEN, L. E. & MERIGAN, T. C. (1978). Selective impairment of lymphocyte reactivity to varicella zoster virus antigens among untreated patients with lymphoma. *Journal of Infectious Diseases*, **137**, 531–40.

BEAN, W. B. (1980). Nail growth. Thirty five years of observation. *Archives of Internal Medicine,* **140,** 73–6.

BLACK, F. L. (1966). Measles endemicity in insular populations: critical community size and its evolutionary implication. *Journal of Theoretical Biology,* **11,** 207–11.

BOULAN, E. R. & SABATINI, D. D. (1978). Asymmetric budding of viruses in epithelial monolayers. Model system for study of epithelial polarity. *Proceedings of the National Academy of Sciences, USA,* **75,** 5071–5.

COLEMAN, D. V., WOLFENDALE, M. R., DANIEL, R. A., DHANJAL, N. K., GARDNER, S. D., GIBSON, P. E. & FIELD, A. M. (1980). A prospective study of human polyomavirus infection in pregnancy. *Journal of Infectious Diseases,* **142,** 1–8.

DUNLOP, M. B. C. & BLANDEN, R. V. (1977). Mechanisms of suppression of cytotoxic T cell responses in murine lymphocytic choriomeningitis virus infection. *Journal of Experimental Medicine,* **145,** 1131–43.

EDINGTON, G. M. & WHITE, H. A. (1972). The pathology of lassa fever. *Transactions of the Royal Society of Tropical Medicine and Hygiene,* **66,** 381–9.

GEE, S. R., CLARK, D. A. & RAWLS, W. E. (1979). Differences between syrian hamster strains in natural killer cell activity induced by infection with Pichinde virus. *Journal of Immunology,* **123,** 2618–26.

HOPE-SIMPSON, R. E. (1965). The nature of herpes zoster: a long term study and a new hypothesis. *Proceedings of the Royal Society for Medicine,* **58,** 9–20.

HUDDLESTONE, J. R., LAMPERT, P. W. & OLDSTONE, M. B. A. (1980). Virus–lymphocyte interactions; infection of T_G and T_M subsets by measles virus. *Clinical Immunology and Immunopathology,* **15,** 502–9.

KNOX, G. E., PASS, R. F., REYNOLDS, D. W., STAGNO, S. & ALFORD, C. A. (1979). Comparative prevalence of subclinical cytomegalovirus and herpes simplex virus infections in the genital and urinary tracts of low-income, urban women. *Journal of Infectious Diseases,* **140,** 419–22.

KORSAGER, B., SPENCER, E. S. & MORDHORST, C. H. (1975). Herpesvirus hominis infections in renal transplant recipients. *Scandinavian Journal of Infectious Diseases,* **7,** 11–19.

LIEW, F. C. & RUSSELL, S. M. (1980). Delayed-type hypersensitivity to influenza virus. *Journal of Experimental Medicine,* **151,** 799–814.

MANDEL, T. E., PHIPPS, R. P., ABBOT, A. & TEW, J. G. (1980). The follicular dendritic cell: long term antigen retention during immunity. *Immunological Reviews,* **53,** 29–60.

MENSER, M. A., FORREST, J. M. & BRANSBY, R. D.(1978). Rubella infection and diabetes mellitus. *Lancet;* **i,** 57–60.

MILLER, A. E. (1980) Selective decline in cellular immune response to varicella-zoster in the elderly. *Neurology,* **30,** 582–7.

MIMS, C. A. (1966). Immunofluorescent study of the carrier state and mechanism of vertical transmission in lymphocytic choriomeningitis virus infection in mice. *Journal of Pathology and Bacteriology,* **91,** 395–402.

MIMS, C. A. (1974). Factors in the mechanism of persistence of viral infections. *Progress in Medical Virology,* **18,** 1–14.

MIMS, C. A. (1978). General features of persistent virus infections. *Postgraduate Medical Journal,* **54,** 581–6.

MIMS, C. A. (1981). Viral interference with the immune response. In *Microbial Perturbation of Host Defences,* ed. F. O'Grady & H. Smith, pp. 211–25. Academic Press, London & New York.

MIMS, C. A. & GOULD, J. (1978). Splenic necrosis in mice infected with cytomegalovirus. *Journal of Infectious Diseases,* **137,** 587–91.

MIMS, C. A. & TOSOLINI, F. A. (1969). Pathogenesis of lesions in lymphoid tissue of mice infected with lymphocytic choriomeningitis (LCM) virus. *British Journal of Experimental Pathology*, **50**, 584–92.

MORGAN-CAPNER, P., ROBINSON, P., CLEWLEY, G., DARBY, A. & PETTINGALE, K. (1981). Measles antibody in Paget's disease. *Lancet*, **i**, 733.

MURPHY, F. A. (1977). Rabies pathogenesis – brief review. *Archives of Virology*, **54**, 279–97.

MURPHY, F. A., BUCHMEIER, M. J. & RAWLS. W. E. (1977). The reticuloendothelium as the target in a virus infection: Pichinde virus pathogenesis in two strains of hamsters. *Laboratory Investigation*, **37**, 502–25.

NASH, A. A. & GELL, P. G. H. (1981). Cell-mediated immunity in herpes simplex virus infected mice: suppression of delayed hypersensitivity by an antigen-specific B lymphocyte. *Journal of General Virology*, **48**, 359–64.

NOTKINS, A. L., MERGENHAGEN, S. & HOWARD, R. (1970). Effect of virus infections on the function of the immune system. *Annual Reviews of Microbiology*, **24**, 525–38.

OLDSTONE, M. B. A. & DIXON, F. J. (1967). Lymphocytic choriomeningitis: production of anti-LCM antibody by 'tolerant' LCM-infected mice. *Science*, **158**, 1193.

OLDSTONE, M. B. A. & DIXON, F.J. (1969). Pathogenesis of chronic disease associated with persistent lymphocytic choriomeningitis viral infection. I. Relationship of antibody production to disease in neonatally infected mice. *Journal of Experimental Medicine*, **129**, 483–99.

OLDSTONE, M. B. A. & DIXON, F. J. (1971). Immune complex disease in chronic viral infections. *Journal of Experimental Medicine*, **134**, 32–40s.

PORTER, D. D., LARSEN, A. E. & PORTER, H. G. (1980). Aleutian disease of mink. *Advances in Immunology*, **29**, 261–86.

REBEL, A., BASLE, M., POUPLARD, A., KOUYOUMDJIAN, S., FILMON, R. & LEPATEZOUR, A. (1980). Viral antigens in osteoclasts from Paget's disease of bone. *Lancet*, **ii**, 344–6.

SAWYER, W. A. (1931). Persistence of yellow fever immunity. *Journal of Preventive Medicine*, **5**, 413.

STAGNO, S., REYNOLDS, D., TSIANTOS, A., FUCILLO, D. A., SMITH, R., TILLER, M. & ALFORD, C. A. (1975). Cervical cytomegalovirus excretion in pregnant and non-pregnant women: suppression in early gestation. *Journal of Infectious Diseases*, **132**, 522–7.

TEW, J. G., PHIPPS, R. P. & MANDEL, T. E. (1980). The maintenance and regulation of the humoral immune response: persisting antigen and the role of follicular antigen-binding dendritic cells as accessory cells. *Immunological Reviews*, **53**, 175–201.

WALKER, D. H., WULFF, H., LANGE, J. V. & MURPHY, F. A. (1975). Comparative pathology of Lassa virus infection in monkeys, guinea-pigs and *Mastomys natalensis*. *Bulletin World Health Organization*, **52**, 523–34.

WIKTOR, T., DOHERTY, P. C. & KOPROWSKI, H. (1977). Suppression of cell-mediated immunity by street rabies virus. *Journal of Experimental Medicine*, **145**, 1617–28.

YOON, J. W., AUSTIN, M., ONODERA, T. & NOTKINS, A. L. (1979). Virus induced diabetes mellitus: isolation of a virus from the pancreas of a child with diabetic ketoacidosis. *New England Journal of Medicine*, **300**, 1173–9.

PERSISTENCE OF INFLUENZA IN A POPULATION

M. S. PEREIRA

Virus Reference Laboratory, Central Public Health Laboratory, Colindale, London NW9 5HT, UK

The persistence of epidemic virus disease in man depends upon the presence of enough susceptibles to maintain the virus in circulation, and since virus infections are normally followed by an immune response in the host which prevents re-infection, virus survival must depend upon balancing the efficiency of infectivity against the availability of susceptible hosts. Few viruses spread so efficiently that the supply of hosts is exhausted, but the more easily a virus spreads the more it will need an equally efficient means to ensure survival. For the most easily transmissible agents, as measured by the proportion of young children found to have been infected in the first years of life, various mechanisms appear to operate. The most frequent is incorporation into the cells of the host itself and the establishment of a latent state. This is observed with the endemic adenoviruses which colonize the reticulo-endothelial system, with the papovavirus BK which probably becomes established in the urinary tract, and with the herpes viruses. Herpes simplex virus and EB virus are possibly slightly less easily transmissible than those already mentioned as they appear to transfer more effectively in the conditions of close contact of larger groups of children such as found in a crèche or day nursery rather than in the domestic situation. There is some evidence that parainfluenza viruses may also survive by means of a latent phase in patients with chronic pulmonary disease (Gross, Green & McCrea-Curnen, 1973) and latency seems to be the only explanation for the findings of Muchmore *et al.* (1981) who demonstrated the presence of virus in both clinical and sub-clinical infections throughout a period of $8\frac{1}{2}$ months among a small population living at the South Pole and in complete social isolation.

For measles virus, latency does not seem to be responsible for its epidemicity, and the two-year cycle of epidemics which in pre-vaccine days was observed in the northern hemisphere was explained by the high infectivity of the virus and the need for a fresh susceptible population to be born in the community before an epidemic could be repeated. In tropical countries, however, such a pattern has not been observed and it has been suggested that sub-clinical infection takes place when the virus is encountered on successive occasions and thus helps maintain a waning immunity (Black & Rosen, 1962; Brown, Gajdusek & Tsai, 1969).

An inadequate or short-lived immune response to virus infection seems to be important for the persistence of respiratory syncytial virus, again a highly infectious agent in infancy. It appears that spread of infection to infants is mediated through the frequent re-infections that take place in the older child or adult (Chanock *et al.*, 1970).

All these viruses are maintained by some such mechanism in the population and survive without difficulty as human pathogens although the perfect answer would appear to have been found by those viruses which after their first attack on a host and despite the presence of circulating neutralizing antibody are able to maintain themselves indefinitely, and in clinical silence in the body, and to be reactivated in response to various stimuli from time to time, thus reaching fresh hosts if these happen to be available.

An alternative or extra means to ensure a continuing source of susceptible hosts is exemplified by the influenza viruses. With this group there is a mechanism by which the virus can overcome the immunity of a host by altering its own antigenic structure in such a way that antibody to previous infection no longer matches the changed virus. That this method is highly successful is shown by the fact that influenza epidemics continue to occur all over the world, affect all age groups, and have resisted all efforts of control by vaccines.

Influenza viruses all have similar physical and biological properties but they can be divided into three types, A, B, C, on the basis of their characteristic ribonucleoprotein. These three types behave very differently epidemiologically, perhaps reflecting the degree of antigenic change they undergo. Influenza C appears to circulate fairly continuously among the young causing such mild illnesses that these are rarely investigated. However, those viruses that have been isolated have been found to be antigenically closely related with only a minor degree of antigenic variation demonstrable and it is possible that immunity is solid enough that second attacks in adult life are unlikely to occur. Influenza B similarly attacks the young predominantly but in this case the illness may be indistinguishable from that produced by classical type A influenza and is detected easily because of the occurrence of multiple cases particularly in close communities such as schools. Again, as with influenza C, antigenic variation is of a modest degree and all known viruses show some relationship one with another thus apparently conferring an immunity which limits infection in adult life. Influenza type A is responsible for pandemic influenza and is the cause of the regular epidemics which occur world wide and which produce a heavy toll both in morbidity and mortality. This virus undergoes very considerable antigenic variation which may be of such a major nature that either one or both of the surface antigens (the haemagglutinin and neuraminidase) becomes completely different from those of earlier viruses. These major changes are referred to by the term 'antigenic shift'. This has occurred only infrequently in the period since influenza viruses were first identified in 1933 (Smith, Andrewes & Laidlaw, 1933): in 1957 when the subtype known as H2N2 arose and replaced the earlier H1N1 subtype and in 1968 when subtype H3N2 arose and replaced the H2N2 subtype. Antigenic shift has not so far occurred among influenza B or C viruses.

The mechanism for the creation of a new subtype is thought to depend upon two factors. One is the presence of influenza A viruses in several animal species besides man, and the other is the possession by influenza viruses of a segmented genome which allows re-assortment of the RNA segments if a cell is infected simultaneously by two different viruses from different species. This has been shown to occur with high frequency in the experimental situation (Tumova & Pereira, 1965; Webster, Campbell & Granoff, 1973). The lack of antigenic shift and separate subtypes of influenza B and C can be explained by the fact that although both have the same sort of segmented genome as influenza A neither has been found in any species but man.

'Antigenic drift' is the term used to describe changes of a less complete nature which occur repeatedly during the period of prevalence of each subtype. It appears to depend upon a completely different circumstance from antigenic shift and current evidence suggests that this sort of minor variation may arise from small mutational changes in the haemagglutinin (or neuraminidase) antigen which result in a virus with sufficient antigenic difference that it may escape from an immune host and spread to contact susceptibles. This sort of selection under immunological pressure has been demonstrated experimentally by the isolation of mutant strains if a

Table 1. *Hosts of the different haemagglutinin (H) subtypes*

H1	Human Swine Avian	H7	Equine Avian
H2	Human Avian	H8	Avian
H3	Human Equine Avian	H9	Avian
H4	Avian	H10	Avian
H5	Avian	H11	Avian
H6	Avian	H12	Avian

Table 2. *Hosts of the different neuraminidase (N) subtypes*

N1	Human Swine Avian	N5	Avian
		N6	Avian
N2	Human Avian	N7	Equine Avian
N3	Avian	N8	Equine Avian
N4	Avian		

virus is grown in the presence of a limiting concentration of antibody (Archetti & Horsfall, 1950).

Both these antigenic changes have a profound effect on the epidemiology of influenza in man and in themselves would be sufficient to maintain the virus indefinitely in the human host. However, there are several curious features of influenza epidemiology which suggest that there are other factors which also play a part, and since influenza A viruses infect certain species other than man the study of how the virus behaves in animals could well shed light on the persistence of the disease in man.

The main animal species in which influenza viruses have been found are pigs, horses and birds and, until recently, the viruses themselves were classified according to their natural host. However with the demonstration of antigenic similarities in the surface glycoproteins in the viruses isolated from different hosts this classification has been abandoned and now all influenza viruses are numbered in series according to the singularity of their haemagglutinins and neuraminidases as shown in Tables 1 and 2. This sharing of antigens between man and other animal species gives added support to the suggestions for the mechanism of antigenic shift.

SWINE INFLUENZA

It was realized for the first time that influenza might affect pigs when outbreaks of acute respiratory illness occurred among swine in the Midwest of the USA at the height of the human pandemic of 1918. The similarity of the disease with that observed concurrently among humans was sufficiently striking for the disease to be named swine influenza. The epizootics at that time were extensive and severe with millions of pigs becoming sick and many thousands dying.

From the time interval of the outbreaks in swine it was thought most likely that infection had spread from man to pigs rather than that pigs were the source of human disease.

The viral aetiology was established by Shope (1931) in studies over the next years when he was able to transfer infection from pig to pig by means of filtered material from lungs and lymph nodes of diseased animals. He also demonstrated the mildness of illness when the filtrate was administered on its own and the typical severe illness which was produced when the filtrate was mixed with *Haemophilus influenzae suis*, an organism regularly encountered in the respiratory tract of sick pigs but which alone did not induce the disease.

At that time a virus had not been identified from the human disease and it was only some years later when influenza virus was isolated from man (Smith *et al.*, 1933) that the two agents could be compared and were found to be the same.

In further studies Shope (1941, 1943) came to the conclusion that the virus in pigs was closely associated with the life cycle of the swine lungworm, a nematode which parasitizes pigs and which when coughed up and swallowed in its larval form passes through the gut to be deposited in faeces in the soil. From there it may be eaten directly by other pigs or be engulfed by earthworms where it may remain for up to four years and then regain a swine host when the earthworm is eaten. Virus carried in this way may thus survive for several seasons.

One feature of considerable interest was that pigs known to be harbouring the virus remained symptomless until they were placed under stress of some kind, such as the administration of *H. influenzae suis*. This was followed rapidly by the onset of typical swine influenza. It was also observed that in these circumstances pigs were usually refractory to the induction of illness during the summer months and it was during the winter months that activation could be induced most readily. Shope (1955) proposed that virus in such an occult form could be extensively seeded among swine and be reactivated on a wide geographical scale by changes in the environmental conditions. Following this lead he looked for other forms of stress more likely to occur in natural conditions and found that when pigs were fed on lungworms from influenza virus-infected pigs (that is given so-called 'masked' virus) and were then submitted to adverse weather conditions, such as low temperatures or a cold and wet environment, some of the animals would develop full-blown typical swine influenza, not at all like the mild 'filtrate' disease which followed the simple administration of virus.

His interpretation of these experiments was that much greater pulmonary involvement occurred in pigs in which virus activation was provoked than in pigs infected intranasally with virus. He also observed that some of the pigs provoked by stress did not become ill at all but had unrecognized infections only shown up by serological tests.

Ordinarily it appears that swine influenza epizootics occur each year in the Midwest of the USA. They begin in the late autumn with the onset of bad weather. The young pigs born since the previous year have meanwhile been infested with lungworms, containing masked influenza virus, taken up in earthworms in the pastures. With the sudden cold wet weather, the masked virus is activated and spreads among the susceptible animals.

However, doubts have been expressed as to whether this mechanism really is how influenza is perpetuated in nature, and evidence has been accumulating which shows that the lungworm cycle may not be an essential feature in the maintenance of swine influenza. An opportunity arose in Hawaii (Wallace, 1979) to study the natural spread of influenza in an area where pigs were not infested with lungworms and where the temperature range was of a different order from the continental climate of the central States of the USA. In these circumstances it was found, in agreement with Shope's findings, that outbreaks occurred in herds in different areas simultaneously without any apparent introduction or transmission from an outside source. But lungworms were not involved nor did outbreaks result from a stressful climatic change. What happened to the virus between the epizootics was not discovered; transmission to other species such as rodents or birds did not appear to occur and no other reservoir of the virus was found. However, another possible way in which influenza might be maintained is by vertical transmission from sows to their young. This would presuppose a carrier state in the sow and would explain the observation by Wallace & Elm (1979) that in one herd of pigs in Hawaii where most pigs had been infected in an epizootic 18 months previously, a considerable number of young pigs aged between 4 and 6 months were found to have significantly raised antibody to swine influenza virus. These authors were only partially convinced by their studies which they had hoped would confirm earlier observations of Mensik (1962) who demonstrated swine influenza virus in foetal membranes and uterine mucous membranes collected at three successive parturitions from one sow exposed to virus in its first pregnancy.

Whatever the mode of preservation of virus, the pattern of epizootics in Hawaii is somewhat different from that found in the USA perhaps because the breeding routine is different. In Hawaii, breeding animals are replaced every 2–4 years by new susceptible animals which, in consequence, results in a low herd immunity.

More extensive studies have now shown, according to Nakamura, Easterday, Ronaldean & Walker (1972), that swine influenza not only occurs regularly every year in the USA as outbreaks in the colder months of the year, but also that the virus can be isolated throughout almost the entire 12 months. They found that infection was not always accompanied by the typical disease and that only a small proportion (25–40%) of the animals in a herd might have overt disease. Whether virus becomes truly latent in pigs or is maintained by a slow spread is still unclear but some serological evidence that the levels of antibody in breeding stocks are consistently high may indicate a continuous antigenic stimulus from persisting virus.

Although the 1918 influenza pandemic in man spread world wide it was apparently only in the USA that the swine population became infected and has continued to be infected with the swine influenza virus. In 1976, after the detection of human cases of infection with this virus in an outbreak in the USA, an international surveillance programme was begun to determine whether pigs in other countries were similarly infected. Serological studies however indicated that in most countries pigs had no antibody to this virus at all. The few exceptions seemed most likely to have originated from pedigree pigs obtained from the USA for breeding purposes (Nardelli, Pascucci, Gualandi & Loda, 1978; Sugimura *et al.*, 1980) although it has been recently reported (Shortridge & Webster, 1979) that pigs in China also carry the swine influenza virus and thus may constitute another reservoir of this virus in the world.

While these investigations into the prevalence of the swine virus in pigs were being pursued it was found that another influenza A virus had apparently passed to pigs and in 1970 Kundin isolated the human H3N2 virus from pigs in Taiwan shortly after an outbreak of infection with this virus in the human population (Kundin, 1970). The evidence indicated that man was infected first, as studies done a few months previously showed pigs then were completely free of antibody.

Evidence of the spread of the H3N2 virus to pigs has now been obtained in nearly every country where it has been sought, either by the demonstration of antibody or by the isolation of the virus itself (Harkness, Schild, Lamont & Brand, 1972; Yamane *et al.*, 1978; Chapman, Lamont & Harkness, 1978). In studies in Hawaii although infection by the H3N2 virus was clearly demonstrated the virus apparently did not persist in the pig population (Wallace, 1979).

In Hong Kong, however, continuing surveillance has now shown (Shortridge, Webster, Butterfield & Campbell, 1977) that the original H3N2 virus A/Hong Kong/68 has persisted in pigs in some areas and been recovered long after the original virus has been replaced in man. Besides this, some of the subsequent variants which have spread in the human population, such as A/England/42/72, A/Port Chalmers/1/73 and A/Victoria/3/75, have apparently also been transmitted to pigs, the evidence being derived from both virus isolation and serology. These viruses may have spread among pigs, or multiple transmissions could have occurred from man to pig. Similar studies by Chapman *et al.* (1978) on serum samples collected from pigs in the UK from 1971 to 1977 have demonstrated the continuing infection of pigs with the H3N2 virus. These authors suggest that there are both man-to-pig and pig-to-pig transmission, as if there were only interspecies transmission, antibody prevalence would correlate more closely with outbreaks of influenza in humans unless events in the two populations were linked in that the same factors control the development of epidemics of influenza in both species.

Evidence that other subtypes of influenza A infect pigs was obtained in Czechoslovakia, USA and Germany (Kaplan & Payne, 1959) with the H2N2 subtype. It is possible that, as in Hawaii with the H3N2 virus, these viruses did not persist. Blakemore & Gledhill (1941) isolated from pigs in the UK viruses which were serologically like the H1N1 viruses which were circulating at that time. There is no information that this virus persisted either.

As far as the H1N1 virus is concerned the possibility exists that this subtype which returned in 1977 after a period of absence of over 20 years may have come from reactivated virus from a swine source. Little evidence has accumulated to support this but D. J. Alexander (personal communication) reports that antibody titres of 1/20 or more to A/USSR/0098/77 (H1N1) have recently been found in pigs in the UK so this time round the virus has apparently passed to the swine population. It remains to be seen if any evidence of long-term carriage in pigs can be obtained.

Thus, to summarize, it is known that influenza A virus spreads freely from man to pigs, where it may or may not be maintained for long periods. How this is achieved is still not certain but there is the possibility of long-term survival of virus either in the lungworm–earthworm cycle or by some other unknown means whereby pigs are permanently or chronically infected by virus which can be subsequently reactivated by various stimuli. It is also possible that the virus is maintained by a continuous slow or inapparent spread from pig to pig with occasional sharp outbreaks of illness when spread becomes rapid and apparent.

EQUINE INFLUENZA

This first indication that influenza viruses infect horses came from serological evidence in Sweden (Heller, Espmark & Viriden, 1956) where outbreaks of respiratory illness occurred in the autumn of 1955 affecting particularly horses in racing stables. Sometime afterwards influenza A virus was isolated from horses in similar outbreaks of respiratory disease in Czechoslovakia (Sovinova, Tumova, Pouska & Nemec, 1958) followed some years later by the isolation of a second distinct variety in the USA (Waddell, Teigland & Siegel, 1963). These two subtypes known as A/equi/Prague/1/56 and A/equi/Miami/1/63 have been found either separately or together in outbreaks of horse influenza in many countries of the world. The viruses are not apparently those common to man, as with pigs, although as shown in Table 1 one of them does share a common surface antigen with the human H3N2 virus. Both viruses have been encountered repeatedly since their first discovery but the mechanism by which they are maintained is by no means clear. In the UK for example confirmed influenza outbreaks occurred on only four occasions in 13 years (Thomson et al., 1977) although other outbreaks of similar illness were known to have occurred but were not investigated during this period. Many outbreaks may be mild and perhaps not recognized as influenza. But in a susceptible population spread of infection can be rapid with an attack rate of up to 98% (Klingeborn, Rockborn & Dinter, 1980), not surprising perhaps in the close community of a stable in view of the explosive coughing which is a prominent feature of the disease.

Spread of infection from stable to stable follows the movement of animals to race meetings and other equestrian events. As these may be attended by horses from overseas countries as well, there is full scope for the encounter of susceptible animals. Efforts to avoid the spread of infection are complicated by the fact that the disease may be inapparent (Lief & Cohen, 1966).

So far only the two serotypes have been defined but antigenic drift has been demonstrated with both A/equi/1/56 virus (Powell et al., 1974; Tumova et al., 1980) and with the A/equi/2/63 virus (Pereira et al., 1972; Thomson et al., 1977; Klingeborn et al., 1980).

When outbreaks occur the attack rate will depend on previous experience of the horses to that particular serotype and in stables with many young horses immunity will be low or absent. Where horses of mixed ages are stabled together and the older animals may have been infected previously spread may be slow.

Whatever the means, equine influenza virus clearly survives satisfactorily between epizootics with the help of only modest antigenic drift. Latency has not been demonstrated to account for the long intervals between recognizable outbreaks but the epidemic pattern seems to demand a population of fully susceptible young animals for apparent infection to be noted. In between such episodes one can only postulate slow asymptomatic infections for the maintenance of the virus in the horse population.

AVIAN INFLUENZA

Fowl plague is an acute disease of chickens which in its classical form causes a heavy mortality among affected birds. Although the disease has long been recognized and since 1900 even known to be provoked by a virus it was only in 1955 that Schäfer established that the virus was influenza A (Schäfer, 1955). Since then, influenza A viruses have been isolated from a variety of other domestic birds such as ducks, geese, turkeys, quails and pheasants, from several cage birds such as parrots and parakeets, as well as from a wide range of wild birds such as terns, shearwaters, ducks, geese, gulls and guillemots. In domestic and cage birds the disease ranges from the severe with a high mortality rate to the relatively benign. In wild birds the severity of disease is less easy to evaluate but influenza viruses have been isolated from terns dying in large numbers in colonies off the coast of South Africa as well as from apparently completely normal healthy shearwaters in Australia.

The viruses which have been identified so far fall into several different antigenic groups none of which is confined to one particular bird species; a particular haemagglutinin may be associated with a neuraminidase similar to those from other avian strains or even from viruses of swine, horse or man. Less often, the haemagglutinin of an avian virus has been found similar to that of a horse or human virus (Butterfield & Campbell, 1978b). However, the evidence so far does not suggest that man has transmitted influenza viruses to birds as seems to have happened with pigs. Nor is there evidence that birds are directly the source of infections in man. What is clear is that an enormous reservoir of influenza viruses exists among birds and since it has been shown by Webster & Laver (1971) that influenza viruses from birds and mammals can recombine *in vivo* when simultaneously introduced into the same animal it is not surprising that in nature such a heterogeneous collection of viruses has been found. The natural spread of avian influenza viruses has not been well defined but these viruses are certainly widely distributed in the world and apparently circulating constantly. The conditions under which domestic birds are kept, particularly in battery farms, are highly favourable to the virus with large numbers of birds in close contact giving the maximum opportunity for virus spread.

Hwang, Lief, Miller & Mallinson (1970) describe an outbreak of influenza among domesticated ducks where the principal symptom was sneezing. The young birds appeared to withstand the illness well but 10% of the older ducks died. A source of infection was not found but the ducks had access to an area holding turkeys and geese as well as wild ducks which mixed freely with the domesticated birds.

In some outbreaks among turkeys (Lang & Wills, 1966; Smithies et al., 1969a, b; Kleven et al., 1970) the disease varied from mild to moderate with mortality rates between 1% to 10%. There was evidence in some cases of the spread of infection through human handlers, usually insemination crews who may visit several farms within a short space of time. The original source of infection has usually not been found but, as infection may be inapparent despite all care in disinfection, virus could be carried mechanically. Besides this possibility, as the act of insemination is a stress, latent virus could be activated.

What happens when the birds recover is still uncertain. The sharp fall in egg laying at the onset of illness is generally restored after a short period and returns to normal levels although some flocks never return to full production. It seems that birds may be persistently infected and virus has been recovered for several months after infection suggesting that the immune chicken may become a virus carrier (Butterfield & Campbell, 1978a). There is also some suggestion that vertical infection may occur via the egg but direct demonstration has not yet been achieved; progeny from affected flocks monitored serologically have not shown evidence of infection throughout the growing period.

The natural history of influenza among wild birds is more difficult to follow and presumably the way viruses spread and survive will depend to some extent on the nesting habits, breeding and migration patterns of the avian hosts. A significant feature is the finding that in wild ducks, even in symptomless birds, virus multiplies in the gut, is shed in faeces (Webster et al., 1978) and apparently survives for considerable periods in the water of ponds and lakes. This in itself gives abundant opportunity for the spread and interchange of viruses between birds of the same or different species. There is some evidence that infection or virus excretion is commoner during the breeding season and some seasonal variation has been suggested but Czechoslovakia, where a continuous search was maintained among aquatic birds, viruses were found throughout the year.

Hinshaw, Webster & Turner (1980) studied the prevalence of influenza viruses in wild ducks over a 3-year period and concluded that these viruses were maintained by infection of young birds when they congregate, with transmission occurring during the subsequent migration, faecal contamination of lake water playing its part in the infectious cycle.

CONCLUSIONS REGARDING INFLUENZA PERSISTENCE IN ANIMALS AND BIRDS

If one looks at the way influenza is maintained in the various animal species described above there does not appear to be a common pattern. Pigs seem to be curiously and highly susceptible to human influenza and to have no particular viruses of their own. The virus which has persisted in some way in swine since 1918 has shown only a moderate degree of antigenic drift in that long period and transmission back to man has been infrequent and then usually to single close contacts who had some underlying disease problem affecting the immune system. The only demonstration of spread in man was short-lived and confined to a few cases in a large military community in the USA (World Health Organization, 1976).

It seems likely that the virus can become latent in swine but equally it could persist by slow smouldering infections maintained by the particular mode of husbandry where pedigree herds of pigs are carefully preserved for long-term production.

In contrast to pigs, horses seem to be completely insusceptible to natural infection with human viruses despite their close contact with man; even though the haemagglutinin of the H3N2 virus of man is close to that of the horse, equally man is apparently insusceptible to the two equine serotypes even in the environment of stables where the concentration of virus during an epizootic can be considerable. How the virus survives in nature is not clear, particularly as the two serotypes show only moderate antigenic drift and latency has not been demonstrated. It could be that infection goes on continuously at low or undetectable levels with the wide movement of animals both nationally and internationally providing the necessary contacts between susceptibles and only when these reach a certain proportion does the virus begin to spread fast to produce the characteristic outbreaks of influenza.

Avian influenza viruses have not been found in man although several avian isolates have been found to have one or other of the haemagglutinin or neuraminidase antigens of the human influenza

·viruses. However, the human influenza virus H2N2 has been isolated from domestic ducks (Shortridge, 1979). The spread and maintenance of the avian viruses among domestic birds must be strongly influenced by the methods of husbandry and by the intestinal multiplication of virus and its apparent survival in the excreta. Among wild birds influenza viruses circulate constantly with the free movement of birds, particularly migratory birds, allowing the exchange of viruses not only between other avian species but also between other animals. That this is a frequent event has been shown by Hinshaw, Webster & Rodriguez (1981) who described the variety of combinations of antigens which have been found in influenza viruses in nature. The possibility for re-assortment of genetic material in these circumstances is considerable and it is doubtful if there would ever be a shortage of susceptible hosts.

Despite the similarity of the virus causing the disease it is hard to see many parallels in the epidemiological impact of influenza in any of the three major animal hosts and it would appear that in none of them is the pattern quite the same as in man.

HUMAN INFLUENZA

An accurate virological history of human influenza is available from 1933 onwards and since that time only three distinct subtypes have circulated in the world. The first, now known as H1N1, was found up to 1957 when the H2N2 appeared and took its place, to be followed and substituted in 1968 by the H3N2 which still circulates to the present time, 1981. The replacement of one subtype by another has been thought to be absolute and sharply demarcated in time and until recently there had been only occasional exceptions to this such as that reported by Isaacs, Hart & Law, (1962) where an H1N1 virus was isolated three years after this subtype had apparently disappeared and as reported by Napiorkowski & Black (1974) who demonstrated an H2N2 virus response among Amazonian Indians two years after the H3N2 subtype had replaced it in the world.

The unexpected return of the H1N1 subtype as a pandemic virus in 1977 seemed almost a freak event and indeed this subtype did not replace the H3N2 subtype but circulated freely with it often even in the same community without either predominating or replacing it.

During the periods of prevalence of the three subtypes repeated significant antigenic drift has been demonstrated and epidemics and outbreaks of influenza have occurred regularly every winter with only infrequent exceptions.

As with subtypes the same replacement of drifted viruses by the latest variant to appear has also been noted, but the time lag with a mixture of old and new is comparatively extended and overlaps of one or two seasons have been found. Longer periods, however, have been rare and once the first drifted variant has begun to circulate the prototype is rarely seen again. One exception to this reported by Moore *et al.* (1981) has been of the isolation in 1979 of a virus biochemically and antigenically similar to the A/Aichi/2/68 virus which had not been isolated for nearly ten years.

The appearance of variant viruses of a new subtype was thought at one time to follow the saturation of the population with a specific immunity to the new virus. It seems, however, that this selective pressure in nature does not necessarily begin to operate only when all the susceptibles are exhausted and as early as 1969, during only the second year of prevalence of the H3N2 virus, significantly drifted viruses were detected. At the same time in both the H2N2 and H3N2 periods the prototype pandemic viruses were each responsible for three major epidemics during the first four years and it was only then that the pandemic strains disappeared completely and were replaced by variant viruses. The factor, or degree of drift, needed to ensure spread of a variant is still unknown. Many variants have only limited or local spread whereas others become the main influenza virus encountered world wide for one or two seasons till they disappear and are replaced by another.

The changes in the viruses which have undergone antigenic drift are usually in the first instance detected by haemagglutination-inhibition tests when a virus fails to be inhibited to the usual titre by an antiserum prepared to the currently circulating viruses. Such variants examined at the molecular level are found to have changes in the base sequence for the amino acids of the polypeptides of the haemagglutinin sub-unit; changes may be slight with perhaps only a single alteration but with sequentially isolated variants more and more changes become apparent (Laver & Webster, 1968; Laver, Air, Dopheide & Ward, 1980). It is not yet known whether variants showing the same antigenic drift appear spontaneously in several unconnected areas or whether, as with a new pandemic strain, the spread of outbreaks of influenza in the world follows lines of communication from a single point of origin.

The number of variant viruses with both detectable antigenic changes and an epidemic potential have been more frequent during the H3N2 period of prevalence than during the H2N2 decade. The longest interpandemic period yielding drifted strains within a sub-type was during the H1N1 period, which probably began some years before the first influenza viruses were isolated in man in 1933 and extended through to 1957 with such antigenically modified viruses that until recently these variants were considered to be different subtypes (Schild et al., 1980).

So far an explanation for this disappearance and replacement of subtypes or variants has not been found and that they do so, particularly in the case of drifted variants when many susceptibles remain, does not support the suggestion that sequential drift occurs in response to an urgent need for new hosts.

A further unexplained feature of the behaviour of the influenza virus is its seasonal appearance and disappearance and the unpre-dictable nature of the epidemic waves it produces. Regular and detailed surveillance has shown that in the UK no winter during the H3N2 period has been entirely free of influenza viruses (Pereira & Chakraverty, 1977) although marked differences in the impact has been noted irrespective of the proportion of susceptibles available.

Where the virus lies between epidemics is unknown: nor why while a new virus can spread round the world within a matter of months, within a family group when all are apparently equally susceptible, it is the exception for all members to be infected (Hope-Simpson, 1979; Pereira et al., unpublished data).

Hope-Simpson (1979) has suggested that an explanation for this feature of influenza epidemiology would be accounted for by the existence of a latent phase for the virus. He has proposed that the virus having caused illness in a patient becomes latent somewhere in the human host and undetectable by the usual methods employed by the laboratory. Subsequently under the influence of some unspeci-fied seasonal stimulation latent virus reactivates and spreads to the non-immune contacts. The appearance of drifted variants would be the rule since the immune state of the host would allow the escape only of mutant viruses. In this situation as the immune status at any one time would be equal throughout the world the antigenicity of mutants would tend to be the same. This hypothesis has some attractive features but so far direct evidence of latency is entirely lacking and one can only speculate about the seasonal factors which might produce the necessary activation.

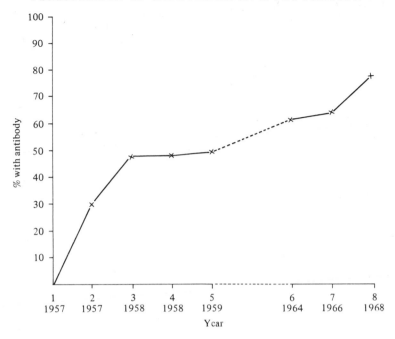

Fig. 1. Acquisition of antibody to the H2N2 subtype of influenza A virus during its 11 year period of circulation.

In an attempt to understand the natural behaviour of influenza in a normal non-institutionalized population a long-term study was begun in 1957 involving 25 unconnected families living in a London suburb (Pereira *et al.*, unpublished data). The objective was to follow the effects of a new pandemic virus on a fully susceptible group of people by regular blood sampling for serology and by the collection of clinical information. The study continued for 13 years covering the whole period of prevalence of the H2N2 subtype and the first two years of the H3N2 subtype.

The rate at which antibody to the new virus was acquired is illustrated in Fig. 1. There was a sharp increase in the first epidemic winter when 48% of the group became infected, half of them sub-clinically. In subsequent years as variants began to circulate these caused either primary infection in those who had missed infection with the prototype, again with half of them becoming clinically ill, or as second infections in those whose antibody to the prototype was apparently inadequate to protect against the drifted variant. In none of these second infections was the illness severe

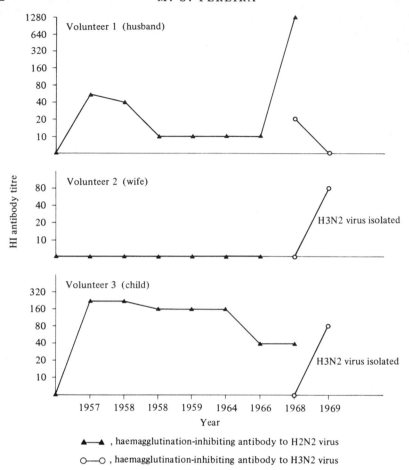

Fig. 2. Acquisition of antibody to the H2N2 and H3N2 influenza subtypes in one family 1957–69.

enough to require bed-rest as had happened with the first infections, and in most cases the volunteer denied any illness at all.

By the end of the H2N2 period there were still 11 people out of the original 56 (20%) who had apparently escaped infection and were without antibody. That it was not a question of an absolute resistance to infection by influenza virus was demonstrated by the infection of half of these by the H3N2 virus in the next two years. In these families, in all of which were at least two children, only infrequently did both parents become ill in the same epidemic and the more common picture was of one parent and one or more children in one year and another member of the family a year or so later. A characteristic pattern in one family is shown in Fig. 2. Right through the period infections were apparently random and how they were acquired it is impossible to say, whether introduced by child or adult, from school or place of work. The opportunities to pick up the virus when it is circulating in the community during the winter months are there for each member of a household. It would be impossible to prove or disprove the possibility that the virus had been activated from a latent infection in any one person to spread within the family.

Where the virus lies between one winter season and the next is unclear but there are several possibilities. The virus could fail to find fresh hosts and die out, to be re-introduced the following year by normal population movements from the alternative hemisphere. Against this, it seems that there are apparently always susceptible hosts still available when the viruses cease to be detected, and in some winters the epidemics come to an end as early as the second month of the year with the winter seasonal characteristics still operating, as happened in 1969–70 and 1972–73. There has often been evidence of the introduction of virus from the southern hemisphere to the northern when children return to boarding schools in the UK from their homes abroad to begin the school year in September. Outbreaks have certainly been initiated from this source but spread from such close communities to the outside has been infrequent. Perhaps the most likely mechanism for virus persistence is by inapparent spread. As already mentioned, sub-clinical infections during epidemics always form a significant proportion of those affected by the virus and it is possible that the disease could go completely undetected between epidemics. However searches for such infections have been unrewarding and the occasional cases which have been detected have been no more than would occur by the introduction of infection by persons coming into the country from the other hemisphere. Serological evidence has similarly been unconvincing and such low grade and slowly evolving infections if they exist must not only occur without provoking symptoms but also without providing an immune response. Another possibility is that latency does in fact exist although where the virus lies in the body and what the factors are that activate it are still unknown. If it were found to exist it would certainly provide answers to some of the unexplained epidemiological features of the disease.

REFERENCES

ARCHETTI, I. & HORSFALL, F. L. (1950). Persistent antigenic variation of influenza A viruses after incomplete neutralisation in ovo with heterologous immune serum. *Journal of Experimental Medicine*, **92**, 441–62.

BLACK, F. L. & ROSEN, L. (1962). Patterns of measles antibody in residents of Tahiti and their stability in the absence of re-exposure. *Journal of Immunology*, **88**, 725–31.

BLAKEMORE, F. & GLEDHILL, A. W. (1941). Discussion on swine influenza in the British Isles. *Proceedings of the Royal Society of Medicine*, **34**, 611–18.

BROWN, P., GAJDUSEK, D. C. & TSAI, T. (1969). Persistence of measles antibody in the absence of circulatory natural virus five years after immunisation of an isolated virgin population with Edmonston B vaccine. *American Journal of Epidemiology*, **90**, 514–18.

BUTTERFIELD, W. K. & CAMPBELL, C. H. (1978a). Vaccination of chickens with influenza A/turkey/Oregon/71 virus and immunity challenge exposure to five strains of fowl plague virus. Proceedings. *Eighty-second Annual Meeting of the United States Animal Health Association*, Buffalo, New York, pp. 320–4.

BUTTERFIELD, W. K. & CAMPBELL, C. H. (1978b). Identification of non-avid influenza A viruses containing human subtype of haemagglutinin and neuraminidase isolated from poultry in Hong Kong. Proceedings. *Eighty-second Annual Meeting of the United States Animal Health Association*, Buffalo, New York, pp. 325–31.

CHANOCK, R. M., KAPIKIAN, A. Z., MILLS, J., KIM, H. W. & PARROTT, R. H. (1970). Influence of immunological factors in respiratory syncytial disease of the lower respiratory tract. *Archives of Environmental Health*, **21**, 347–55.

CHAPMAN, M. S., LAMONT, P. H. & HARKNESS, J. W. (1978). Serological evidence of continuing infection of swine in Great Britain with an influenza A virus (H3N2). *Journal of Hygiene*, **80**, 415–22.

GROSS, P. A., GREEN, R. H. & McCREA-CURNEN, M. G. (1973). Persistent infection with parainfluenza type 3 virus in man. *Annual Review of Respiratory Disease*, **108**, 894–8.

HARKNESS, J. W., SCHILD, G. C., LAMONT, P. H. & BRAND, C. M. (1972). Studies on relationships between human and porcine influenza. *Bulletin of the World Health Organization*, **40**, 709–19.

HELLER, L., ESPMARK, A. & VIRIDEN, P. (1956). Immunological relationships between infectious coughs in horses and human influenza A. *Archiv für die Gesamte Virusforschung*, **7**, 1204.

HINSHAW, V. S., WEBSTER, R. G. & RODRIGUEZ, R. J. (1981). Influenza A viruses: combinations of hemagglutinin and neuraminidase subtypes isolated from animals and other sources. *Archives of Virology*, **67**, 191–201.

HINSHAW, V. S., WEBSTER, R. G. & TURNER, B. (1980). The perpetuation of orthomyxoviruses and paramyxoviruses in Canadian waterfowl. *Canadian Journal of Microbiology*, **26**, 622–9.

HOPE-SIMPSON, R. E. (1979). Epidemic mechanisms of type A influenza. *Journal of Hygiene*, **83**, 11–26.

HWANG, J., LIEF, F. S., MILLER, C. W. & MALLINSON, E. T. (1970). An epornitic of type A influenza virus infection in ducks. *Journal of the American Veterinary Medical Association*, **157**, 2106–8.

ISAACS, A., HART, R. J. C. & LAW, V. G. (1962). Influenza viruses 1957–60. *Bulletin of the World Health Organization*, **26**, 253–9.

KAPLAN, M. M. & PAYNE, A. M. (1959). Serological survey in animals for type A influenza in relation to the 1957 pandemic. *Bulletin of the World Health Organization*, **20**, 465–88.

KLEVEN, S. H., NELSON, R. C., DESHMUKH, D. R., MOULTHROP, J. I. & POMEROY, B. S. (1970). Epidemiologic and field observations on avian influenza in Minnesota turkeys. *Avian Diseases*, **14**, 153–66.

KLINGEBORN, B., ROCKBORN, G. & DINTER, Z. (1980). Significant antigenic drift within the influenza equi-2 subtype in Sweden. *Veterinary Record*, **106**, 363–4.

KUNDIN, W. D. (1970). Hong Kong A-2 influenza virus infection among swine during a human epidemic in Taiwan. *Nature*, **228**, 857.

LANG, G. & WILLS, C. G. (1966). Wilmot virus: a new influenza A virus infecting turkeys. *Archiv für die Gesamte Virusforschung*, **19**, 81–90.

LAVER, W. G., AIR, G. M., DOPHEIDE, T. A. & WARD, C. W. (1980). Amino acid sequence changes in the haemagglutinin of A/Hong Kong (H3N2) influenza virus during the period 1968–77. *Nature*, **283**, 454–7.

LAVER, W. G. & WEBSTER, R. G. (1968). Selection of antigenic mutants of influenza viruses. Isolation and peptide mapping of their haemagglutinating proteins. *Virology*, **34**, 193–202.

LIEF, F. S. & COHEN, D. (1966). Equine influenza. Studies of the virus and of antibody patterns in convalescent, interepidemic and post-vaccination sera. *American Journal of Epidemiology*, **82**, 225–46.

MENSIK, J. (1962). Experimental infection of pregnant sows with influenza suis virus. I. Proof of virus in placental tissue and in organs of newborn piglets. *Vědecké Práce Výzkumného ústavo Vet Lek Brné*, **2**, 31–47.

MOORE, B. W., WEBSTER, R. G., BEAN, W. J., VAN WYKE, K. L., LAVER, W. G., EVERED, M. G. & DOWNIE, J. C. (1981). Reappearance in 1979 of a 1968 Hong Kong-like influenza virus. *Virology*, **109**, 219–22.

MUCHMORE, H. G., PARKINSON, A. J., HUMPHRIES, J. E., SCOTT, E. N., MCINTOSH, D. A., SCOTT, L. V., COONEY, M. K. & MILES, J. A. R. (1981). Persistent parainfluenza virus shedding during isolation at the South Pole. *Nature*, **289**, 187–9.

NAKAMURA, R. M., EASTERDAY, B. C., RONALDEAN, P. & WALKER, G. L. (1972). Swine influenza: epizootiological and serological studies. *Bulletin of the World Health Organization*, **47**, 481–7.

NAPIORKOWSKI, P. A. & BLACK, F. L. (1974). Influenza A in an isolated population in the Amazon. *Lancet*, **2**, 1390–1.

NARDELLI, L., PASCUCCI, S., GUALANDI, G. L. & LODA, P. (1978). Outbreaks of classical swine influenza in Italy in 1976. *Zentralblatt für Veterinärmedicin*, **25B**, 853–7.

PEREIRA, M. S. & CHAKRAVERTY, P. (1977). The laboratory surveillance of influenza epidemics in the United Kindom 1968–1976. *Journal of Hygiene*, **79**, 77–87.

PEREIRA, H. G., TAKIMOTO, S., PIEGAS, N. S. & RIBEIRO DE VALLE, L. A. (1972). Antigenic variation of equine (Heq2Neq2) influenza virus. *Bulletin of the World Health Organization*, **47**, 465–9.

POWELL, D. G., THOMPSON, G. R., SPOONER, P., PLOWRIGHT, W., BURROWS, R. & SCHILD, G. C. (1974). The outbreak of equine influenza in England April/May 1973. *Veterinary Record*, **94**, 282–7.

SCHÄFER, W. (1955). Vergleichende sero-immunologische untersuchungen uber die Viren der Influenza und Klassischen Geflugelpest. *Zeitschrift für Naturforschung*, **10b**, 81–91.

SCHILD, G. C., NEWMAN, R. W., WEBSTER, R. G., MAJOR, D. & HINSHAW, V. S. (1980). Antigenic analysis of influenza A virus surface antigens: considerations for the nomenclature of influenza virus. *Archives of Virology*, **63**, 171–84.

SHOPE, R. E. (1931). Swine influenza. III. Filtration experiments and etiology. *Journal of Experimental Medicine,* **54,** 373–85.

SHOPE, R. E. (1941). The swine lungworm as a reservoir and intermediate host for swine influenza virus. II. The transmission of swine influenza virus by the swine lungworm. *Journal of Experimental Medicine,* **74,** 49–68.

SHOPE, R. E. (1943). The swine lungworm as a reservoir and intermediate host for swine influenza virus. III. Factors influencing transmission of the virus and the provocation of influenza. *Journal of Experimental Medicine,* **77,** 111–26.

SHOPE, R. E. (1955). The swine lungworm as a reservoir and intermediate host for swine influenza virus. V. Provocation of swine influenza by exposure of prepared swine to adverse weather. *Journal of Experimental Medicine,* **102,** 567–72.

SHORTRIDGE, K. F. (1979). H2N2 influenza viruses in domestic ducks. *Lancet,* **1,** 439.

SHORTRIDGE, K. F. & WEBSTER, R. G. (1979). Geographical distribution of swine (HSw1N1) and Hong Kong (H3N2) influenza virus variants in pigs in South-east Asia. *Intervirology,* **11,** 9–15.

SHORTRIDGE, K. F., WEBSTER, R. G., BUTTERFIELD, W. K. & CAMPBELL, C. H. (1977). Persistence of Hong Kong influenza virus variants in pigs. *Science,* **196,** 1454–5.

SMITH, W., ANDREWES, C.H. & LAIDLAW, P.P. (1933). A virus obtained from influenza patients. *Lancet,* **2,** 66.

SMITHIES, L.K., RADLOFF, D.B., FRIEDELL, R.W., ALBRIGHT, G.W., MISNER, V.E. & EASTERDAY, B.C. (1969a). Two different type A influenza virus infections in turkeys in Wisconsin. I. 1965–66 outbreak. *Avian Diseases,* **13,** 603–6.

SMITHIES, L. K., EMERSON, F. G., ROBERTSON, S. M. & RUEDY, D. D. (1969b). Two different type A influenza virus infections in turkeys in Wisconsin. II. 1968 outbreak. *Avian Diseases,* **13,** 606–10.

SOVINOVA, O., TUMOVA, B., POUSKA, F. & NEMEC, J. (1958). Isolation of a virus causing respiratory disease in horses. *Acta Virologica,* **2,** 52.

SUGIMURA, T., YONEMOCHI, H., OGAWA, T., TANAKA, Y. & KUMAGAI, T. (1980). Isolation of a recombinant influenza virus (HSw1N2) from swine in Japan. *Archives of Virology,* **60,** 271–4.

THOMSON, G. R., MUMFORD, J. A., SPOONER, P. R., BURROWS, R. & POWELL, D. G. (1977). The outbreak of equine influenza in England, January 1976. *Veterinary Record,* **100,** 465–8.

TUMOVA, B. & PEREIRA, H. G. (1965). Genetic interaction between influenza A viruses of human and animal origin. *Virology,* **27,** 253–61.

TUMOVA, B., STUMPA, A., ZAKOPAL, J., VEZNIKOVA, D. & MENSIK, J. (1980). Persistence in nature of influenza virus A/eq/Praha/56 (Heq1Neq1). *Acta Virologica,* **24,** 63–7.

WADDELL, G. H., TEIGLAND, M. B. & SIEGEL, M. M. (1963). A new influenza virus associated with equine respiratory disease. *Journal of the American Veterinary Association,* **143,** 587.

WALLACE, G. D. (1979). Natural history of influenza in swine in Hawaii. Swine influenza virus (HSw1N1) in herds not infected with lungworms. II. Prevalence of infection with A/Hongkong (H3N2) subtype virus and its variants 1974–7. *American Journal of Veterinary Research,* **40,** 1159–68.

WALLACE, G. D. & ELM, J. L. (1979). Transplacental transmission and neonatal infection with swine influenza virus (HSw1N1) in swine. *American Journal of Veterinary Research,* **40,** 1169–72.

WEBSTER, R. G., CAMPBELL, C. H. & GRANOFF, A. (1973). The 'in vivo' production of 'new' influenza viruses. III. Isolation of recombinant influenza viruses under simulated conditions of natural transmission. *Virology,* **51,** 149–62.

WEBSTER, R. G. & LAVER, W. G. (1971). Antigenic variation in influenza virus biology and chemistry. *Progress in Medical Virology,* **13,** 271–83.
WEBSTER, R. G., YAKHNO, M., HINSHAW, V. S., BEAN, W. J., JR & MURTI, K. G. (1978). Intestinal influenza: replication and characterisation of influenza viruses in ducks. *Virology,* **84,** 268–78.
WORLD HEALTH ORGANIZATION (1976). *Weekly Epidemiological Record,* No. 14.
YAMANE, N., ARIKAWA, J., ODAGIRI, T., KUMASAKA, M. & ISHIDO, N. (1978). Distribution of antibodies against swine and Hong Kong influenza viruses among pigs in 1977. *Tohoku Journal of Experimental Medicine,* **126,** 199–200.

PERSISTENCE OF HEPATITIS B VIRUS IN THE POPULATION

ARIE J. ZUCKERMAN

*Department of Medical Microbiology and WHO Collaborating
Centre for Reference and Research on Viral Hepatitis, London
School of Hygiene and Tropical Medicine (University of London),
London WC1E 7HT, UK*

INTRODUCTION

Human viral hepatitis is caused by at least four different viruses, hepatitis A (infectious or epidemic hepatitis), hepatitis B (referred to in the past as serum hepatitis) and the more recently identified type of hepatitis, non-A, non-B hepatitis which is caused by more than two viruses and probably by several different viruses.

Acute viral hepatitis is a generalized infection with particular emphasis on the liver and all known types are endemic throughout the world. Inapparent or subclinical infections and infections without jaundice are common. However, the clinical picture of the illness ranges from an asymptomatic infection, a mild anicteric illness, acute disease with jaundice, severe prolonged jaundice to acute fulminant hepatitis. While hepatitis A virus does not persist in the host nor is there evidence of progression to chronic liver damage, hepatitis B and non-A, non-B hepatitis may be associated with persistent infection, prolonged carrier state and progression to chronic liver disease, which may be severe. In addition, there is now substantial evidence of a close and indeed aetiological association between hepatitis B virus and primary hepatocellular carcinoma.

Hepatitis A and hepatitis B can now be differentiated by sensitive laboratory tests for specific antigens and antibodies and the viruses have been characterized. On the other hand, there are as yet no precise virological criteria nor specific laboratory tests for non-A, non-B hepatitis.

Survival and persistence of hepatitis B virus in the population on a global scale is ensured by a huge reservoir of carriers, estimated conservatively to number between 175 and 200 million; prolonged 'shedding' of the virus by a proportion of carriers; varied mechanisms and routes of transmission including perinatal infection and relative stability of the virus in the environment.

The discovery of Australia antigen in 1965 and the demonstration by B. S. Blumberg and his colleagues of its association with hepatitis B led to rapid and unabated progress in the understanding of this complex infection, resulting in the publication of over 5000 papers during the last two years alone. The subject has been amply reviewed recently elsewhere (Zuckerman, 1979a; Zuckerman & Howard, 1979; Sherlock, 1980; Bianchi, Gerok, Sickinger & Stalder, 1980 and others).

Serological markers of hepatitis B virus

Infection with the hepatitis B virus leads to the appearance in the plasma during the incubation period of a specific antigen, hepatitis B surface antigen (originally referred to as Australia antigen) about 2–8 weeks before biochemical evidence of liver dysfunction or the onset of jaundice. This antigen persists during the acute illness and is usually cleared from the circulation during convalescence. Next to appear in the circulation is a specific viral DNA polymerase associated with the core or nucleocapsid of the virus and at about the same time another antigen, the e antigen, becomes detectable, again preceding serum aminotransferase elevations. The e antigen is a distinct soluble antigen which is located within the core and correlates closely with the number of virus particles and relative infectivity. Antibody to the hepatitis B core antigen is found in the serum 2–4 weeks after the appearance of the surface antigen, and it is always detectable during the early acute phase of the illness. Core antibody of the IgM class becomes undetectable within some months of the onset of uncomplicated acute infection, but IgG core antibody persists after recovery for many years and possibly for life. The next antibody to appear in the circulation is directed against the e antigen, and there is evidence that, in general terms, anti-e indicates relatively low infectivity of serum. Antibody to the surface antigen component, hepatitis B surface antibody, is the last marker to appear late during convalescence. More recently, precipitating antibodies reacting with specificities on the complete virus particle have been described. These antibodies may be relevant to the clearance of circulating hepatitis B virions and the termination of infection and their absence in patients with chronic active hepatitis may explain why the infection persists in such patients (Alberti et al., 1978). Cell-mediated immunity also appears to be important in terminating hepatitis B infection and, under certain circumstances, in promoting liver damage and in the genesis of autoimmunity (World Health Organization, 1977; Zuckerman, 1978, 1979b).

Hepatitis B virus

Examination by electron microscopy of serum containing hepatitis B surface antigen reveals the presence of small spherical particles measuring about 22 nm in diameter, tubular forms of varying length but with a diameter close to 22 nm and large double-shelled or solid particles approximately 42 nm in diameter (Fig. 1). The large particles contain a core or nucleocapsid about 28 nm in diameter. The 42 nm particle is the hepatitis B virus, whereas the small particles and the tubules are non-infectious surplus virus coat protein.

The core of the virus contains a DNA-dependent DNA polymerase, closely associated with a DNA template. Double-stranded circular DNA has been isolated from circulating virus and also from cores extracted from the nuclei of infected hepatocytes. The molecular weight of the DNA is about 2.3×10^6 and the DNA is approximately 3600 nucleotides in length, containing a single-stranded gap varying from 600–2100 nucleotides. The endogenous DNA polymerase reaction appears to repair the gap.

The morphological complexity of hepatitis virus is surpassed by the antigenic heterogeneity of the surface antigen reactivities. Careful serological analysis has shown that the hepatitis B surface antigen particles share a common group-specific antigen *a* and generally carry at least two mutually exclusive subdeterminants, *d* or *y* and *w* or *r*. The subtypes are the phenotypic expressions of distinct genotype variants of hepatitis B virus. Four principal phenotypes are recognized, *adw*, *adr*, *ayw* and *ayr*, but other complex permutations of these subdeterminants and new variants have been described, all apparently on the surface of the same physical particles. The major subtypes have differing geographical distribution. For example, in northern Europe, the Americas and Australia subtype *adw* predominates. Subtype *ayw* occurs in a broad zone which includes northern and western Africa, the eastern Mediterranean, eastern Europe, northern and central Asia and the Indian subcontinent. Both *adw* and *adr* are found in Malaysia, Thailand, Indonesia and Papua New Guinea, while subtype *adr* predominates in other parts of south-east Asia including China,

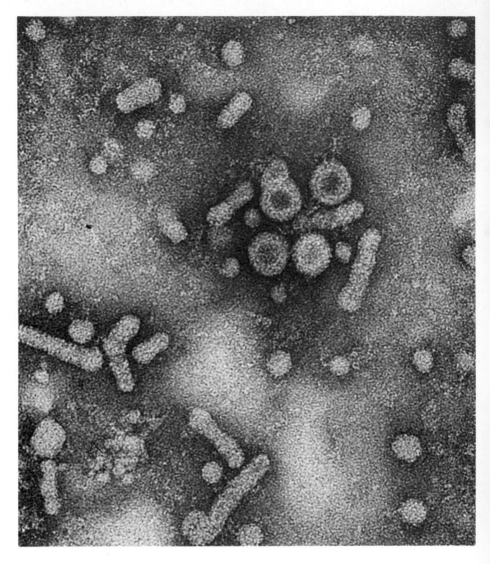

Fig. 1. Electron microscopy of serum containing hepatitis B virus and surface antigen. Three distinct morphological entities are present: small pleomorphic spherical particles (20–25 nm diameter), tubular forms of varying length, and large spherical particles (approximately 42 nm diameter), some of which are penetrated by negative stain to reveal an inner core of 28 nm diameter.

Negative stain, ammonium molybdate: final magnification × 250 000. (Reproduced with permission from A. J. Zuckerman, *Human Viral Hepatitis,* 1975.)

Japan and the Pacific Islands. The subtypes provide useful epidemiological markers of hepatitis B virus.

Epidemiology of hepatitis B

In the past, hepatitis B was diagnosed on the basis of infection occurring about 60–180 days after the injection of human blood or plasma fractions or the use of inadequately sterilized syringes and needles. The development of specific laboratory tests for hepatitis B confirmed the importance of the parenteral routes of transmission and infectivity appears to be especially related to blood. However, a number of factors have altered the epidemiological dogma that hepatitis B is spread exclusively by blood and blood products. These include the observations that under certain circumstances the virus is infective by mouth, that it is endemic in closed institutions and institutions for the mentally-handicapped, that it is more prevalent in adults in urban communities and in poor socio-economic conditions, that there is a huge reservoir of carriers of markers of hepatitis B virus in the human population and that the carrier rate and age distribution of the surface antigen varies in different regions.

There is much evidence for the transmission of hepatitis B by intimate contact and by the sexual route. The sexually promiscuous, particularly male homosexuals, are at very high risk. Hepatitis B surface antigen has been found in blood and in various body fluids such as saliva, menstrual and vaginal discharges, seminal fluid, colostrum and breast milk and serous exudates and these have been implicated as vehicles of transmission of infection. The presence of the antigen in urine, bile, faeces, sweat and tears has been reported at times, but not confirmed. It is not surprising therefore that contact-associated hepatitis B is of major importance. Transmission of the infection may result from accidental inoculation of minute amounts of blood or fluids contaminated with blood such as may occur during medical, surgical and dental procedures, immunization with inadequately sterilized syringes and needles, intravenous and percutaneous drug abuse, acupuncture, laboratory accidents and accidental inoculation with razors and similar objects which have been contaminated with blood. Additional factors may be important for the transmission of hepatitis B infection in the tropics and in hot-climate countries. These include traditional tattooing and scarification, blood letting, ritual circumcision and repeated biting by blood-sucking arthropod vectors. Results of investigations into the role which biting insects may play in the spread of hepatitis B are conflicting. Hepatitis B surface antigen has been detected in several species of mosquitoes and in bed-bugs which have either been trapped in the wild or fed experimentally on infected blood, but no convincing evidence of replication of the virus in insects has been obtained. Mechanical transmission of the infection, however, is a possibility.

Clustering of hepatitis B also occurs within family groups, but on the whole it is not related to genetic factors and does not reflect maternal and venereal transmission. The mechanisms of intrafamilial spread of the infection are not known.

THE CARRIER STATE

A carrier state of hepatitis B virus, which may be life-long, becomes established in approximately 5–10% of infected adults. The carrier state is more likely to follow infection acquired in childhood. Evidence that a prolonged carrier state may result in some patients was obtained many years before specific serological tests became available and before viral antigens were localized in the liver. Experimental transmission of hepatitis to human volunteers during World War II and in the early 1950s revealed viraemia in hepatitis type B up to 87 days before the onset of illness (Neefe, Stokes, Reinhold & Lukens, 1944; Paul, Havens, Sabin & Philip, 1945; Havens, 1946). Stokes *et al.* (1954), Neefe *et al.* (1954) and Murray *et al.* (1954) provided evidence that a prolonged carrier state may become established for as long as five years in some patients, with or without signs of liver disease.

Zuckerman & Taylor (1969) detected hepatitis B surface antigen in the serum of a former blood donor PJG, who was identified as a carrier in 1951 (Stokes *et al.*, 1954). This donor was implicated in 1950 in three cases of homologous serum jaundice, one of whom died 73 days after transfusion. There was no suggestion in his past medical history that he had ever suffered from hepatitis. Ten months after the first incident of infection occurred in a recipient of his blood, a sample from PJG was injected in the USA into five volunteers. At least one of the volunteers developed hepatitis with jaundice 59–62 days after intramuscular inoculation. Hepatitis B surface antigen was detected in fresh serum collected from PJG in 1968, 1969, 1970 and 1971, and furthermore, the antigen was found by electron microscopy in the original 1950–51 serum (MacCallum, 1972). This is a well-documented example of apparently healthy carriage of hepatitis B antigen for over 20 years (Zuckerman, 1979a).

Giles, McCollum, Berndtson & Krugman (1969) tested for hepatitis B surface antigen in serial samples of serum obtained before and up to 200 days after experimental infection of volunteer patients with serum containing the MS-2 strain of type B hepatitis. The antigen was detected in all patients after infection with MS-2, frequently before there was any clinical evidence of infection. In approximately half the patients the antigen persisted in the serum for 200 days and subsequently for at least three years after infection.

These observations confirmed the long-held epidemiological and clinical findings that a prolonged carrier state of hepatitis B virus may exist with persistent viraemia.

On the basis of longitudinal studies of patients with hepatitis B, the persistent carrier state has been defined as the presence of hepatitis B surface antigen in blood for more than six months (World Health Organization, 1977). Such a carrier state may be associated with liver damage ranging from minor changes in the nucleus of the hepatocytes to chronic persistent hepatitis, chronic active hepatitis and cirrhosis, with persistence of the viral components in the hepatocytes (Nazarewicz-de Mezer, Krawczynski, Michalak & Nowoslawski, 1980).

Extensive serological surveys in many parts of the world have shown that the prevalence of hepatitis B surface antigen in apparently healthy persons varies from 0.1 to 20%. Europe may be roughly divided into three regions, with a low prevalence of about 0.1% or less in northern countries, 0.1 to 3% in some central and eastern European countries, and about 5% and even higher in some countries bordering the Mediterranean. The prevalence in North America and Australia is 0.1% or less, 3–5% or higher in Asia and 15–20% or more in several tropical countries. It should be noted, however, that the global distribution of hepatitis B surface antigen is by no means complete and, furthermore, standardized techniques and reagents have not been uniformly used.

A number of factors have been identified which increase the risk of developing the carrier state. It is more common in males, more likely to follow infections acquired in childhood than those acquired in adult life, and more likely to occur in patients with natural or acquired immune deficiencies.

In countries in which infection with hepatitis B virus is relatively uncommon, the highest prevalence of the surface antigen is found in the 20–40 year age group. The prevalence of surface antibody increases steadily with age. In countries where infection with hepatitis B virus is common, the highest prevalence of the surface antigen is observed in children 4–8 years old, with declining rates among older age groups. The decline in antigen carriage rates with age suggests that the carrier state is not invariably life-long.

The carrier state of hepatitis B is characterized serologically by persistence of hepatitis B surface antigen, with and without detectable complete virion (the Dane particle), and generally in the absence of measurable surface antibody. Core antibody is present, often in high titre, and in some persistent carriers IgM core antibody remains detectable. In some carriers, hepatitis B DNA polymerase is found, often with fluctuating levels, and hepatitis B e antigen persists. In other carriers, specific DNA polymerase is not detectable and anti-e is present. Hepatitis B e antigen has been found more commonly in young than in adult carriers, while the prevalence of anti-e seems to increase with age. These observations suggest that young carriers may be the most infective.

Sex differences in the carrier state

An excess of males with hepatitis B surface antigen was noted in early extensive family studies carried out in the islands of Cebu in the Philippines and in Bougainville, New Guinea (Blumberg et al., 1969). A higher incidence of the surface antigen was also found among male patients with Down's syndrome resident in institutions, among blood donors, and among patients with acute hepatitis B. London et al., (1977a) also reported that males were more likely to become persistent carriers when treated by maintenance haemodialysis and females were more likely to develop surface antibody. Robertson & Sheard (1973) reported that the sex ratio at birth in a small town in Lincolnshire in England was disturbed in 1969. There were 43 live female births but only 21 live male births and one male stillbirth. There was no increase in the incidence of spontaneous abortions in the town. It was pointed out that the ratio of boys to girls could be affected by any factor which impaired the viability of sperms bearing the Y chromosome, by factors which gave some advantage in penetration of mucus or ovum to X-bearing sperm, or, after conception, by factors which selectively impaired the viability or survival of the male embryo. A striking excess of female births occurred twice before in the same town in 1955 and 1967. The most notable feature of 1967 in the town was a severe outbreak of viral hepatitis and it was considered that subclinical and anicteric hepatitis may have persisted in the town until 1970. An excess of males among siblings of carriers of hepatitis B surface antigen in a Melanesian population on the island of Santa Cruz in the British Solomon Islands has also been reported and it has been suggested that this alteration in the sex ratio might be the result of selection by an infectious disease.

Hesser, Blumberg & Drew (1976) reported that matings with a parent with surface antigen resulted in an increased sex ratio among males (64%) in a Greek population compared with matings of negative parents (53% males). In Melanesian populations, however, the sex ratio was decreased when the mother was positive for surface antigen and increased when the father was positive. Thus the presence of the surface antigen was associated with alterations in the live birth sex ratio, and altered sex ratios in turn affect population reproduction rates.

The interactions between hepatitis B virus and sex were studied further. London et al. (1977b) found that among patients treated by haemodialysis, males had a 68% chance of remaining persistent carriers of the surface antigen once infected with hepatitis B virus, whereas females had only a 33% chance. Conversely, females had a 55% chance of developing surface antibody once infected, whereas males had only a 30% chance. In addition, the response of renal graft recipients to infection with hepatitis B virus before transplantation and the sex of their kidney donors was related to the duration of graft survival. Kidneys from HLA non-identical male donors which were transplanted into male or female patients with surface antibody survived only a few months. However, survival of grafts from male donors was significantly longer in both uninfected patients and in persistent surface antigen carriers. There were no differences in graft survival among the relatively few patients who received grafts from female donors. The reason why patients with surface antibody have accelerated rejection of kidney transplants is not known. This may be related to an early recognition of HLA antigens, but the observation that the greatest risk of graft rejection occurred when the recipient had surface antibody and the kidney donor was a male is not explained by the hypothesis. Another hypothesis is that Y-linked histocompatibility antigens influenced the host response to both the surface antigen and the HLA antigens.

After further studies on the relation of hepatitis B infection to the sex ratio in a Greek village, Drew *et al.* (1978) suggested that there may be a male-associated antigen. If the surface antigen cross-reacts with a male-associated antigen, males would be more likely to recognize the surface antigen as 'self' and therefore would persistently carry the antigen. Females, however, would be more likely to recognize the surface antigen as 'foreign' and produce surface antibody. In renal transplant patients, tolerance to the surface antigen would result in relative tolerance of male tissues, whereas surface antibody in the recipient would react with male antigens or renal allografts from male donors leading to early rejection of the grafts. Similarly, it is speculated that tolerance to the surface antigen in carrier pregnant women would result in lack of sensitization against male tissues, and therefore good survival of male foetuses. Surface antibody, on the other hand, would react with male antigens and perhaps hinder fertilization by sperm bearing a Y chromosome or increase the probability of spontaneous abortion of male foetuses. Male carriers of surface antigen would have the antigen in their semen which could perhaps protect Y-bearing sperm from surface antibody in the reproductive tract of their spouses. The effect of foetal loss in antigen-carrying women could be explained by replication of hepatitis B virus in male foetuses, and this could result in a greater loss of male than female embryos. Further studies of these observations are in progress.

Genetic susceptibility and the carrier state

In order to explain the geographical variation in the prevalence of asymptomatic carriers of hepatitis B antigens, Blumberg *et al.* (1969) postulated that persistence of hepatitis B surface antigen was dependent not only on infection by hepatitis B virus but also on the presence in the homozygous state (Au^1/Au^1) of an autosomal recessive gene which conferred the ability to maintain the antigen in an individual acquiring it. This gene was considered to be common in tropical areas but rare in temperate zones. It was also suggested that individuals with such an inherited susceptibility do not usually display overt manifestations of hepatitis but nevertheless remain carriers. The interpretation of previous genetic analyses was based upon the assumption of total exposure of the population to the virus, since only then could the effect of the gene be demonstrated.

Although there are many reports of familial clustering, critical matings and other data indicate that genetic factors are not necessarily involved since perinatal transmission from mother to child may also take place. The alternative hypothesis to the genetic factor is that familial clustering may be entirely a function of an increased opportunity for environmental exposure to the virus. Furthermore, it stands to reason that if a parent is excreting the virus, this will augment the chance that any of the offspring may acquire it, particularly if the mother is the transmitter. It is, however, accepted that the genetic composition of the host, for example histocompatibility antigens, may influence the response of the host to infection (reviewed by Zuckerman, 1975 and Zuckerman & Howard, 1979).

Perinatal transmission of hepatitis B

Transmission of hepatitis B virus from carrier mothers to their babies can occur during the perinatal period and appears to be an important factor in determining the prevalence of infection in some regions. The risk of infection may reach 50–60%, although it varies from country to country and appears to be related to ethnic group. The risk is greatest if the mother has a history of transmission to previous children and has a high titre of hepatitis B surface antigen and/or e antigen. There is also a substantial risk of perinatal infection if the mother has acute hepatitis B in the second or third trimester of pregnancy or within two months after delivery. Although hepatitis B virus can infect the foetus *in utero*, this appears to be rare. The mechanism of perinatal infection is uncertain, but it probably occurs during or shortly after birth as a result of a leak of maternal blood into the baby's circulation, its ingestion or inadvertent inoculation. Most children infected during the perinatal period become persistent carriers (reviewed by Boxall, 1980).

Hepatitis B in non-human primates

Hepatitis B surface antigen and surface antibody have been detected in 6–12% of captive chimpanzees when tested by relatively insensitive techniques. Surface antibody was found in a significant proportion of captive non-human primates when sensitive serological techniques such as passive haemagglutination and radioimmunoassay were used. Hepatitis B surface antibody was found in the chimpanzee, orangutan, gibbon, baboon, Celebes ape, patas monkey, vervet, several species of macaque, mangebey and langur and in several species of New World monkey. Surface antibody was found in approximately 50% of chimpanzees examined but in less than 10% of most Old World and New World monkeys (World Health Organization, 1973). As in the human population, the prevalence of surface antibody in chimpanzees also increases with age.

An interesting example of natural infection with hepatitis B virus among the chimpanzees and other apes in the care of the Zoological Society of London was described by Zuckerman *et al.* (1978). In 1976, a serum specimen was examined from a female chimpanzee receiving treatment in the London Zoo Hospital for a chronic dermatosis of unknown aetiology. Electron microscopy revealed a large number of all three morphological forms of hepatitis B. The titre of the surface antigen in the serum determined by reverse passive haemagglutination was greater than 1 : 131 072, and specific hepatitis B DNA polymerase activity was elevated. Surface antigen was also detected repeatedly in the saliva, but not in urine or faeces. There was no past history of infection in this chimpanzee, which was born at the Zoo, and as far as was known she had not received human blood or blood products. Serological survey of the apes kept at the London Zoo revealed that surface antigen was present in very high titre ($>$1 : 1 048 756 by reverse passive haemagglutination) in another four out of eight chimpanzees living as a group, and surface antibody was detected in the remaining four chimpanzees. The antigen was not found in the sera from two gorillas, eleven orangutans and two gibbons, although surface antibody was present in the serum of one gorilla and two orangutans.

Repeat serum samples from the chimpanzees confirmed these findings and established that four chimpanzees were persistent hepatitis B carriers. All three morphological forms of the antigen particles were found in their serum, the titres of the surface antigen were extremely high, ranging by reverse passive haemagglutination from $>$1 : 262 144 to $>$1 : 1 048 756, elevated specific DNA polymerase activity was found in all four chimpanzees, and *e* antigen was detected in each case. Surface antigen was also detected in the saliva of each chimpanzee, often to high titre, but the antigen was not found in the urine or faeces. The subtype of the surface antigen was *adw2*. Hepatitis B surface antibody was never found by radioimmunoassay in any of the serial samples collected over a period of two years and immune complexes were not seen by electron microscopy. Hepatitis B core antibody was present in each case.

Serum aspartic and alanine aminotransferase levels were within the normal range for the species. The histological changes in liver biopsy specimens were consistent with chronic persistent hepatitis in two chimpanzees and mild mesenchymal reaction with occasional inflammatory changes in the other two chimpanzees. Orcein staining revealed scattered liver cells with cytoplasmic staining for hepatitis B surface antigen.

The source of the outbreak at the London Zoo could not be traced. Carriers were not detected among current members of staff. It should be noted, however, that hepatitis B surface antigen was found in two frozen serum samples from chimpanzees which died in 1966 and 1972 and these, or indeed others, may have introduced the original infection. It is interesting that the subtype of the surface antigen in all the cases was identical.

Another observation which is of interest is that three of the carrier chimpanzees born at the Zoo were the offspring of either a carrier mother or father. Furthermore, the implicated carrier mother and father were captured from the wild, the female in 1948 and the male sometime before 1964. Some years ago it was common practice for animal-catchers and dealers to inoculate newly-captured chimpanzees with pooled human blood for 'protection against human disease' by passive transfer of antibodies. Since the blood for such inoculations was obtained in areas with high prevalence of endemic hepatitis B, it is possible that the chimpanzees were infected by this route. It is, however, not yet known whether hepatitis B infection may also be acquired in the wild. Another point of interest is the possibility of perinatal transmission of hepatitis B virus in at least three of the chimpanzee matings, leading ultimately to the persistent carrier state, as in human beings.

Shouval et al. (1980) found hepatitis B viral DNA in the nucleic acid extract of percutaneous liver biopsy specimens from each of five chimpanzee carriers of hepatitis B virus. Hybridization of biopsy DNA on nitrocellulose filters with ^{32}P-labelled plasmid hepatitis B virus probe permitted the detection of as little as 1 pg of hepatitis B viral DNA sequence. The viral DNA was found in two major molecular size classes, of about 3 400 and about 4 100 base pairs. Although the specific function of these DNAs has not been established, neither appeared to be integrated into the host genome. The technique used by Shouval et al. (1980) is approximately 10^3

times more sensitive than any known serological test for hepatitis B, and therefore when serological markers of infection with this virus are absent, hybridization analysis could be used to identify viral DNA in the liver. Such studies have not been reported in patients with unexplained chronic hepatitis or cryptogenic cirrhosis, but information on the size of viral DNA molecules, copy number per cell and state of integration would be valuable for following the progression of all stages of infection with hepatitis B virus from the onset of acute hepatitis to chronic liver disease. It should be noted, however, that several reports have recently been published on the presence of integrated hepatitis B virus DNA sequences in the surface antigen-secreting PLC/PRF/5 cell line derived from a patient with primary hepatocellular carcinoma (Summers *et al.*, 1978; Brechot *et al.*, 1980; Edman *et al.*, 1980; Chakraborty, Ruiz-Opazo, Shouval & Shafritz, 1980; Marion, Salazar, Alexander & Robinson, 1980b, Shafritz & Kew, 1981).

VIRUSES WHICH ARE PHYLOGENETICALLY RELATED TO HEPATITIS B VIRUS IN ANIMALS OTHER THAN PRIMATES

For over twenty years R. L. Snyder of the Philadelphia Zoo has been studying the eastern woodchuck or groundhog (*Marmota monax*) and has noted the presence of liver cancer in 2 out of 76 animals which lived longer than four years in an established colony. Furthermore, the lesions of chronic active hepatitis and sometimes cirrhosis were usually found in the non-tumour liver tissue (Snyder, 1968). Examination by electron microscopy of sera collected from the captive woodchucks revealed virus particles closely resembling human hepatitis B virus, thereby establishing a second member of this remarkable class of viruses (Summers, Smolec & Snyder, 1978). Human hepatitis B virus and the woodchuck hepatitis virus share the following characteristics: infection with either virus results in the accumulation in blood of large amounts of excess virus coat protein in the form of spherical and tubular particles measuring 20–25 nm in diameter, 40–50 nm double-shelled or solid particles with a nucleo-capsid or core containing double-stranded circular DNA with a gap, a viral DNA polymerase, and both viruses are associated with chronic hepatitis and primary hepatocellular carcinoma. In addi-tion, Werner, Smolec, Snyder & Summers (1979) identified a high degree of antigenic cross-reactivity between the cores of the two viruses but only minor common antigenic determinants on the virus surface protein. A small region of 100–150 base pairs of nucleotide sequence homology, as measured by liquid hybridization, was found in the genomes of the two viruses. It seems likely that this 3–5% of homology represents one or two regions of nearly identical nucleotide sequence. This degree of nucleic acid homology has been detected among papovaviruses, for example between SV40 and BK virus. More recently, Cummings *et al.* (1980) cloned the DNA of human hepatitis virus and the DNA of woodchuck hepatitis virus in the vector λgtWES and subcloned into the kanamycin-resistant plasmid pAO1. Comparison of the recombinant DNAs with authentic virus DNAs by specific hybridization, size and restriction enzyme analysis indicated that the recombinants contained the complete genome of each virus. The nucleotide sequence homology between the two viral DNAs was confirmed with the cloned DNAs. Thus the woodchuck hepatitis virus and the human hepatitis B virus are phylogenetically related. The analogy between the two viruses is even closer when judged by their adaptation in their respective hosts causing persistent infection and close association with chronic liver disease and primary liver cancer.

Yet another virus which is related to human hepatitis B has been described in Beechey ground squirrels (*Spermophilus beecheyi*) in one region in northern California (Marion *et al.*, 1980a). Common features with the human hepatitis B virus include virus morphology, size and structure of the viral DNA, a virion DNA polymerase which repairs a single-stranded region in the double-stranded circular genome, cross-reacting surface viral antigens, antigen-antibody systems similar to hepatitis B e antigen and the core antigen, and persistent infection with viral antigen present continuously in the blood. Because the ground squirrels were only bled and then released, it has not been possible to observe a disease accompanying this virus in the ground squirrel.

Evidence suggesting a fourth member of this class of viruses has been described by Snyder (1979) in black-tailed prairie dogs (*Cynonys ludoviccianus*) of the family Sciuridae, a close relative of the woodchuck. Hepatitis and hepatocellular carcinoma were found in prairie dogs, and positive orcein staining of liver cells implies a surface antigen similar to human hepatitis B virus. Virological studies, however, have not yet been completed. A fifth member of this virus group is being investigated in domestic ducks in the People's Republic of China and in the United States.

Since the human hepatitis B virus has not been grown in tissue culture, these phylogenetically related animal viruses may provide valuable models for the study of viral replication, the mechanisms of persistence of the infection and the pathogenesis of chronic liver disease and primary hepatocellular carcinoma.

The work on viral hepatitis at the London School of Hygiene and Tropical Medicine is generously supported by grants from the Medical Research Council, the Wellcome Trust, the Department of Health and Social Security, the World Health Organization, the National Research Development Corporation and Organon International, Holland.

REFERENCES

ALBERTI, A., DIANA, S., SCULLARD, G. H., EDDLESTON, A. L. W. F. & WILLIAMS, R. (1978). Detection of a new antibody system reacting with Dane particles in hepatitis B virus infection. *British Medical Journal*, **2**, 1056–8.

BIANCHI, L., GEROK, W., SICKINGER, K. & STALDER, G. A. (eds.) (1980). *Virus and the Liver*. Lancaster: MTP Press.

BLUMBERG, B. S., ALTER, H. J. & VISNICH, S. (1965). A 'new' antigen in leukemia sera. *Journal of the American Medical Association*, **191**, 541–6.

BLUMBERG, B. S., FRIEDLANDER, J. S., WOODSIDE, A., SUTNICK, A. I. & LONDON, W. T. (1969). Hepatitis and Australia antigen: autosomal recessive inheritance of susceptibility to infection in humans. *Proceedings of the National Academy of Sciences, USA*, **62**, 1108–15.

BOXALL, E. H. (1980). Maternal transmission of hepatitis B. In *Recent Advances in Clinical Virology*, Vol. 2, ed. A. P. Waterson, pp. 17–29. Edinburgh: Churchill Livingstone.

BRECHOT, C., POURCEL, C., LUOISA, A., RAIN, B. & TIOLLAIS, P. (1980). Presence of integrated hepatitis B virus DNA sequences in cellular DNA of human hepatocellular carcinoma. *Nature*, **286**, 533–5.

CHAKRABORTY, P. R., RUIZ-OPAZO, N., SHOUVAL, D. & SHAFRITZ, D. (1980). Identification of integrated hepatitis B virus DNA and expression of viral RNA in an HBsAg-producing hepatocellular carcinoma cell line. *Nature*, **286**, 531–3.

CUMMINGS, I. W., BROWNE, J. K., SALSER, W. A., TYLER, G. V., SNYDER, R. L., SMOLEC, J. M. & SUMMERS, J. (1980). Isolation, characterisation and comparison of recombinant DNAs derived from genomes of human hepatitis B virus and woodchuck hepatitis virus. *Proceedings of the National Academy of Sciences, USA*, **77**, 1842–6.

DREW, J. S., LONDON, W. T., LUSTBADER, E. D., HESSER, J. E. & BLUMBERG, B. S. (1978). Hepatitis B virus and sex ratio of offspring. The response to hepatitis B virus infection in parents is related to the sex ratio of their children. *Science*, **201**, 687–92.

EDMAN, J. C., GRAY, P., VALENZUELA, P., RALL, L. B. & RUTTER, W. J. (1980).

Integration of hepatitis B virus sequences and their expression in a human hepatoma cell. *Nature*, **286**, 535–8.

GILES, J. P., McCOLLUM, R. W., BERNDTSON, L. W. JR & KRUGMAN, S. (1969). Viral hepatitis: relationship of Australia/SH antigen to the Willowbrook MS-2 strain. *New England Journal of Medicine*, **281**, 119–22.

HAVENS, W. P. JR (1946). The period of infectivity of patients with homologous serum jaundice and routes of infection in this disease. *Journal of Experimental Medicine*, **83**, 441–7.

HESSER, J. E., BLUMBERG, B. S. & DREW, J. S. (1976). Hepatitis B surface antigen, fertility and sex ratio: implications for health planning. *Human Biology*, **48**, 73–81.

LONDON, W. T., DREW, J. S., BLUMBERG, B. S., GROSSMAN, R. A. & LYONS, P. J. (1977a). Association of graft survival with host response to hepatitis B infection in patients with kidney transplants. *New England Journal of Medicine*, **296**, 241–4.

LONDON, W. T., DREW, J. S., LUSTBADER, E. D., WERNER, B. G. & BLUMBERG, B. S. (1977b). Host responses to hepatitis B infection in patients in a chronic hemodialysis unit. *Kidney International*, **12**, 51–9.

MACCALLUM, F. O. (1972). Hepatitis. *American Journal of Diseases of Childhood*, **123**, 332–5.

MARION, P. L., OSHIRO, L., REGNERY, D. C., SCULLARD, G. H. & ROBINSON, W. S. (1980a). A virus in Beechey ground squirrels which is related to hepatitis B virus of man. *Proceedings of the National Academy of Sciences, USA*, **77**, 2941–5.

MARION, P. L., SALAZAR, F. H., ALEXANDER, J. J. & ROBINSON, W. S. (1980b). State of hepatitis B viral DNA in a human hepatoma cell line. *Journal of Virology*, **33**, 795–806.

MURRAY, R., DIEFENBACH, W. C. L., RAINER, F., LEONE, N. C. & OLIPHANT, J. W. (1954). Carriers of hepatitis virus in the blood and viral hepatitis in whole blood recipients. 2. Confirmation of carrier state by transmission experiments in volunteers. *Journal of the American Medical Association*, **154**, 1072–4.

NAZAREEWICZ-DE MEZER, T., KRAWCZYNSKI, K., MICHALAK, T. & NOWOSLAWSKI, A. (1980). Intracellular localization of HB antigens in liver tissue. In *Virus and the Liver*, ed. L. Bianchi, W. Gerek, K. Sichinger & G. A. Stalder, pp. 85–95. Lancaster: MTP Press

NEEFE, J. R., NORRIS, R. F., REINHOLD, J. G., MITCHELL, C. B. & HOWELL, D. S. (1954). Carriers of hepatitis virus in the blood and viral hepatitis in whole blood recipients. 1. Studies on donors suspected as carriers of hepatitis virus and as sources of post-transfusion viral hepatitis. *Journal of the American Medical Association*, **154**, 1066–71

NEEFE, J. R., STOKES, J. JR, REINHOLD, J.G. & LUKENS, F. D. W. (1944). Hepatitis due to injection of homologous blood products in human volunteers. *Journal of Clinical Investigation*, **23**, 836–55.

PAUL, J. R., HAVENS, W. P. JR, SABIN, A. B. & PHILIP C. B. (1945). Transmission experiments in serum jaundice and infectious hepatitis. *Journal of the American Medical Association*, **128**, 911–15.

ROBERTSON, J. S. & SHEARD, A. V. (1973). Altered sex ratio after an outbreak of hepatitis. *Lancet*, **1**, 532–4.

SHAFRITZ, D. A. & KEW, M. C. (1981). Identification of integrated hepatitis B virus DNA sequences in human hepatocellular carcinomas. *Hepatology*, **1**, 1–8.

SHERLOCK, S. (ed.) (1980). Virus hepatitis. In *Clinics in Gastroenterology*, vol. 9. London: W. B. Saunders.

SHOUVAL, D., CHAKRABORTY, P. R., RUIZ-OPAZO, N., BAUM, S., SPIGLAND, I., MUCHMORE, E., GERBER, M. A., THUNG, S. N., POPPER, H. J. & SHAFRITZ, D. A.

(1980). Chronic hepatitis in chimpanzee carriers of hepatitis B virus: morphologic, immunologic and viral DNA studies. *Proceedings of the National Academy of Sciences, USA,* **77,** 6147–51.

SNYDER, R. L. (1968). Hepatomas of captive woodchucks. *American Journal of Pathology,* **52,** 32.

SNYDER, R. L. (1979). Hepatitis and hepatocellular carcinoma in captive prairie dogs (*Cynomys ludovicianus*). Sonderdruck aus Verhandlungsbericht des XXI. *International Symposiums über die Erkrankungen der Zootiere,* pp. 325–8. Berlin: Akademie Verlag.

STOKES, J. JR, BERK, J. E., MALAMUT, L. L., DRAKE, M. E., BARONDESS, J. A., BASH, W. J., WOLMAN, I. J., FARQUHAR, J. D., BEVAN, B., DRUMMOND, R. J., MAYCOCK, W. D'A., CAPPS, R. B. & BENNETT, A. M. (1954). The carrier state in viral hepatitis. *Journal of the American Medical Association,* **154,** 1059–65.

SUMMERS, J., O'CONNELL, A., MAUPAS, P., GOUDEAU, A., COURSAGET, P. & DRUCKER, J. (1978). Hepatitis B virus DNA in primary hepatocellular carcinoma. *Journal of Medical Virology,* **2,** 207–14.

SUMMERS, J., SMOLEC, M. J. & SNYDER, R. L. (1978). A virus similar to human hepatitis B virus associated with hepatitis and hepatoma in woodchucks. *Proceedings of the National Academy of Sciences, USA,* **75,** 4533–7.

WERNER, B. G., SMOLEC, J. M., SNYDER, R. L. & SUMMERS, J. (1979). Serological relationship of woodchuck hepatitis virus to human hepatitis B virus. *Journal of Virology,* **32,** 314–22.

WORLD HEALTH ORGANIZATION (1973). Viral hepatitis. *Report of a WHO Scientific Group. World Health Organization Technical Report Series, No. 512.* Geneva.

WORLD HEALTH ORGANIZATION (1977). Advances in viral hepatitis. *Report of the WHO Expert Committee on Viral Hepatitis. World Health Organization Technical Report Series, No. 602.* Geneva.

ZUCKERMAN, A. J. (1975). Genetic factors in viral hepatitis. In *Human Viral Hepatitis,* 2nd edition, pp, 254–63. Amsterdam: North Holland/American Elsevier.

ZUCKERMAN, A. J. (1978). The three types of human viral hepatitis. *Bulletin of the World Health Organization,* **56,** 1–20.

ZUCKERMAN, A. J. (1979a). The chronicle of viral hepatitis. *Abstracts on Hygiene,* **54,** 1113–35.

ZUCKERMAN, A. J. (1979b). Specific serological diagnosis of viral hepatitis. *British Medical Journal,* **2,** 84–6.

ZUCKERMAN, A. J. & HOWARD, C. R. (1979). Genetic susceptibility to hepatitis B. In *Hepatitis Viruses of Man,* pp. 137–43. London: Academic Press.

ZUCKERMAN, A. J. & TAYLOR, P. E. (1969). Persistence of the serum hepatitis (SH-Australia) antigen for many years. *Nature,* **223,** 81–2.

ZUCKERMAN, A. J., THORNTON, A., HOWARD, C. R., TSIQUAYE, K. N., JONES, D. M. & BRAMBELL, M. R. (1978). Hepatitis B outbreak among chimpanzees at the London Zoo. *Lancet,* **2,** 652–4.

PERSISTENCE OF INSECT VIRUSES

H. F. EVANS AND K. A. HARRAP

*Natural Environment Research Council, Institute of Virology,
Mansfield Road, Oxford OX1 3SR, UK*

INTRODUCTION

One of the features which influences the interaction of insects and their viruses is the discontinuity of the host population. Most insects have a dormant phase where metabolic activity is extremely low and this stage is frequently unavailable for virus infection. In such circumstances the virus has to persist in an infective state until the host is once more available for infection.

Persistence of insect viruses therefore means their survival in the natural environment as in most instances insect viruses do not require intimate host-to-host transfer to retain viability. The means by which insect viruses can persist in nature are important both in the initiation and maintenance of infection in a population. The two key features of this persistence are the preservation of virus infectivity and the ways in which transmission to a susceptible host can occur. These are influenced by many factors and their inter-relationship is examined in this chapter.

The quantity of virus persisting between insect host generations and initiating new infections is a function of a complex interaction between the virus, the host insect and the environment. This is outlined diagramatically in Fig. 1. Initiation of infection in an insect population and the subsequent multiplication of virus are two criteria which measure the success of virus persistence. They are considered here as the beginning and end of the virus persistence cycle, and data will be presented to outline the wide range of susceptibilities of some virus–host interactions. However, the inter-dependence of the various factors must be regarded as central to a consideration of insect virus persistence even though its individual components will be considered separately.

Insect viruses are well known for the very high yield of virus often produced in infections. That apart, the most striking feature of many known insect viruses is the inclusion body. Both of these properties are significant in the persistence of virus in the natural environment. The inclusion body, polyhedron or capsule, certainly

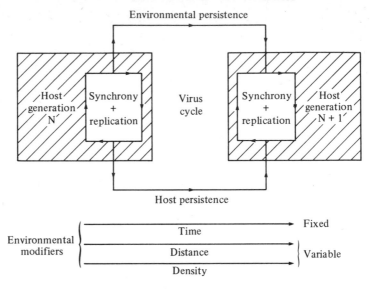

Fig. 1. Schematic representation of factors influencing insect virus persistence.

has a vital role in preserving the infectivity of the virus particles embedded within it. Particular emphasis is given to inclusion body viruses in the larval feeding stage of phytophagous insects, mainly because they have received most attention. Some of these viruses are proving useful as insecticides and their persistence has obvious implications in such developments.

THE PROPERTIES OF INSECT VIRUSES

Structural properties

The majority of the viruses described from insect hosts are of the inclusion body type and this is undoubtedly because the presence of such inclusion bodies in the infected insect has been easily detectable under the light microscope. Most baculoviruses (nuclear polyhedrosis and granulosis viruses), cytoplasmic polyhedrosis viruses and entomopoxviruses are found in proteinaceous pseudo-crystalline inclusions. These are rigid solid structures that are extremely stable. Around 400 isolates of baculoviruses alone have been isolated from insects of different species (Martignoni & Iwai, 1975) though it is not clear from characterization studies how many of them might be identical. As with almost all insect viruses each

Table 1. *Groups of insect viruses*

Group name	Dimensions (nm)	Abbreviation	Insect host order	Nucleic acid	Symmetry of particle	Inclusion body	Cryptogram
Baculoviruses (Nuclear polyhedrosis viruses)	40–140 × 250–400[a]	NPV	Lepidoptera Hymenoptera Diptera Coleoptera	dsDNA	Helical	+	D/2:80/10–15:Ue, o/(E):I/O
Granulosis virus	30–50 × 220–320[b]	GV				+	
Non-occluded nuclear viruses)	120 × 220	—				—	
Cytoplasmic polyhedrosis	48–69	CPV	Lepidoptera Hymenoptera Diptera	dsRNA	Isometric	+	R/2:Σ13–Σ18/25–30:So/S:I/O
Entomopoxviruses	320–400 × 230–250	EPV	Lepidoptera Orthoptera Coleoptera Diptera	csDNA	'Brick-shaped'	+	D/2:140–240/5–6:X/∗:I/O
Iridoviruses	130–180	IV	Lepidoptera Hymenoptera Coleoptera Diptera Ephemeroptera Hemiptera	dsDNA	Isometric	—	D/2:130/15–20:S/S:I/I
Parvoviruses	20–24	DNV	Lepidoptera	ssDNA	Isometric	—	D/1:1.5–2.2/19–32:S/S:I/O
Small RNA viruses	27–39[c]	—	Lepidoptera Hymenoptera Coleoptera Orthoptera	ssRNA	Isometric	—	Various
Rhabdoviruses	140–180 × 70	σ	Diptera	ssRNA	Helical	—	∗/∗:∗/∗:Ue/E:I/C

[a] With envelope.   [b] Without envelope.   [c] Except 'Mini' viruses 13–17 nm.

isolate assumes the name of the host insect from which it was originally isolated. It should not be forgotten that insects harbour other types of viruses where incorporation into an inclusion body is not a feature of infection. The iridoviruses, parvoviruses and small isometric RNA viruses of insects fall into this category. Increasing numbers of the latter group have been reported in recent years and it is likely that many viruses of this type exist that have not been immediately obvious and remain to be described. The various groups of insect viruses with their host range and principal characteristics are shown in Table 1. The properties of insect viruses, in particular those with inclusion bodies, have a significant bearing on their persistence in nature. These properties will be stressed as the viruses are considered group by group. A more detailed analysis of the molecular properties of insect viruses can be found in reviews by Tinsley & Harrap (1978) and Harrap & Payne (1979).

Baculoviruses

The nuclear polyhedrosis viruses (NPV) and granulosis viruses (GV) have inclusion bodies that are either polyhedral or capsular (granule-like) in shape according to type. A further subgroup typified by the virus of the rhinoceros beetle, *Oryctes rhinoceros*, are not found in inclusion bodies. In NPV several virus particles are occluded within the polyhedron whereas in GV one virus particle only is contained in the capsule. The inclusion bodies range in diameter from 0.3μm to 3μm, are refractile, and usually visible under the light microscope. The matrix of the inclusion body is a protein (polyhedrin or granulin) of around 30 000 daltons arranged in a cubic lattice structure clearly visible in electron microscope thin section (Harrap, 1972a). In NPV the virus particles are occluded in this matrix in a random manner but do not affect the regularity of the lattice structure. Inclusion bodies contain protein, DNA, lipid and a number of trace elements (Wellington, 1954; Shapiro & Ignoffo, 1971). Because of their resistance to environmental conditions polyhedra and capsules will retain their infectivity for long periods, for example 20 years stored as dry powders or in flame-sealed glass tubes (Bergold, 1963).

Virus particles released from the inclusion body by treatment with alkali are rod-shaped and the nucleocapsid is enveloped either singly or in groups or 'bundles' of up to 20 or so. Singly enveloped virus particles are about 60–70 nm wide by 260–300 nm in length. The 'bundled' nucleocapsids can be significantly wider than this depending on the number enveloped. The virus envelope has a trilaminar structure typical of a unit membrane despite the fact that it is usually acquired *de novo* in the nucleus. The nucleocapsid shows subunit structure of regular periodicity (Harrap, 1972b; Beaton & Filshie, 1976).

The virus particles contain double-stranded DNA with an average size of around 80×10^6 daltons. Restriction endonuclease analysis of *Baculovirus* DNA is now used to compare and identify different isolates.

Baculovirus proteins have been studied serologically and electrophoretically. The molecular weights of the virus structural polypeptides range from 9 to 160×10^3 and around 30 or so can now be detected. Some polypeptides have been shown to be components of the virus envelope and others of the nucleocapsid. The detailed properties of *Baculovirus* DNA and proteins have been reviewed by Harrap & Payne (1979).

Cytoplasmic polyhedrosis viruses

Over 150 cytoplasmic polyhedrosis virus (CPV) isolates have been described from different insect species, though it is likely that a number of these are the same virus. The polyhedra formed in CPV infections can vary widely in size from less than 0.5μm to several micrometres in diameter. The virus particles are occluded in the polyhedra in large numbers and may often be found close to or on the surface of the inclusion body. In ultrathin section a cubic lattice pattern can be demonstrated which is very similar to that seen in baculoviruses. The major protein of the inclusion body matrix (polyhedrin) is a polypeptide with a molecular weight in the range $25-30 \times 10^3$ (Payne & Rivers, 1976). The virus particles are spherical with diameters in the range $50-65\mu$m and bear large spikes at the icosahedral vertices. Virus particles contain 25–30% double-stranded RNA in 10 pieces ranging in size from $0.3-2.6 \times 10^6$ daltons, total RNA size usually being in the range $12-16 \times 10^6$ daltons. Three to six proteins are found of molecular weights $28-163 \times 10^3$. The virus particles also contain several enzyme activities. A detailed review of the proteins and nucleic acids of cytoplasmic polyhedrosis viruses can be found in Harrap & Payne (1979).

Entomopoxviruses

Insect poxviruses (EPV) are generally found in crystalline proteinaceous inclusion bodies (spherules) which can be 10–20 μm in diameter and are analogous to those of baculoviruses and cytoplasmic polyhedrosis viruses. There is some evidence of a definite arrangement of the virus particles within the inclusion body in contrast to nuclear and cytoplasmic polyhedrosis viruses. Inclusion body protein molecular weights have been reported of between 90 and 112×10^3 (Langridge & Granados, 1978). The morphology of the virus particles is similar to that of vertebrate poxviruses with a surface 'mulberry' appearance, one or two lateral bodies, and 'ropelike' internal structures all being described (Bergoin & Dales, 1971). The virus particles contain large double-stranded DNA molecules ranging in average size from 134.7×10^6 to 200.4×10^6 daltons depending on the isolate. There are reports of up to 40 structural polypeptides and a number of enzyme activities analogous to those found in vertebrate poxviruses (Harrap & Payne, 1979).

Iridoviruses
The iridoviruses or insect iridescent viruses (IV) are large viruses with a distinctive morphology causing infection in insects without the production of proteinaceous inclusion bodies. Virus is found in the cytoplasm of many tissues but a main site of replication is often the fat body. Virus particles occur freely in the tissues, often in such concentration that a blue to yellow-green iridescence is produced that typifies the infection. In some hosts the quantity of virus produced can be as much as 25% of the dry weight of the whole larva (Williams & Smith, 1957). Surface subunit structure has been demonstrated on certain insect iridescent virus isolates and arrangements involving around 1 500 subunits have been proposed (Wrigley, 1969, 1970; Manyakov, 1977). The virus particles have an internal lipid-containing membrane possibly surrounding the core.

The virus particles contain between 10.5 and 19% DNA, the molecule being linear and double-stranded. Many of the iridescent viruses have more than 20 structural proteins (Elliott, Lescott & Kelly, 1977; Kelly *et al.*, 1979; Carey *et al.*, 1978) though the larger size group appears to have only nine (Wagner, Paschke, Campbell & Webb, 1974). In general the sizes of the structural proteins fall in the range 11 000–220 000 daltons. A number of enzymes have been found associated with the smaller iridescent viruses.

Small isometric viruses
Increasing numbers of small isometric viruses have been reported from insects in the last 15 years. Such viruses contain either DNA

(insect parvoviruses) or RNA. The insect parvoviruses are assigned to the *Densovirus* genus of the Parvoviridae (Matthews, 1979). This genus contains the densonucleosis virus (DNV 1) of the greater wax moth (*Galleria mellonella*), the related viruses of *Junonia coenia* (DNV 2), and *Agraulis vanillae* (DNV 3) (Longworth, 1978). There is a striking difference in host range between DNV 1 and DNV 2. *In vivo* DNV 1 is restricted to *G. mellonella* whereas DNV 2 will infect several lepidopterous larvae but not *G. mellonella*. DNV 1 is very virulent and many host tissues are affected; the larvae being killed in 4–6 days at 28 °C.

Longworth (1978) has divided the small isometric RNA viruses into seven subgroups according to their biophysical and biochemical properties. The pathology of the infections caused by small RNA viruses of insects is inadequately described. The viruses develop in the cytoplasm and in many instances the gut is the principal target tissue, though nervous (including brain) and muscle tissue can also be involved. In gut infections vomiting and anal discharge of virus certainly aid its dispersal. Some of the viruses, for example bec viruses, are rather host specific (Bailey, 1976) whereas others have a broad host range, for example cricket paralysis virus (Longworth, 1978) and the specific insect name adopted for the virus is misleading. Documented reactions between small insect RNA viruses and naturally occurring antibodies in vertebrate sera (Longworth *et al.*, 1973) perhaps give a hint of the ubiquitous nature of these viruses or their close similarity to viruses commonly found in vertebrates. However, details of the ecology of infections by the viruses are lacking, possibly because of the difficulties of rapid diagnosis and identification. Detailed biophysical and biochemical data are now available on a number of the isolates so ecological studies should become feasible as identification methods are developed. It is no purpose of this chapter to review the detailed chemical properties of small isometric insect viruses. For such information the reader should consult the reviews of Longworth (1978) and Harrap & Payne (1979).

Routes of infection

In nature the insect larva may acquire virus by mouth, transovarially, or through injury. The most usual route of infection is by mouth, usually as a result of contamination of the insect's food source with virus. In the laboratory successful infection is often achieved by injection even though this is not likely to be a common route of entry except, for example, by the occasional activities of parasites. In this section what is known about the fate of virus after ingestion will be considered. The dynamics of how the virus persists prior to ingestion and the vertical transmission of virus from generation to generation will be considered in later sections.

All occluded insect viruses (and certain of the viruses not forming inclusion bodies) can successfully infect the insect host via the mouth. The inclusion bodies of baculoviruses, cytoplasmic polyhedrosis viruses and entomopoxviruses dissociate in the gut lumen both because of the alkaline conditions there and as a result of enzyme action, particularly alkaline protease associated with the inclusion bodies. The virus particles are released and can be found in association with the absorptive microvilli of the gut columnar cells (Stoltz & Hilsenhoff, 1969; Harrap, 1970; Granados, 1973; Kawanishi, Summers, Stoltz & Arnott, 1972). With baculoviruses and entomopoxviruses fusion of the microvillus membrane with the virus envelope has been demonstrated and the virus nucleocapsid or core appears to 'migrate' along the microvillus into the cell so initiating infection. In granulosis virus infections naked nucleocapsids have been observed situated 'end-on' to nuclear pores and, as full and partly empty nucleocapsids were found in this position, it was postulated that the DNA content of the nucleocapsid was injected into the nucleus through the nuclear pore (Summers, 1969, 1971). In baculovirus infections of Hymenoptera and cytoplasmic polyhedrosis virus infections where the gut is the principal target tissue, histopathological events can be seen leading to the formation of polyhedra containing virus particles. In *Baculovirus* infection the nucleus enlarges and an unusually dense network of granular nucleoplasm forms, the virogenic stroma, from which nucleocapsids are derived. These become enveloped either singly or in bundles within the nucleus and polyhedrin crystallizes leading to the production of polyhedra (Harrap, 1972c). Eventually the cell may lyse and polyhedra and other virus components are liberated. With granulosis virus infection the cytoplasm is also a site of virus replication. In cytoplasmic polyhedrosis virus infections the virogenic areas occur in regions of the cytoplasm which become devoid of cell organelles and contain only areas of viroplasm and the virus particles (Stoltz & Hilsenhoff, 1969). The virus particles become occluded by crystalline polyhedrin to form the polyhedra.

In Lepidoptera, haemocytes, tracheal cells, fat body and

hypodermis are all infected by baculoviruses. The gut is not a primary target tissue though a localized infection involving the nucleus does occur at the site of virus entry though polyhedra are not found (Harrap & Robertson, 1968). Enveloped virus can be found at the base of the columnar cells and external to these cells in the basal lamina and adjacent to tracheal epithelium cells in which conventional *Baculovirus* morphogenesis with polyhedron formation occurs. A variation of this pathway of infection has been proposed (Granados & Lawler, 1981) which envisages two populations of virus particles; those that replicate in the columnar cell nucleus and those that cross the cell cytoplasm directly from the microvillus to initiate infection in the tracheal cells.

In nature it has been difficult to substantiate that iridescent virus infection is acquired by mouth. In the laboratory intrahaemocoelic injection is the infection technique of choice though mosquito larvae are usually infected from an aqueous virus-contaminated environment. Stoltz & Summers (1971) tried to follow the infection pathway of an ingested iridescent virus in mosquito larvae but found that virus was often degraded in the midgut and failed to penetrate the peritrophic membrane and come into contact with the gut cells. Accidental injury or predatory behaviour may be important in iridescent virus gaining access to susceptible tissues but it is perhaps less likely than cannibalism or feeding on infected cadavers (Linley & Nielsen, 1968; Carter, 1973).

In nature parvoviruses and the small RNA viruses of insects almost certainly initiate infection after oral ingestion and in many cases infection can be established by this means in the laboratory. However, there is no detailed information on the uptake of virus in the insect gut for these two groups of viruses. Intrahaemocoelic injection is often a much more effective way of infecting a laboratory culture of insects. Incubation time is often shorter and a more dilute inoculum virus suspension can be used than is the case when the virus is fed to the insect.

In summary then, it is clear that in nature virus acquisition by insects occurs most commonly through the mouth, usually of the feeding larva. In the laboratory *per os* infection is still the method of choice for all viruses occluded in polyhedra or capsules as these must dissociate in the gut lumen to provide infective virus. With non-occluded viruses, or with preparations of virus particles released from the inclusion bodies or obtained directly from infected cell cultures, intrahaemocoelic injection is the method of choice, since circumventing the gut lumen increases the speed of action of the virus and a more dilute inoculum virus suspension will bring about lethal infection.

DYNAMICS OF VIRUS PERSISTENCE IN INSECT POPULATIONS

Synchrony of host and virus

Most information on susceptibility of insects to pathogenic viruses relates to the baculoviruses. A common feature of NPV infection in the Lepidoptera is the wide range of susceptibility between the first and final instars. Thus, for the NPV of cabbage moth, *Mamestra brassicae*, LD_{50} dosages range from 7 polyhedra in the first instar to approximately 240 000 polyhedra in the fifth instar, a factor of 34 000 times in range of susceptibility (Evans, 1981). The susceptibility range of Hymenoptera is much narrower and factors as low as 10 from first to fifth instar have been observed for *Gilpinia hercyniae* (P. F. Entwistle, P. H. W. Adams & H. F. Evans, unpublished information).

The modifying influence of larval body weight on dosage responses is commonly observed. For example 90% of the observed variation in susceptibility of *M. brassicae* to its NPV was removed when results were expressed as LD_{50}/mg body weight (Evans, 1981; see also Allen & Ignoffo, 1969, NPV of *Heliothis* sp. and Wigley, 1976, NPV of *Operophthera brumata*). Work in our Institute on the LD_{50} responses of *G. hercyniae* suggests that sawflies also appear to have weight-related changes in response despite having virus replication confined to the midgut. The increasing maturation resistance with larval age during a host generation reduces the effectiveness of a constant pool of inoculum which therefore has to be augmented by release of secondarily produced virus. This stage is important in defining the population of virus available both within and, more significantly, at the end of a host generation.

Fully mature lepidopterous larvae infected with NPV produce up to 1.5×10^{10} polyhedra at death compared with a peak of 2×10^8 in Hymenoptera at an equivalent stage, while in both orders production is related directly to larval weight. Productivity of CPV polyhedra is generally lower than NPV, probably as a result of infection being confined to the gut.

Release of virus into the environment depends on the primary site of infection. In lepidopterous larvae NPV replicates in the majority of body tissues except the gut (see section on 'Routes of infection') and the insect eventually consists of a fragile 'bag' of virus which ruptures easily, releasing the liquefied body contents containing polyhedra. In Hymenoptera the gut only is infected and polyhedra may be released gradually in frass as the gut epithelial cells break down. Virus-killed larvae generally stick to their feeding substrate and are broken down slowly, releasing polyhedra over a long time period.

The total virus produced from a given level of infection in a single host generation depends on the interaction of a number of variables; for example, the initial quantity of virus present, the age structure and density of the host, and any factors which reduce the effectiveness of released virus. Studies which estimate some or all of these parameters, and hence allow a quantitative assessment of the amounts of inoculum persisting, are rare.

Detailed investigations of an epizootic of NPV in spruce sawfly, G. hercyniae, populations in mid-Wales forests have included measurement of many of these parameters which govern virus persistence in an insect population (Evans & Entwistle, 1982). The relationship between inoculum produced in one year and virus present at the start of the next was curvilinear with a decreasing rate of virus retention at higher inoculum concentrations. This may reflect a relative saturation of the carrying capacity of foliage for infective virus, particularly since virus showed greater retention on Norway spruce than on Sitka spruce.

The question remains as to what effect the overwintered inoculum had on the spruce sawfly population on the trees. This was tested by regular quantitative sampling of the host population to establish the rate of build-up of infection during the larval period (Entwistle, Evans & Adams, 1980). The results established a link between virus production in one year and primary infection in the next. The changing patterns of response reflected the declining host populations as the virus epizootic progressed resulting in less virus persisting between generations and fewer larvae available to ingest the virus.

Although the data have yet to be analysed fully, a few simple calculations based on known characteristics of the spruce sawfly epizootic in Wales serve to illustrate the true potential of available inoculum for intergeneration persistence. Although in excess of 10^{14}

polyhedra were produced per hectare of forest, extrapolation from known lethal dosages and observed infection levels indicated that 0.00025% only of these were available during the following year. Clearly, virus persistence in terms of polyhedra effectively utilized is extremely low. However, as we have shown above, the true measure of success is the rate of infection induced in the new host population. It is, therefore, pertinent to discuss in some detail the underlying mechanisms which govern the amount of virus persisting in the environment. These will be divided into a number of interrelated categories including environmental factors, secondary hosts, specialized environment and host-mediated factors.

Environmental factors

Here we consider the environmental attributes which affect the persistence of extra-host virus both within and between generations. Of prime importance is the immediate fate of virus following replication in the host. Virus may be lost immediately from the host environment, for example by being removed with crop plants or by deposition outside the normal sphere of the host, when the persistence probability is extremely low. At the other extreme virus may be protected from environmental degradation as soon as it is released. For example, the larvae of wattle bagworm, *Kotochalia junodi*, (Ossowski, 1957) and of *Dasychira basiflava* (Kaya & Anderson, 1976) die from NPV in larval feeding nests which provide a protective site for both host and virus. Virus persistence between these extremes is a function of the food source and of the feeding habits of the host insects.

Both time and space considerations modify the probability of virus persisting. Time will have importance particularly in an unstable or open environment where virus degradation may proceed rapidly and synchrony of virus and host is short-lived. Space also influences persistence since one of the longer-term features of virus epizootics in insects is a decline in the numbers of susceptible individuals at the epicentre of disease. For survival the virus has to both persist and disperse efficiently to exploit the changing host population. The immediate environment of the virus produced is therefore a major factor in persistence and features such as host plant physico-chemical characteristics and growth as well as the nature of the soil beneath the plants must be considered.

Table 2. *Gross host plant characteristics influencing potential persistence of insect viruses*

Host plant characteristic	Stability	Plant structure for virus persistence
Evergreen tree	} permanent	} foliage, bark & epiphytes
Evergreen shrub		
Deciduous tree		} bark & epiphytes
Deciduous shrub		
Perennial crop	} semi-permanent rotation	} foliage, bark & epiphytes
Annual crop	} temporary	} none
Grass	} permanent	} foliage

Host plant characteristics

Plants are extremely complex sites for virus persistence. Gross characteristics relevant to virus retention are outlined in Table 2 where we have considered the permanency of the plant and its constituent parts and related these to long-term persistence of virus outside the living host. The presence or absence of leaves, or of the entire host plant, has a major influence on persistence and the micro-environment of the plant itself may be equally important. Degree of protection from ultra-violet light, surface temperature, surface structure and chemical nature are primary components of this micro-environment.

Presence of foliage throughout the year provides the best examples of long-term retention of virus on the plant itself. Studies of forest pests, particularly on conifers, have shown that viable inoculum may be retained on foliage between larval generations, a period which may extend to several months; for example, *Orgyia leucostigma* NPV on balsam fir foliage (Elgee, 1975) and the NPV of *G. hercyniae* on spruce foliage in Wales (Entwistle & Adams, 1977). However, Bird (1961) in Canada, also studying the NPV of *G. hercyniae*, stated that no virus was retained on foliage between host generations. A major source of foliar inoculum may be the disintegration of virus-killed larvae (Entwistle & Adams, 1977) and the poor persistence of *G. hercyniae* NPV in Canada could be attributable to the absence of virus-killed larvae at the start of the new larval season (Bird, 1961). Other factors including the different spruce species, geographical locations and the enzootic NPV status in Canada probably contributed to this.

Both impure and pure preparations of NPV polyhedra will adhere strongly to coniferous foliage. Evans & Entwistle (1982) dipped Norway spruce foliage in pure and impure polyhedral suspensions and counted polyhedra on needles over an extended time period. Counts showed an exponential rate of loss with half-lives for pure deposits ranging from 23 days in summer to 38 days in winter while significantly longer retention (half-life, 55 days) was observed for impure suspensions in winter. The disintegration of virus-killed larvae contributes to a reservoir of inoculum adhering directly to foliage and is evidenced by the demonstration of the movement of polyhedra between branches on spruce trees (Evans & Entwistle, 1982).

Physical presence of polyhedra may not necessarily imply that activity has been retained. A series of experiments involving bioassay of foliage dipped in impure virus suspensions of G. hercyniae NPV indicated that the increased mortality observed in test larvae fell to background levels in approximately 16 weeks. Although the sets of data on physical presence and infectivity are not comparable directly the latter experiment suggests that a high proportion of polyhedra adhering to spruce foliage retain activity.

Despite the loss of potential sites for virus retention at leaf-fall there is evidence for good virus persistence on deciduous trees. One of the earliest examples was provided by Clark (1956) who demonstrated that the NPV of *Malacosoma fragile* persisted overwinter on the bark of quaking aspen. Recently Podgwaite, Shields, Zerillo & Bruen (1979) showed that bark was an important site for overwinter persistence of *Lymantria dispar* NPV on oak and red maple. Further protection of virus may be provided by epiphytic mosses (NPV of *O. brumata* on oak, Wigley, 1976), by debris mats resulting from the host feeding (NPV of *L. dispar*, Doane, 1970) or by feeding nests containing living and virus-killed larvae (NPV of *K. junodi* on wattle, Ossowski, 1957; NPV of *D. basiflava* on oak, Kaya & Anderson, 1976).

Demonstration of long-term foliar persistence seems to be an exception and virus generally retains its activity on foliage for short periods only (for review of early literature see Tanada, 1971). More recently, in studying the NPV of *Orgyia pseudotsugata* on douglas-fir, Thompson & Scott (1979) showed that although larvae died and released virus onto foliage, its activity was soon lost. The authors assumed this was the result of inactivation by solar radiation.

In contrast to the situation on a tree, virus is rapidly inactivated on the foliage of most crop plants which, compounded with the temporary nature of the plant itself, precludes long-term persistence at the host feeding site. The major mechanisms for intergeneration virus persistence in these cases are retention of virus in the surface layers of soil and host-mediated transmission. We are, therefore, more concerned here with intrageneration persistence on the plants themselves which is significant in the initiation of primary and secondary infection and virus multiplication.

Jaques (1975) has reviewed most of the early work on virus persistence in agro-ecosystems and concludes that 'deposits of NPV and GV on foliage in a field environment retain activity for relatively short periods, usually less than two weeks'. Paramount in this rapid inactivation are the effects of sunlight, especially its ultra-violet light component, on the virus deposits.

Deposits of virus adhere strongly to plant surfaces and on cabbage plants have been shown to be remarkably resistant to rain and physical abrasion (David & Gardiner, 1966). However, the NPV of *Trichoplusia ni* and the GV of *Pieris rapae* lost 50% of the infectivity within two days and were almost entirely inactivated by 10 days (Jaques, 1975). When virus-contaminated collard plants were grown in a dark room the viruses remained active for at least 20 days, confirming that sunlight was a significant factor in field crops (Jaques, 1967, 1972).

Loss of activity was rapid for the NPVs of *Heliothis* sp. on cotton plants (Bullock, 1967), corn silks (Ignoffo *et al.*, 1973) and soybean plants (Ignoffo, Hostetter & Pinnell, 1974). Elnagar & Abul Nasr (1980) showed, by bioassay, that purified *Spodoptera littoralis* NPV sprayed onto cotton plants in Egypt lost over 50% of its activity within 24 hours while impure and shaded preparations retained between 60% and 80% activity.

Protection of virus from sunlight may be achieved in a number of ways, all related to provision of shade. Reed (1971) demonstrated the presence of *Phthorimaea operculella* GV capsules in the stomatal cavities of potato plants in Australia, while the complex structure of sorghum heads was said to provide protection for *Heliothis armigera* NPV (Roome & Daoust, 1976). Selective placement of *H. armigera* NPV on the underside of cotton leaves gave increased persistence (Stacey, Young & Yearian, 1977a).

There has been considerable experimentation to identify the components of sunlight which inactivate occluded viruses. Much of this work has been done '*in vitro*' using deposits of purified virus exposed to artificial light or sunlight. The peak wavelength for virus inactivation is in the far UV around 2500Å, a wavelength emitted by bactericidal lamps. The major component of solar radiation penetrating the atmosphere is in the near ultraviolet (3000–3800Å), although a small amount of far UV (1900–3000Å) may be present intermittently (Barker, 1968). Witt & Stairs (1975) showed that the continuous presence of near UV may have a cumulative effect on virus, with most rapid inactivation occurring between 10 and 30 hours, contrasting markedly with the extremely short duration of virus exposed to far UV. Ignoffo et al. (1977) demonstrated that although the half-lives of various insect viruses were less than 4 hours when exposed to UV light at 2540Å, they could be ranked EPV > NPV = CPV > GV in order of stability.

Sunlight was implicated as the major component in inactivation of Heliothis NPV on cotton (McLeod, Yearian & Young, 1977) but was complicated by leaf exudates which in themselves had deleterious effects on NPV deposits (Young, Yearian & Kim, 1977). Plant exudates as modifiers of virus persistence are receiving more attention in the light of work on the cotton/NPV system. Cotton leaves produce dew which occasionally has a pH in excess of 9.3 and during drying on the plant causes substantial inactivation of virus (McLeod et al., 1977). These authors postulated that high pH alone was not necessarily the reason for this and that the observed inactivation resulted more from exposure to high concentrations of metallic ions.

These factors would depend on the nature and duration of rainfall, relative humidity and age of the plant (see Godfrey (1976) for discussion of plant microclimate in relation to microbial adhesion).

Soil persistence

Persistence of insect viruses in soil was mentioned as a source of inoculum by Komarek & Breindl (1924) and Prell (1926) who suggested that viruses in soil could initiate epizootics in insect pests. Since that time many workers have speculated on the potential of soil as a major reservoir of virus. Tanada (1971) and Jaques (1975) have reviewed the earlier work carried out in this area.

It is generally true that the ultimate fate of viruses of phytophagous insects is to accumulate in the soil. This is probably the case regardless of the life style of the host or of the type of plant on which it feeds.

A characteristic of soil is that it provides a medium for strong adsorption of virus which remains in high concentration near the soil surface; for example, *T. ni* NPV in cabbage plots (Jaques, 1975). Laboratory experiments with columns of soil confirmed that very little leaching of virus from soil occurs (Jaques, 1969, *T. ni* NPV; David & Gardiner, 1967, *Pieris brassicae* GV) while Harcourt & Cass (1968) showed that soil disturbed by cultivation retained less virus as a result of mixing of surface with deeper soil, although discing of soil had little effect on retention of *T. ni* NPV and *P. rapae* GV in the upper layer of soil (Jaques, 1970).

Soil pH can have a significant effect on the length of time viruses persist. Lower pH values inactivated *T. ni* NPV more rapidly than higher values (Thomas, Reichelderfer & Heimpel, 1973). However, Jaques & Harcourt (1971) concluded that pH had little effect on *T. ni* viruses but *P. rapae* GV persisted better in soils of pH 5.1–6.0 than in more alkaline soils. Hukuhara & Wada (1972) studying the CPV of *Bombyx mori* showed that more polyhedra adsorbed to soil at acid pHs (pH 4.9–5.6) than at higher values. These apparently contradictory findings highlight the difficulty of comparing physical presence of virus inclusion bodies in soil and their infectivity in, probably, greatly different soils. Evans, Bishop & Page (1980), in developing a method for counting NPV polyhedra accurately in soil, showed, by bioassay against standard virus preparations, good correlations between polyhedra counted and their infectivity.

The mechanism of adsorption of virus to soil is poorly understood. Jaques (1975) assumed that polyhedra were adsorbed to clay particles, while Hukuhara & Wada (1972) showed that the amount of adsorption of CPV polyhedra to soil increased with decreasing pH. It is more difficult to remove NPV polyhedra from acid forest soils than from agricultural soils of a higher pH but this can be reversed partially by buffering the soil prior to processing (H. F. Evans, unpublished data).

Conclusive evidence that virus adsorbed to soil initiates infection in host insects is dependent on the system studied. Its role is most clearly demonstrated in grassland where the relatively undisturbed environment increases the effectiveness of virus produced on death of host insects. Hurpin & Robert (1972) showed that an Entomopoxvirus of the cockchafer *Melolontha melolontha* persisted well from season to season. Crawford & Kalmakoff (1977) demonstrated the importance of a stable soil habitat in enhancing the effect of NPV on *Wiseana* sp.

Soil has good potential as a virus reservoir in agricultural crops, especially where plant leaves may be in direct contact with it. Soil acted as a source of infection for the NPVs of *T. ni* (Jaques, 1975), *Pseudoplusia includens* (Young & Yearian, 1979), *P. operculella* (Reed, 1971, Reed & Springett, 1971) and a variety of alfalfa pests (Tanada & Omi, 1974). *M. brassicae* NPV accumulates in the top 10 cm of soil directly below cabbage plants carrying infected larvae and up to 90% of the virus produced can be accounted for by this route (H. F. Evans, unpublished information).

Virus persistence in soil of stable habitats, not liable to cultivation and disturbance each year, has been described by several authors. Examples include fall webworm, *Hyphantria cunea* NPV (Hukuhara, 1973; Hukuhara & Namura, 1972); *L. dispar* NPV (Podgwaite *et al.*, 1979) and *O. pseudotsugata* NPV (Thompson & Scott, 1979). Kaupp (1980) counted the numbers of polyhedra of *Neodiprion sertifer* entering soil beneath pine trees and recovered up to 9×10^7 polyhedra from a 10 cm diameter collection pot.

In general, it is not clear whether, or how much, virus in soil is available for initiating infection in forest pests. Foliage is located progressively further from the virus source and transport mechanisms have to be postulated to explain movement of virus back to the trees. These would include the host insect itself, small mammals, birds, invertebrate parasites and predators, wind blow of dry soil and rain splash at canopy edges. Conclusive evidence for the existence of such mechanisms is scant.

Secondary hosts

This mode of virus persistence includes any other host, passive or active, which provides a means of protection for the virus while it is outside its primary host. In this context the availability of virus carried by the secondary host is a significant factor since it may be removed from the immediate environment of the primary host. Thus time and distance constraints are involved which will ultimately determine the contribution made by secondary host inoculum.

Three modes of secondary host passage of virus can be identified.

Virus replicates in the secondary host and is transmissible back to the primary host. The essential components of this mode are synchrony of both hosts in space and time and no modification of the virus.

Cunningham (1968) showed that the baculoviruses of a number of nymphalid butterflies on nettle near Oxford were serologically related and speculated that they had acquired a single baculovirus as a result of feeding on the same host plant. However, most studies indicating potential cross-transmission and persistence in alternate hosts have been carried out experimentally under laboratory conditions. Examples include cross-transmission of CPV from *L. dispar* to 10 species of Lepidoptera (Aratake & Kayamura, 1973), transmission of *Autographa californica* NPV to at least 15 species of Lepidoptera in six families (Vail & Jay, 1973; Entwistle, 1978) and, by injection, transmission of *Chilo* iridescent virus to non-insect arthropods (Ohba & Aizawa, 1979). Tanada (1971) demonstrated cross-transmission of field-collected GV and NPV between several alfalfa pests in California and stated 'the presence of alternate hosts enables the virus to persist longer and more effectively in the insect ecosystem'.

Assuming that cross-transmissions occur in the field, virus in the secondary host may be subject to the same problems of long-term persistence as the first. However, it is possible that the life style of the secondary host could confer significant advantages for virus persistence if, for instance, it overwinters as an infected larval stage, via the secondary host adult stage or has more generations per year, hence allowing greater virus production and persistence.

On present evidence, reciprocal transmission between hosts remains more a potential than an actual means of virus persistence. A major problem in citing reciprocal transmission is the lack of sound identification of both the primary and secondary viruses involved (see section on 'Host-mediated features'). Tinsley (1979), in reviewing the potential of insect pathogenic viruses as pesticidal agents, advocates strongly the wider use of reliable diagnostic techniques in studies of virus specificity.

Secondary hosts ingest virus which remains unaffected in the gut and is passed into the environment in an infective state. This is a relatively short-term persistence mechanism and provides a means for rapid transfer of virus during the period when susceptible primary hosts are present.

Early examples include passage of infective NPV through guts of the predators *Rhinocoris annulatus* (Franz, Krieg & Langenbruch, 1955) and *Ephippiger bitterensis* (Vago, Fosset & Bergoin, 1966). Birds have been strongly implicated in transmission of the NPV of

G. hercyniae in spruce forests in Wales (Entwistle, Adams & Evans, 1977a, b).

Lautenschlager & Podgwaite (1979) demonstrated passage of infective *L. dispar* NPV by five mammal and three bird species. Entwistle, Adams & Evans (1978) demonstrated prolonged passage of *G. hercyniae* NPV (up to seven days) by blue tits fed infected larvae. Passage of NPV through mammalian guts does not always take place and Smirnoff & MacLeod (1964) showed that the NPV of *Neodiprion swainei* was destroyed in mouse gut.

Secondary hosts are surface contaminated with virus which is transmitted by direct or indirect contact with the primary host. Excellent examples are provided by insect parasitoids which may oviposit in an infected host and pass virus via the ovipositor to other susceptible hosts. *Apanteles melanoscelus* can transmit NPV between *L. dispar* larvae during oviposition (Raimo, 1975; Raimo, Reardon & Podgwaite, 1977) and *Campoletis sonorensis* transmits NPV between *Heliothis virescens* larvae (Irabagon & Brooks, 1974). Other examples are cited by Vinson & Iwantsch (1980) who reviewed host suitability for insect parasitoids.

Thus in the short term, parasitoids, by virtue of transmission directly to the primary host, are potentially highly efficient agencies of virus persistence. However, the relative rates of development of parasitoid and virus in the same host individual are critical factors in the interaction. If the parasitoid develops faster than the virus the amount of inoculum generated may be small. This occurs with *Hyposoter exiguae*, an ichneumonid parasitoid of *T. ni* which has an accelerated rate of development in infected hosts (Beegle & Oatman, 1975). However, despite the lower immediate yield of NPV in *T. ni*, it was shown that 40% of female parasites developing from infected larvae transmitted infective dosages to 65% of healthy larvae which they encountered. This would compensate for reduced virus yield in the original host.

Evidence for large-scale transmissions of virus in field populations of insects is often circumstantial. Reardon & Podgwaite (1976) have shown strong correlations between NPV incidence in *L. dispar* populations and the presence of the parasites *A. melanoscelus* and *Parasetigena silvestris*. They suggest that the broad geographical range and multivoltine characteristics of *A. melanoscelus* would provide a means of increasing virus incidence in widely dispersed host populations.

Insects in specialized environments

Specialized environments can be defined broadly as those attributable to human activities and which do not occur naturally. We include stored product warehouses, greenhouses, museums, etc., under this heading although examples of virus–host interactions are found mainly in the first category.

Persistence of insect viruses in specialized environments is dependent on two main factors, namely the temporal stability of the food source and the seasonal presence of the host insect. In stored product warehouses the length of storage time of the various commodities will depend on supply and demand which may have a cycle many times longer than the generation time of storage pests. In such situations virus released by death of hosts may remain viable in the food medium for several host generations. The moth, *Plodia interpunctella*, was controlled by application of GV to stored grain (McGaughey, 1975) and persisted in an active state for up to 12 months at normal storage temperatures (Kinsinger & McGaughey, 1976).

The mechanism of virus persistence in stored grain is unknown but, shielded from UV inactivation and kept at relatively constant temperatures and relative humidity, the environment would appear to be ideal for retaining virus infectivity. Infection in the host will therefore depend more on the distribution of host and virus than on absolute amounts persisting in the stored commodity. Thus, McGaughey (1975) found that treatment of grain with GV to a depth of 100 mm gave control equal to treating the entire stock.

The extremely low economic threshold for crops grown in greenhouses would appear to preclude the use of viruses for control of pests in this environment. However, as pointed out by Franz (1971), the increasing resistance of some greenhouse pests to chemical insecticides might make the use of viruses a viable proposition in the future.

Host-mediated features

Vertical transmission
Virus may persist in an insect population by vertical transmission from generation to generation either by transovarian transmission from mother to egg within the ovary or by transovum transmission outside the ovary, generally during oviposition (Martignoni & Milstead,

1962). In transovum transmission the larva acquires virus as it chews through the egg case to hatch, and there are numerous examples of this for insect viruses, the evidence for which often comes from the elimination of virus transmission between generations by surface sterilizing the eggs usually with sodium hypochlorite or formaldehyde. A particularly convincing example can be found with NPV of *Colias eurytheme*, the alfalfa caterpillar (Martignoni & Milstead, 1962) in which the female genitalia were deliberately contaminated with virus. Other examples of transovum transmission of NPV are in *Heliothis zea*, the corn earworm, (Hamm & Young, 1974) and in *N. swainei* a hymenopteran (Smirnoff, 1962). Transovum transmission of GV has been reported in the codling moth, *Laspeyresia pomonella* (Etzel & Falcon, 1976), in the large cabbage white butterfly, *Pieris brassicae* (David & Taylor, 1976), and in the corn borer, *Sesamia nonagroides* (Melamed-Madjar & Raccah, 1979). There are several instances reported on transovum transmission of CPV. For example, in the tobacco budworm, *H. virescens* (Sikorowski, Andrews & Broome, 1973; Simmons & Sikorowski, 1973; Mery & Dulmage, 1975) and in the pink bollworm, *Pectinophora gossypiella* (Bullock, Mangum & Guerra, 1969). With non-occluded insect viruses transovum transmission has been claimed with mosquito iridescent virus in the mosquito, *Aedes taeniorhyncus* (Linley & Nielsen 1968), and with a small isometric RNA virus of the field crickets, *Teleogryllus oceanicus* and *Teleogryllus commodus* (Reinganum, O'Loughlin & Hogan, 1970).

Reports of transovarian transmission, that is of virus within the egg, are fewer and less convincing. Indeed, Longworth (1973) in a review of the topic could find two instances only where transovarian transmission seemed to have been demonstrated convincingly. One example described by Rivers (in Smith, 1967 and personal communication) was with larvae of the atlas moth, *Attacus atlas*, which failed to emerge from the egg and contained large numbers of nuclear polyhedra when examined. In the other example surface sterilization of the cotton leafworm, *Prodenia litura*, eggs from two lines, one contaminated with NPV, did not reduce the disease incidence in the developing larvae (Harpaz & Ben Shaked, 1964). Nevertheless, David (1978) still feels that there is no unequivocal evidence that any *Baculovirus* is transmitted transovarially. Other claims for transovarian transmission of viruses of insects are by Swayne (1966) with *Spodoptera exempta* NPV and Hukuhara (1962) and Aruga & Nagashima (1962) with *B. mori* CPV. Sikorowski,

Andrews & Broome (1971) found that CPV virus particles were present in the haemolymph of infected *H. virescens* larvae and adults so that an egg from the time of its primary differentiation of a female germ cell might be under the influence of haemolymph containing virus particles. In another instance, the non-occluded *Baculovirus* of *O. rhinoceros*, the rhinoceros beetle, genital tissue has been shown to be infected with virus (Monsarrat, Duthoit & Vago, 1974) but no transovarial transmission seems to occur (Bedford, 1980). There is, however, one insect virus infection where vertical transmission is not in doubt. A characteristic feature of sigma virus infection in *Drosophila melanogaster* is oocyte infection in virus-infected females and hereditary transmission involving uninterrupted association of the virus with the germ cells (L'Heritier, 1970).

Latency and resistance

Latent virus infections are notoriously common in insects though the topic is still largely an enigma. However, vertical transmission and more particularly transovarian transmission, is potentially important in giving rise to and maintaining latent infection in populations. The actual mechanism by which latency arises in insect virus infections is not understood (Smith, 1976). Most cases of latency might be explained as being inapparent infections where the host does not die or show clinical signs of infection until subjected to some stress factor such as overcrowding, unsatisfactory diet, incorrect temperature or humidity, or the feeding of chemicals or heterologous virus. Such an explanation might be particularly relevant to latent infections by cytoplasmic polyhedrosis virus which in the authors' experience are very common. In such instances a form of insect immune response might be involved which prevents the usual acute infection. Certainly a virus infection in an insect may be inapparent when the initial dose is small and if it is given late in larval development (Smith, 1967). Such infections might easily go unnoticed for the effects of insect viruses are generally scored solely on the basis of larval mortality and infection as such is often not recorded or indeed investigated. It is true that invertebrate animals do not seem to possess an adaptive immune system like that of vertebrates or humoral factors similar to vertebrate antibodies. The immune system of invertebrates seems to depend largely on the phagocytic activity of circulating blood cells which recognize and ingest foreign materials (Lafferty & Crichton, 1973). However, such a system could be quite effective in spite of the fact that it is simpler than the immune response system found in vertebrates. Salt (1970) has pointed out that insects could not have been so dominant if their defence reactions had been anything other than fully effective.

Some cases of virus latency in insects may be more subtle however and it is possible that with some viruses and some hosts genetic interaction of host cell and virus genome may occur giving rise to a state of latency in the strictest sense. As with an inapparent infection some external stress factor activates the virus so that a frank infection results.

Some examples of latent infection in insects and insect cells are worthy of mention. Longworth & Cunningham (1968) recorded a high NPV mortality when larvae of *Aglais urticae* and *Porthetria dispar* were each fed the NPV of the other species. The two viruses are morphologically and serologically distinct, and occult viruses were considered to have been activated in both species rather than cross-transmission having occurred. McKinley, Brown, Payne & Harrap (1981) studied both cross-infectivity and activation with four baculoviruses isolated from different noctuid insects. In this work specific biophysical and serological identification criteria were used and progeny virus was shown to be different from that fed to the insects as inoculum. Activation was much more common than cross-infection. The mechanism of activation has been ascribed to the polyhedron protein of the heterologous virus (Longworth & Cunningham, 1968; Grace, 1962). Activation of baculoviruses in apparently healthy noctuid insects has also been reported by Burgerjon, Biache & Chaufaux (1975), Burgerjon (1977) and Maleki-Milani (1978). In a comparable study Jurkovicova (1979) used the exacting identification technique of restriction endonuclease cleavage of the *Baculovirus* genome to demonstrate that cross-infection did not occur when larvae of *Adoxophyes orana*, a tortricid, and larvae of *Barathra brassicae*, a noctuid, were fed on polyhedra of the reciprocal species. *Eco*Rl cleavage patterns of DNA isolated from the virus used as inoculum and the progeny virus showed that latent viruses must have been activated in both insects. It was also found that a CPV from *B. brassicae* would activate *A. orana* latent NPV and as cross-inoculation with an extract prepared from healthy larvae would also cause activation the activating agent did not have to be a component of the polyhedra.

Progress on understanding the mechanism of activation or the form in which the virus is latent is most likely to come from studies of the phenomenon in cell culture but such an approach has been thwarted by the lack of a suitable model system. However, Granados, Nguyen & Cato (1978) found a *Baculovirus*-like particle persistently infecting an established lepidopteran (*H. zea*) cell line. The virus caused a cytopathic effect in less than 1% of the cells as measured by phase-contrast and electron microscopy. However, the virus would replicate efficiently in other cell lines, in particular the *T. ni* cell line where 90–100% infection was achieved. Kelly *et al.* (1981) observed what are probably the same *Baculovirus*-like particles in the same *H. zea* cell line as a result of attempts to infect a variety of insect cell lines with singly enveloped baculoviruses. Attempted infections of the cell line with both singly and multiply enveloped baculoviruses were found to induce the *Baculovirus*-like particle. These workers also reported experiments using metabolic inhibitors and inactivated inducing virus to show that it is probably a structural component of the activating virus, most likely a protein, that is responsible for inducing the *Baculovirus*-like particle. This situation mimics at least in part the activation of baculoviruses in insects by feeding heterologous or 'foreign' polyhedra and as such it is likely to prove a useful model for investigating the molecular basis of insect virus latency.

Vector insects

The bulk of the information presented in this chapter concerns the persistence of viruses that are pathogenic for their insect host. However, it must not be forgotten that many other viruses which show little, if any, pathogenicity for the insect but are pathogens of other hosts, can persist in natural habitats in insects which act as vectors of the virus. A survey of the persistence of viruses in their insect vectors is beyond the scope of this chapter. Mention of it is made here for completeness, but detailed accounts can be found elsewhere, for example in Gibbs (1973). Plant viruses can be transmitted by aphids, leafhoppers, mites, whiteflies, mealybugs, thrips, treehoppers, grasshoppers and beetles, but it is those vector insects such as aphids and leafhoppers, in which certain plant viruses replicate or remain capable of transmission, that have a particular significance in virus persistence (Sylvester, 1980). Several other insects, in particular mosquitoes and ticks, have a similarly important role in the spread and persistence of virus diseases of vertebrate animals (Fallis, 1980), including man. Arthropod-borne viruses (arboviruses) multiply in the insect vector and transovarian transmission of virus is recognized in ticks and with certain viruses in mosquitoes. In this way many important disease-causing viruses can persist both in the vector insect population and in secondary hosts from which the vector insect will also take a blood meal. The pivotal role of the vector insect in which the virus multiplies is intriguing. The virus is disseminated by the insect in the natural environment and the persistence of the virus is greatly assisted, and in some cases guaranteed, by the activity of the insect. For such viruses it is tempting to ask whether in evolutionary terms the virus is a virus of a plant or vertebrate animal in which it may cause a dramatic disease or truly a virus of an insect where it may go unnoticed.

IMPLICATIONS AND CONCLUSIONS

Knowledge gained on the nature and role of insect virus persistence in populations can be used in explaining the appearance and maintenance of virus epizootics in the natural environment and can aid the development and utilization of microbial agents in pest control.

Virus persistence and epizootics in the natural environment

Insect pests in stable environments commonly exhibit cycles of population fluctuation which may result periodically in severe defoliation of their food plants. Virus epizootics have been cited as major factors in limiting pest outbreaks. Examples include NPV epizootics of *Choristoneura murinana* (Bucher, 1953), *M. fragile* (Clark, 1956), *O. pseudotsugata* (Morris, 1963; Dahlsten & Thomas, 1969), *L. dispar* (Doane, 1970), *D. basiflava* (Kaya & Anderson, 1976) and *Lymantria fumida* (Katagiri, 1977). Viral epizootics typically develop two to three years after the outbreak phase of the pest begins and within two years reduce populations to low levels. During the intermediate phases virus is enzootic and persists at a low incidence in the host population.

An explanation of the effects of viruses on cyclic populations was put forward by Doane (1976) for *L. dispar* and its NPV. Virus acts as a direct density-dependent mortality factor but only at high population densities, at other times being relatively independent of density. This results, following a lag phase indicating that density dependence is delayed, in large quantities of virus persisting between host generations. Although host populations would be low as a result of the full development of the virus epizootic they would still be regulated by virus independently of density until the large source of persistent virus reached quantities too low to be encountered by a sparse population. At this stage the host population increases and the outbreak cycle begins again.

Stable environments with mechanisms for long-term virus persistence therefore provide ready explanations for cyclic epizootics. Appearance of epizootics in pests of annual crops does not follow a cyclic pattern, which probably results from a combination of two principal factors. First, even when virus is epizootic as a response to high host density, the host plant is removed hence reducing the persistent virus to low levels. Secondly, constant provision of suitable food plants in large monocultures often results in rapid and destructive pest outbreaks. Economic thresholds are low and any cyclic tendency in the host population will be obscured by the need for rapid control.

Thus the appearance of virus epizootics in, for example, *T. ni* depends heavily on a good source of persistent virus which can exploit the relatively short-lived host population. *T. ni* NPV persists well in the soil of cabbage fields where epizootics are frequently a major regulatory factor (Jaques, 1975). *T. ni* is also a pest of cotton in California and here epizootics are more sporadic, a result of poor persistence of NPV between and within generations (Ehler, 1977).

Virus persistence and pest control

Design and utilization of microbial control programmes against insect pests can benefit from the knowledge gained from the study of natural epizootics. Various approaches can be used depending on the economic thresholds and turn-over time of the host plant or food source involved.

The stability of forests and grasslands and their high economic thresholds make them ideal candidates for ecologically based control measures. Here it is more prudent to look at ways of enhancing long-term virus persistence than to hope for rapid control akin to conventional insecticides. This approach for grassland pests has been advocated by Longworth & Kalmakoff (1977) and Crawford & Kalmakoff (1977). They suggest measures to maintain old pastures by oversowing rather than complete renewal so that *Wiseana* NPV persists and accumulates between generations and is spread within pastures by stock movement, eliminating the need for frequent addition of virus.

Control of forest dwelling pests can be considered in a similar manner. Applications of virus can be timed, not for rapid mortality, but for maximal virus production and persistence. High damage levels in the first year are more than compensated by the control exerted by the persistent virus source. This method works well for *N. sertifer* NPV which persists and spreads in high sawfly populations in young pine forests, eventually reducing the pest to low levels (unpublished information). Thompson & Scott (1979) warn of the dangers of direct control using NPV against *O. pseudotsugata* and suggest applications of moderate dosages in order to encourage accumulation of naturally produced NPV in the host environment. However, this approach is not always successful: dissemination of low dosages of GV against codling moth on apple reduced damage to acceptable levels but although accumulation and spread of virus was noted, very little persisted to control the pest in the following year (Sheppard & Stairs, 1976).

A special case of an ecological approach is provided by the *Baculovirus* of *O. rhinoceros*, a pest of coconut palm in Asia. Based on demonstrated transmission of virus from infected adults to progeny, methods which increased the incidence of infection-bearing adults in natural populations were initiated. These ranged from placing virus-killed larvae in artificial compost heaps and relying on adults visiting them (Bedford, 1976) to, more effectively, releasing laboratory infected adults into the environment (Zelazny, 1977). These methods proved effective in maintaining infection but, more significantly, resulted in spread over long distances; for example virus spread at 3 km per month throughout the island of Tongatapu (Young, 1974). This approach has been reviewed by Bedford (1980) who makes the point that good control by the *Baculovirus* can be expected only if adequate populations of the beetle remain following infection to maintain a source of persistent virus.

In general, research is centred on enhancing the persistence of sprayed deposits of purified virus. This may be achieved in a wide variety of ways, used alone or in combination. Stickers and UV protectants serve to fulfil two of the primary requirements in this respect, maintaining the virus at the optimum site for persistence and serving to reduce UV inactivation of virus. Maksymiuk (1975) and Pinnock (1975) have reviewed many of the methods used to apply insect viruses and emphasize safety considerations in the use of microbial agents.

Another approach is to improve the activity of the virus in order to overcome its poor persistence during the host generation. New formulations of *Heliothis* NPV with greater activity and persistence were described by Ignoffo *et al.* (1976) with resultant increases of cotton yields between 15 and 114%. On the other hand, Andrews, Spence & Miller (1980) were unable to produce *A. californica* NPV with increased virulence in experimental cloning of the virus. Increased persistence by selecting for resistance to UV inactivation was demonstrated by Witt & Hink (1979) for *A. californica* NPV, although this was effective against near UV only and, after repeated selection, the virus proved less virulent than wild-type virus. Development of efficient virus clones is still in its infancy and as yet does not present a viable method of increasing virus persistence or virulence.

The interrelationship between host, virus and food source can be exploited to improve virus effectiveness. Ignoffo, Garcia, Hostetter & Pinnell (1980) described a method termed 'transplanting' which involved dipping seedlings in a formulation of NPV which then persisted as the plants grew and was supplemented by natural NPV as larvae died from the original preparation. NPV preparations mixed with feeding stimulants have proved effective in increasing mortality from virus in *Heliothis* sp. on cotton (McLaughlin, Andrews & Bell, 1971; Stacey *et al.*, 1977b; Bell & Romine, 1980). Virus can be added to soil prior to planting where it remains in the upper layers and is infective during the larval period. *P. includens* was controlled, in cage tests, by applications of NPV to soil below soybean plants, and also persisted to the following year (Young & Yearian, 1979). Similarly, Jaques (1975) showed that NPV applied to soil prior to planting could be used to control *T. ni* on cabbages.

Selective placement of virus on the undersides of leaves and on fruiting bodies of cotton reduced inactivation of *Heliothis* NPV by UV radiation and concentrated virus at the main sites of larval feeding (Stacey *et al.*, 1977a).

We have discussed examples of viruses which persist at a low incidence in insect populations. It is possible to manipulate the host to encourage frank virus infection in these cases. Stress factors applied to a host may induce a greater incidence of infection than normal. High nitrogen and carbohydrate in plants may increase the susceptibility of the host to low virus concentrations (Pimentel & Shapiro, 1962) while chemical sprays may have a similar effect (silica powder for the CPV of *Dendrolimus spectabilis*, Katagiri, 1975; 3-methylcholanthrene for the NPV of *Spodoptera frugiperda*, Reichelderfer & Benton, 1973; sodium silicate and plant ash for the NPV of *L. dispar*, Yadava, 1971).

Application with other pathogens may result in increased viral infection by synergism. This has been discussed by Tanada (1976) who described a number of examples including a factor in the GV of *Pseudaletia unipuncta* which was synergistic for the NPV of the same and other host species. However, pathogens in combination may have an antagonistic effect thus reducing the effectiveness of the favoured pathogen. Fuxa (1979) provided a recent example in describing antagonism of *H. zea* NPV by the fungus *Variomorpha necatrix*, which he ascribed to competition for gut entry sites. Aruga (1968) suggested that, at the cellular level, antagonism resulted from lack of simultaneous infection in the same cell, although Kurstak, Garzon & Onji (1975) showed that it may be interference rather than complete exclusion of the second virus.

Insect populations may themselves be manipulated to increase virus persistence and dissemination. Autodissemination was a term coined by Ignoffo (1978) and illustrated by examples including release of contaminated adults to transmit virus to their progeny (Martignoni & Milstead, 1962), release of infected larvae to induce and spread infection (Ignoffo, 1978) and release of parasites and predators to disseminate viruses (Smirnoff, 1959). A further example is the auto-dissemination of *Oryctes Baculovirus* by adult beetles described above.

In summary, a full understanding of the modes of virus persistence in insect populations can lead to novel methods for enhancing the effect of the virus in the environment. A few of the possible methods have been described; there will undoubtedly be many other approaches as our knowledge of virus epizootiology increases.

REFERENCES

ALLEN, G. E. & IGNOFFO, C. M. (1969). The nucleopolyhedrosis of *Heliothis*: quantitative in vivo estimates of virulence. *Journal of Invertebrate Pathology*, **13**, 378–81.

ANDREWS, R. E. JR, SPENCE, K. D. & MILLER, L. K. (1980). Virulence of cloned variants of *Autographa californica* nuclear polyhedrosis virus. *Applied and Environmental Microbiology*, **39**, 932–3.

ARATAKE, Y. & KAYAMURA, T. (1973). Pathogenicity of a nuclear-polyhedrosis virus of the silkworm, *Bombyx mori*, for a number of lepidopterous insects. *Japanese Journal of Applied Entomology and Zoology*, **17**, 121–6.

ARUGA, H. (1968). Induction of polyhedroses and interaction among viruses in insects. *Proceedings Joint US–Japan Seminar Microbial Control Insect Pests, Fukuoka*, pp. 33–6.

ARUGA, H. & NAGASHIMA, E. (1962). Generation-to-generation transmission of the cytoplasmic polyhedrosis virus of *Bombyx mori* (Linnaeus) *Journal of Insect Pathology*, **4**, 313–20.

BAILEY, L. (1976). Viruses attacking the honey bee. *Advances in Virus Research*, **20**, 271–304.

BARKER, R. E. (1968). The availability of solar radiation below 290 nm and its importance in photomodification of polymers. *Photochemistry and Photobiology*, **7**, 275–95.

BEATON, C. D. & FILSHIE, B. K. (1976). Comparative ultrastructural studies of insect granulosis and nuclear polyhedrosis viruses. *Journal of General Virology*, **31**, 151–61.

BEDFORD, G. O. (1976). Use of a virus against the coconut palm rhinoceros beetle in Fiji. *Pesticides Abstracts and News Summary*, **22**, 11–25.

BEDFORD, G. O. (1980). Biology, ecology and control of palm rhinoceros beetles. *Annual Review of Entomology*, **25**, 309–39.

BEEGLE, C. C. & OATMAN, E. R. (1975). Effect of a nuclear polyhedrosis virus on the relationship between *Trichoplusia ni* (Lepidoptera: Noctuidae) and the parasite, *Hyposeter exiguae* (Hymenoptera: Ichneumonidae). *Journal of Invertebrate Pathology*, **25**, 59–71.

BELL, M. R. & ROMINE, C. L. (1980). Tobacco budworm field evaluation of microbial control in cotton using *Bacillus thuringiensis* and a nuclear polyhedrosis virus with a feeding adjuvant. *Journal of Economic Entomology*, **73**, 427–30.

BERGOIN, M. & DALES, S. (1971). Comparative observations on poxviruses of invertebrates and vertebrates. In *Comparative Virology*, ed. K. Maramorosch & E. Kurstak, pp. 169–205. New York: Academic Press.

BERGOLD, G. H. (1963). The nature of nuclear polyhedrosis viruses. In *Insect Pathology*, vol. 1, ed. E. A. Steinhaus, pp. 413–56. New York: Academic Press.

BIRD, F. T. (1961). Transmission of some insect viruses with particular reference to ovarial transmission and its importance in the development of epizootics. *Journal of Insect Pathology*, **3**, 352–80.

BUCHER, G. E. (1953). Biotic factors of control of the European fir budworm, *Choristoneura murinana* (Hbn). (N. Comb.), in Europe. *Canadian Journal of Agricultural Science*, **33**, 448–69.

BULLOCK, H. R. (1967). Persistence of *Heliothis* nuclear polyhedrosis virus on cotton foliage. *Journal of Invertebrate Pathology*, **16**, 352–6.

BULLOCK, H. R., MANGUM, C. L. & GUERRA, A. A. (1969). Treatment of eggs of the pink bollworm, *Pectinophora gossypiella* with formaldehyde to prevent infection with a cytoplasmic polyhedrosis virus. *Journal of Invertebrate Pathology*, **14**, 271–3.

BURGERJON, A. (1977). Use of specificity difference indices for the identification of nuclear polyhedrosis viruses (*Baculovirus*) of insects. *Entomophaga*, **22**, 187–92.

BURGERJON, A., BIACHE, G. & CHAUFAUX, J. (1975). Recherches sur la spécificité de trois virus à polyhedres nucléaires vis-à-vis de *Mamestra brassicae, Scotia segetum, Trichoplusia ni* et *Spodoptera exigua. Entomophaga*, **20**, 153–60.

CAREY, G. P., LESCOTT, T., ROBERTSON, J. S., SPENCER, L. K. & KELLY, D. C. (1978). Three African isolates of small iridescent viruses: Type 21 from *Heliothis armigera* (Lepidoptera:Noctuidae), Type 23 from *Heteronychus arator* (Coleoptera:Scarabaeidae), and Type 28 from *Lethocerus columbiae* (Hemiptera Heter-optera:Belostomatidae). *Virology*, **85**, 307–9.

CARTER, J. B. (1973). The mode of transmission of *Tipula* iridescent virus. I. Source of infection. *Journal of Invertebrate Pathology*, **21**, 123–30.

CLARK, E. C. (1956). Survival and transmission of a virus causing polyhedrosis in *Malacosoma fragile*. *Ecology*, **37**, 728–32.

CRAWFORD, A. M. & KALMAKOFF, J. (1977). A host-virus interaction in a pasture habitat: *Wiseana* spp. (Lepidoptera:Hepialidae) and its baculoviruses. *Journal of Invertebrate Pathology*, **29**, 81–7.

CUNNINGHAM, J. C. (1968). Serological and morphological identification of some nuclear-polyhedrosis and granulosis viruses. *Journal of Invertebrate Pathology*, **11**, 132–41.

DAHLSTEN, D. L. & THOMAS, G. M. (1969). A nucleopolyhedrosis virus in populations of the douglas-fir tussock moth, *Hemerocampa pseudotsugata*, in California. *Journal of Invertebrate Pathology*, **13**, 264–71.

DAVID, W. A. L. (1978). The granulosis virus of *Pieris brassicae* (L) and its relationship with its host. *Advances in Virus Research*, **22**, 111–61.

DAVID, W. A. L. & GARDINER, B. O. C. (1966). Persistence of a granulosis virus of *Pieris brassicae* on cabbage leaves. *Journal of Invertebrate Pathology*, **8**, 180–3.

DAVID, W. A. L. & GARDINER, B. O. C. (1967). The persistence of a granulosis virus of *Pieris brassicae* in soil and in sand. *Journal of Invertebrate Pathology*, **9**, 342–7.

DAVID, W. A. L. & TAYLOR, C. E. (1976). Transmission of a granulosis virus in the eggs of a virus-free stock of *Pieris brassicae*. *Journal of Invertebrate Pathology*, **27**, 71–5.

DOANE, C. C. (1970). Primary pathogens and their role in the development of an epizootic in the gypsy moth. *Journal of Invertebrate Pathology*, **15**, 21–3.

DOANE, C. C. (1976). Ecology of pathogens of the gypsy moth. In *Perspectives in Forest Entomology*, ed. J. F. Anderson & H. K. Kaya, pp. 285–93. New York & London: Academic Press.

EHLER, L. E. (1977). Natural enemies of cabbage looper on cotton in the San Joaquin Valley. *Hilgardia*, **45**, 73–106.

ELGEE, E. (1975). Persistence of a virus of the white-marked tussock moth on balsam fir foliage. *Bimonthly Research Notes – Canada, Forestry Service*, **31**, 33–4.

ELLIOTT, R. M., LESCOTT, T. & KELLY, D. C. (1977). Serological relationships of an iridescent virus (Type 25) recently isolated from *Tipula* sp. with two other iridescent viruses (Types 2 and 22). *Virology*, **81**, 309–16.

ELNAGAR, S. & ABUL NASR, S. (1980). Effect of direct sunlight on the virulence of NPV (nuclear polyhedrosis virus) of the cotton leafworm, *Spodoptera littoralis* (Boisd.). *Zeitschrift für angewandte Entomologie*, **90**, 75–80.

ENTWISTLE, P. F. (1978). Microbial control of insects and other pests. *Crop Protection: Proceedings of the British Association for the Advancement of Science, University of Bath*, pp. 72–96.

ENTWISTLE, P. F. & ADAMS, P. H. W. (1977). Prolonged retention of infectivity in the nuclear polyhedrosis virus of *Gilpinia hercyniae* (Hymenoptera, Diprionidae) on foliage of spruce species. *Journal of Invertebrate Pathology*, **29**, 392–4.

ENTWISTLE, P. F., ADAMS, P. H. W. & EVANS, H. F. (1977a). Epizootiology of a nuclear-polyhedrosis virus in European spruce sawfly (*Gilpinia hercyniae*): the status of birds as dispersal agents of the virus during the larval season. *Journal of Invertebrate Pathology*, **29**, 354–60.

ENTWISTLE, P. F., ADAMS, P. H. W. & EVANS, H. F. (1977b). Epizootiology of a nuclear-polyhedrosis virus in European spruce sawfly, *Gilpinia hercyniae*: birds as dispersal agents of the virus during winter. *Journal of Invertebrate Pathology*, **30**, 15–19.

ENTWISTLE, P. F., ADAMS, P. H. W. & EVANS, H. F. (1978). Epizootiology of a nuclear polyhedrosis virus in European spruce sawfly (*Gilpinia hercyniae*): the rate of passage of infective virus through the gut of birds during cage tests. *Journal of Invertebrate Pathology*, **31**, 307–12.

ENTWISTLE, P. F., EVANS, H. F. & ADAMS, P. H. W. (1980). The role of foliar residues of nuclear polyhedrosis virus in epizootics in European spruce sawfly, *Gilpinia hercyniae*. *Proceedings of the 16th International Congress of Entomology*. Kyoto, Japan.

ETZEL, L. K. & FALCON, L. A. (1976). Studies of transovum and transstadial transmission of a granulosis virus of the codling moth. *Journal of Invertebrate Pathology*, **27**, 13–26.

EVANS, H. F. (1981). Quantitative assessment of the relationships between dosage and response of the nuclear polyhedrosis virus of *Mamestra brassicae*. *Journal of Invertebrate Pathology*, **37**, 101–9.

EVANS, H. F., BISHOP, J. M. & PAGE, E. A. (1980). Methods for the quantitative assessment of nuclear-polyhedrosis virus in soil. *Journal of Invertebrate Pathology*, **35**, 1–8.

EVANS, H. F. & ENTWISTLE, P. F. (1982). Epizootiology of the nuclear polyhedrosis virus of European spruce sawfly with emphasis on persistence of virus outside the host. In *Microbial Pesticides*, ed. E. Kurstak. New York: Marcel Dekker (in press).

FALLIS, A. M. (1980). Arthropods as pests and vectors of disease. *Veterinary parasitology*, **6**, 47–73.

FRANZ, J. M. (1971). Influence of environment and modern trends in crop management on microbial control. In *Microbial Control of Insects and Mites*, ed. H. D. Burges & N. W. Hussey, pp. 407–44. New York & London: Academic Press.

FRANZ, J., KRIEG, A. & LANGENBRUCH, R. (1955). Untersuchungen über den Einflus der Passage durch den Darm von Raubinsekten und Vögeln auf die Infektiosität insektenpathogener Viren. *Zeitschrift für Pflanzenkrankheiten Pflanzenpathologie und Pflanzenschutz*, **62**, 721–6.

FUXA, J. R. (1979). Interactions of the microsporidium *Vairimorpha necatrix* with a bacterium, virus and fungus in *Heliothis zea*. *Journal of Invertebrate Pathology*, **33**, 316–23.

GIBBS, A. J. (1973). *Viruses and Invertebrates*. Amsterdam: North-Holland.

GODFREY, B. E. S. (1976). Leachates from aerial parts of plants and their relation to plant surface microbial populations. In *Microbiology of Aerial Plant Surfaces*, ed. C. H. Dickinson & T. F. Preece, pp. 433–9. New York & London: Academic Press.

GRACE, T. D. C. (1962). The development of a cytoplasmic polyhedrosis in insect cells grown in vitro. *Virology*, **18**, 33–42.

GRANADOS, R. R. (1973). Entry of an insect poxvirus by fusion of the virus envelope with the host cell membrane. *Virology*, **52**, 305–9.

GRANADOS, R. R. & LAWLER, K. A. (1981). *In vivo* pathway of *Autographa californica* baculovirus invasion and infection. *Virology*, **108**, 297–308.

GRANADOS, R. R., NGUYEN, T. & CATO, B. (1978). An insect cell line persistently infected with a baculovirus-like particle. *Intervirology*, **10**, 309–17.

HAMM, J. J. & YOUNG, J. R. (1974). Mode of transmission of nuclear-polyhedrosis virus to progeny of adult *Heliothis zea*. *Journal of Invertebrate Pathology*, **24**, 70–81.

HARCOURT, D. G. & CASS, L. M. (1968). Persistence of a granulosis virus of *Pieris rapae* in soil. *Journal of Invertebrate Pathology*, **11**, 142–3.

HARPAZ, I. & BEN SHAKED, Y. (1964). Generation-to-generation transmission of the nuclear polyhedrosis virus of *Prodenia litura* (Fabricus). *Journal of Insect Pathology*, **6**, 127–30.

HARRAP, K. A. (1970). Cell infection by a nuclear polyhedrosis virus. *Virology*, **42**, 311–18.

HARRAP, K. A. (1972a). The structure of nuclear polyhedrosis viruses. I. The inclusion body. *Virology*, **50**, 114–23.

HARRAP, K. A. (1972b). The structure of nuclear polyhedrosis viruses. II. The virus particle. *Virology*, **50**, 124–32.

HARRAP, K. A. (1972c). The structure of nuclear polyhedrosis viruses. III. Virus assembly. *Virology*, **50**, 133–9.

HARRAP, K. A. & PAYNE, C. C. (1979). The structural properties and identification of insect viruses. *Advances in Virus Research*, **25**, 273–355.

HARRAP, K. A. & ROBERTSON, J. S. (1968). A possible infection pathway in the development of a nuclear polyhedrosis virus. *Journal of General Virology*, **3**, 221–5.

HUKUHARA, T. (1962). Generation-to-generation transmission of the cytoplasmic polyhedrosis virus of the silkworm, *Bombyx mori* (Linn) *Journal of Insect Pathology*, **4**, 132–5.

HUKUHARA, T. (1973). Further studies on the distribution of a nuclear-polyhedrosis virus of the fall webworm, *Hyphantria cunea*, in soil. *Journal of Invertebrate Pathology*, **22**, 345–50.

HUKUHARA, T. & NAMURA, H. (1972). Distribution of a nuclear-polyhedrosis virus of the fall webworm, *Hyphantria cunea*, in soil. *Journal of Invertebrate Pathology*, **19**, 308–16.

HUKUHARA, T. & WADA, H. (1972). Absorption of polyhedra of a cytoplasmic-polyhedrosis virus by soil particles. *Journal of Invertebrate Pathology*, **20**, 309–16.

HURPIN, B. & ROBERT, P. H. (1972). Comparison of the activity of certain pathogens of the cockchafer, *Melolontha melolontha* in plots of natural meadowland. *Journal of Invertebrate Pathology*, **19**, 291–8.

IGNOFFO, C. M. (1978). Strategies to increase the use of entomopathogens. *Journal of Invertebrate Pathology*, **31**, 1–3.

IGNOFFO, C. M., GARCIA, C., HOSTETTER, D. L. & PINNELL, R. E. (1980). Transplanting: a method of introducing an insect virus into an ecosystem. *Environmental Entomology*, **9**, 153–4.

IGNOFFO, C. M., HOSTETTER, D. L. & PINNELL, R. E. (1974). Stability of *Bacillus thuringiensis* and *Baculovirus heliothis* on soybean foliage. *Environmental Entomology*, **3**, 117–19.

IGNOFFO, C. M., HOSTETTER, D. L., SIKOROWSKI, P. P., SUTTER, G. & BROOKS, W. M. (1977). Inactivation of representative species of entomopathogenic viruses, a bacterium, fungus and protozoan by an ultraviolet light source. *Environmental Entomology*, **6**, 411–15.

IGNOFFO, C. M., PARKER, F. D., BOENING, O. P., PINNELL, R. E. & HOSTETTER, D. L. (1973). Field stability of the *Heliothis* nucleopolyhedrosis virus on corn silks. *Environmental Entomology*, **2**, 302–3.

IGNOFFO, C. M., YEARIAN, W. C., YOUNG, S. Y., HOSTETTER, D. L. & BULL, D. L. (1976). Laboratory and field persistence of new commercial formulations of the *Heliothis* nucleopolyhedrosis virus, *Baculovirus heliothis*. *Journal of Economic Entomology*, **69**, 233–6.

IRABAGON, T. A. & BROOKS, W. M. (1974). Interaction of *Campoletis sonorensis* and a nuclear polyhedrosis virus in larvae of *Heliothis virescens*. *Journal of Economic Entomology*, **67**, 229–31.

JAQUES, R. P. (1967). The persistence of a nuclear polyhedrosis virus in the habitat of the host insect *Trichoplusia ni*. I. Polyhedra deposited on foliage. *Canadian Entomologist*, **99**, 785–94.

JAQUES, R. P. (1969). Leaching of the nuclear-polyhedrosis virus of *Trichoplusia ni* from soil. *Journal of Invertebrate Pathology*, **13**, 256–63.

JAQUES, R. P. (1970). Natural occurrence of viruses of the cabbage looper in field plot. *Canadian Entomologist*, **102**, 36–41.

JAQUES, R. P. (1972). The inactivation of foliar deposits of viruses of *Trichoplusia ni* and *Pieris rapae* and tests on protectant additives. *Canadian Entomologist*, **104**, 1985–94.

JAQUES, R. P. (1975). Persistence, accumulation, and denaturation of nuclear polyhedrosis and granulosis viruses. In *Baculoviruses for Insect Pest Control: Safety Considerations*, ed. M. Summers, R. Engler, L. A. Falcon & P. Vail, pp. 90–101. Washington, DC: American Society for Microbiology.

JAQUES, R. P. & HARCOURT, D. G. (1971). Viruses of *Trichoplusia ni* and *Pieris rapae* in soil in fields of crucifers in southern Ontario. *Canadian Entomologist*, **103**, 1285–90.

JURKOVICOVA, M. (1979). Activation of latent virus infections in larvae of *Adoxophyes orana* (Lepidoptera: Torticidae) and *Barathra brassicae* (Lepidoptera: Noctuidae) by foreign polyhedra. *Journal of Invertebrate Pathology*, **34**, 213–23.

KATAGIRI, K. (1975). Control of forest pest insects by virus. *Proceedings of 1st Intersection Congress. International Association of Microbiological Societies, Tokyo*, **2**, 613–20.

KATAGIRI, K. (1977). Epizootiological studies on the nuclear and cytoplasmic polyhedroses of the red belly tussock moth, *Lymantria fumida* Butler (Lepidoptera:Lymantriidae). (In Japanese.) *Bulletin of the Government Forest Experiment Station, Meguro*, **294**, 85–135.

KAUPP, W. J. (1980). A simple method of assessing the quantities of nuclear polyhedrosis virus entering the soil from diseased populations of European pine sawfly. *Bimonthly Research Notes – Canada, Forestry Service*, **36**, 31.

KAWANISHI, C. Y., SUMMERS, M. D., STOLTZ, D. B. & ARNOTT, H. J. (1972). Entry of an insect virus *in vivo* by fusion of viral envelope and microvillus membrane. *Journal of Invertebrate Pathology*, **20**, 104–8.

KAYA, H. K. & ANDERSON, J. F. (1976). Biotic mortality factors in dark tussock moth populations in Connecticut. *Environmental Entomology*, **5**, 1141–5.

KELLY, D. C., AYRES, M. D., LESCOTT, T., ROBERTSON, J. S. & HAPP, G. M. (1979). A small iridescent virus (Type 29) isolated from *Tenebrio molitor*: a comparison of its proteins and antigens with six other iridescent viruses. *Journal of General Virology*, **42**, 95–105.

KELLY, D. C., LESCOTT, T., AYRES, M. D., CAREY, D., COUTTS, A & HARRAP, K. A. (1981). Induction of a non-occluded baculovirus persistently infecting *Heliothis zea* cells by *Heliothis armigera* and *Trichoplusia ni* nuclear polyhedrosis viruses. *Virology*, **112**, 174–89.

KINSINGER, R. A. & MCGAUGHEY, WM. H. (1976). Stability of *Bacillus thuringiensis* and a granulosis virus of *Plodia interpunctella* on stored wheat. *Journal of Economic Entomology*, **69**, 149–54.

KOMAREK, J. & BREINDL, V. (1924). Die Wipfelkrankheit der Nonne und der Erreger derselben. *Zeitschrift für Angewandte Entomologie*, **10**, 99–162.

KURSTAK, E., GARZON, S. & ONJI, P. A. (1975). Multiple viral infections of insect cells and host pathogenesis: multicomponent viral insecticides. *Proceedings 1st Intersectional Congress, International Association of Microbiological Societies, Tokyo*, **2**, 650–7.

LAFFERTY, K. J. & CRICHTON, R. (1973). Immune responses of invertebrates. In *Viruses and Invertebrates*, chapter 16, ed. A. J. Gibbs, pp. 300–20. Amsterdam: North-Holland.

LANGRIDGE, W. H. B. & GRANADOS, R. R. (1978). Recent developments in entomopoxvirus replication. In *Abstracts of the Fourth International Congress for Virology*, p. 548. The Hague.

LAUTENSCHLAGER, R. A. & PODGWAITE, J. D. (1979). Passage of nucleo-polyhedrosis virus by avian and mammalian predators of the gypsy moth, *Lymantria dispar. Environmental Entomology*, **8**, 210–14.

L'HÉRITIER, P. (1970). *Drosophila* viruses and their role as evolutionary factors. *Evolutionary Biology*, **4**, 185–209.

LINLEY, J. R. & NIELSEN, H. T. (1968). Transmission of a mosquito iridescent virus in *Aedes taeniorhyncus*. II. Experiments related to transmission in nature. *Journal of Invertebrate Pathology*, **12**, 17–24.

LONGWORTH, J. F. (1973). Viruses and Lepidoptera. In *Viruses and Invertebrates*, chapter 20, ed. A. J. Gibbs, pp. 428–441. Amsterdam: North-Holland.

LONGWORTH, J. F. (1978). Small isometric viruses of invertebrates. *Advances in Virus Research*, **23**, 103–57.

LONGWORTH, J. F. & CUNNINGHAM, J. C. (1968). The activation of occult nuclear-polyhedrosis viruses by foreign nuclear polyhedra. *Journal of Invertebrate Pathology*, **10**, 361–7.

LONGWORTH, J. F. & KALMAKOFF, J. (1977). Insect viruses for biological control: an ecological approach. *Intervirology*, **8**, 68–72.

LONGWORTH, J. F., ROBERTSON, J. S., TINSLEY, T. W., ROWLANDS, D. J. & BROWN, F. (1973). Reactions between an insect picornavirus and naturally occurring IgM antibodies in several mammalian species. *Nature (London)*, **242**, 314–16.

McGAUGHEY, W. H. (1975). A granulosis virus for Indian meal moth control in stored wheat and grain. *Journal of Economic Entomology*, **68**, 346–8.

McKINLEY, D. J., BROWN, D. A., PAYNE, C. C. & HARRAP, K. A. (1981). Cross-infectivity and activation studies with four baculoviruses. *Entomophaga*, **26**, 79–90.

McLAUGHLIN, R. E., ANDREWS, G. & BELL, M. R. (1971). Field tests for control of *Heliothis* spp. with nuclear polyhedrosis virus included in a boll weevil bait. *Journal of Invertebrate Pathology*, **18**, 304–5.

McLEOD, P. J., YEARIAN, W. C. & YOUNG, S. Y. (1977). Inactivation of *Baculovirus heliothis* by ultraviolet irradiation, dew and temperature. *Journal of Invertebrate Pathology*, **30**, 237–41.

MAKSYMIUK, B. (1975). Pattern of use and safety aspects in application of insect viruses in agriculture and forestry. In *Baculoviruses for Insect Pest Control: Safety Considerations*, ed. M. Summers, R. Engler, L. A. Falcon & P. Vail, pp. 123–8. Washington DC: American Society for Microbiology.

MALEKI-MILANI, H. (1978). Influence de passages répétés du virus de la polyédrose nucléaire de *Autographa californica* chez *Spodoptera littoralis* (Lep:Noctuidae). *Entomophaga*, **23**, 217–24.

MANYAKOV, V. F. (1977). Fine structure of the iridescent virus Type 1 capsid. *Journal of General Virology*, **36**, 73–9.

MARTIGNONI, M. E. & IWAI, P. J. (1975). A catalogue of viral diseases of insects and mites. *USDA Forest Service General Technical Report PNW-40*.

MARTIGNONI, M. E. & MILSTEAD, J. E. (1962). Trans-ovum transmission of the nuclear polyhedrosis virus of *Colias eurytheme* Boisduval through contamination of the female genitalia. *Journal of Insect Pathology*, **4**, 113–21.

MATTHEWS, R. E. F. (1979). Classification and nomenclature of viruses. Third

Report of the International Committee on Taxonomy of Viruses. *Intervirology*, **12**, 131–296.

MELAMED-MADJAR, V. & RACCAH, B. (1979). The transstadial and vertical transmission of a granulosis virus from the corn borer *Sesamia nongrioides*. *Journal of Invertebrate Pathology*, **33**, 259–64.

MERY, C. & DULMAGE, H. T. (1975). Transmission, diagnosis and control of cytoplasmic polyhedrosis virus in colonies of *Heliothis virescens*. *Journal of Invertebrate Pathology*, **26**, 75–9.

MONSARRAT, P., DUTHOIT, J.-L. & VAGO, C. (1974). Mise en evidence de Virions de type *Baculovirus* dans l'appareil genital du Coleoptere *Oryctes rhinoceros* L. *Comptes Rendus Hebdomadaires des Séances de l'Académie des sciences. Series D*, **278**, 3259–61.

MORRIS, O. N. (1963). The natural and artificial control of the Douglas-fir tussock moth, *Orgyia pseudotsugata* McDunnough, by a nuclear-polyhedrosis virus. *Journal of Insect Pathology*, **5**, 401–14.

OHBA, M. & AIZAWA, K. (1979). Multiplication of *Chilo* iridescent virus in noninsect arthropods. *Journal of Invertebrate Pathology*, **33**, 278–83.

OSSOWSKI, L. L. J. (1957). The biological control of the wattle bagworm (*Kotochalia junodi* Heyl.) by a virus disease. I. Small-scale pilot experiments. *Annals of Applied Biology*, **45**, 81–9.

PAYNE, C. C. & RIVERS, C. F. (1976). A provisional classification of cytoplasmic polyhedrosis viruses based on the sizes of the RNA genome segments. *Journal of General Virology*, **33**, 71–85.

PIMENTEL, D. & SHAPIRO, M. (1962). The influence of environment on a virus–host relationship. *Journal of Insect Pathology*, **4**, 77–87.

PINNOCK, D. W. (1975). Pest populations and virus dosage in relation to crop productivity. In *Baculoviruses for Insect Pest Control: Safety Considerations*, ed. M. Summers, R. Engler, L. A. Falcon & P. Vail, pp. 145–57. Washington DC: American Society for Microbiology.

PODGWAITE, J. D., SHIELDS, K. S., ZERILLO, R. T. & BRUEN, R. B. (1979). Environmental persistence of the nucleopolyhedrosis virus of the gypsy moth, *Lymantria dispar*. *Environmental Entomology*, **8**, 528–36.

PRELL, H. (1926). Die Polyederkrankheiten der Insckten. *Third International Congress of Entomology*, Zurich 1925, **2**, 145–68.

RAIMO, B. (1975). Infecting the gypsy moth, *Porthetria dispar* (L.), with nuclear polyhedrosis virus vectored by *Apanteles melanoscelus* (Ratzeburg). *Journal of the New York Entomological Society*, **83**, 246.

RAIMO, B., REARDON, R. C. & PODGWAITE, J. D. (1977). Vectoring gypsy moth nuclear polyhedrosis virus by *Apanteles melanoscelus* (Hym.: Braconidae). *Entomophaga*, **22**, 207–15.

REARDON, R. C. & PODGWAITE, J. D. (1976). Disease-parasitoid relationships in natural populations of *Lymantria dispar* (Lep.:Lymantriidae) in the Northeastern United States. *Entomophaga*, **21**, 333–41.

REED, E. M. (1971). Factors affecting the status of a virus as a control agent for the potato moth (*Phthorimaea operculella* (Zell.) (Lep. Gelechiidae)). *Bulletin of Entomological Research*, **61**, 207–22.

REED, E. M. & SPRINGETT, B. P. (1971). Large-scale field testing of a granulosis virus for the control of potato moth (*Phthorimaea operculella* (Zell.) (Lep., Gelechiidae)). *Bulletin of Entomological Research*, **61**, 223–33.

REICHELDERFER, C. F. & BENTON, C. V. (1973). The effect of 3-methylcholanthrene treatment on the virulence of a nuclear polyhedrosis virus of *Spodoptera frugiperda*. *Journal of Invertebrate Pathology*, **22**, 38–41.

REINGANUM, C., O'LOUGHLIN, G. T. & HOGAN, T. W. (1970). A non-occluded virus of the field cricket *Teleogryllus oceanicus* and *T. commodus* (Orthoptera:Gryllidae). *Journal of Invertebrate Pathology*, **16**, 214–20.

ROOME, R. E. & DAOUST, R. A. (1976). Survival of the nuclear polyhedrosis virus of *Heliothis armigera* on crops and in soil in Botswana. *Journal of Invertebrate Pathology*, **27**, 7–12.

SALT, G. (1970). The cellular defence reactions of insects. In *Cambridge Monographs in Experimental Biology No. 16*. Cambridge University Press.

SHAPIRO, M. & IGNOFFO, C. M. (1971). Amino-acid and nucleic acid analysis of inclusion bodies of the cotton bollworm *Heliothis zea* nucleo-polyhedrosis virus. *Journal of Invertebrate Pathology*, **18**, 154–5.

SHEPPARD, R. F. & STAIRS, G. R. (1976). Effects of dissemination of low dosage levels of a granulosis virus in populations of the codling moth. *Journal of Economic Entomology*, **69**, 583–6.

SIKOROWSKI, P. P., ANDREWS, G. L. & BROOME, J. R. (1971). Presence of cytoplasmic polyhedrosis virus in the hemolymph of *Heliothis virescens* larvae and adults. *Journal of Invertebrate Pathology*, **18**, 167–8.

SIKOROWSKI, P. P., ANDREWS, G. L. & BROOME, J. R. (1973). Transovum transmission of a cytoplasmic polyhedrosis virus of *Heliothis virescens* (Lepidoptera:Noctuidae). *Journal of Invertebrate Pathology*, **21**, 41–5.

SIMMONS, C. L. & SIKOROWSKI, P. P. (1973). A laboratory study of the effects of cytoplasmic polyhedrosis virus on *Heliothis virescens* (Lepidoptera:Noctuidae). *Journal of Invertebrate Pathology*, **22**, 369–71.

SMIRNOFF, W. A. (1959). Predators of *Neodiprion swainei* Midd. (Hymenoptera: Tenthredinidae). *Canadian Entomologist*, **92**, 957–8.

SMIRNOFF, W. A. (1962). Transovum transmission of virus of *Neodiprion swainei* (Middleton) (Hymenoptera:Tenthredinidae). *Journal of Insect Pathology*, **4**, 192–200.

SMIRNOFF, W. A. & MacLEOD, C. F. (1964). Apparent lack of effects of orally introduced polyhedrosis virus on mice and of pathogenicity of rodent-passed virus for insects. *Journal of Insect Pathology*, **6**, 537–8.

SMITH, K. M. (1967). *Insect Virology*. New York & London: Academic Press.

SMITH, K. M. (1976). *Virus–Insect Relationships*. London: Longman.

STACEY, A. L., YOUNG, S. Y. & YEARIAN, W. C. (1977a). *Baculovirus heliothis*: effect of selective placement on *Heliothis* mortality and efficacy of directed sprays on cotton. *Journal Georgia Entomological Society*, **12**, 167–73.

STACEY, A. L., YEARIAN, W. C. & YOUNG, S. Y. (1977b). Evaluation of *Baculovirus heliothis* with feeding stimulants for control of *Heliothis* larvae on cotton. *Journal of Economic Entomology*, **70**, 779–84.

STOLTZ, D. B. & HILSENHOFF, W. L. (1969). Electron-microscopic observations on the maturation of a cytoplasmic-polyhedrosis virus. *Journal of Invertebrate Pathology*, **14**, 39–48.

STOLTZ, D. B. & SUMMERS, M. D. (1971). Pathway of infection of mosquito iridescent virus. I. Preliminary observations on the fate of ingested virus. *Journal of Virology*, **8**, 900–9.

SUMMERS, M. D. (1969). Apparent in vivo pathway of granulosis virus invasion and infection. *Journal of Virology*, **4**, 188–90.

SUMMERS, M. D. (1971). Electron microscope observations on granulosis virus entry, uncoating and replication processes during infection of the midgut cells of *Trichoplusia ni*. *Journal of Ultrastructure Research*, **35**, 606–25.

SWAYNE, G. (1966). Generation-to-generation passage of the nuclear polyhedral virus of *Spodoptera exempta* (Wlk). *Nature (London)*, **210**, 1053–4.

SYLVESTER, E. S. (1980). Circulative and propagative virus transmission by aphids. *Annual Review of Entomology*, **25**, 257–86.

TANADA, Y. (1971). Persistence of entomogenous viruses in the insect ecosystem. In *Entomological Essays to Commemorate the Retirement of Professor K. Yasumatsu*, pp. 367–79.

TANADA, Y. (1976). Ecology of insect viruses. In *Perspectives in Forest Entomology*, Proceedings of Lockwood Conference on Perspectives of Forest Pest Management, New Haven, Connecticut, XXIII + 428 pp. ed. J. F. Anderson & H. K. Kaya, pp. 265–83. New York & London: Academic Press.

TANADA, Y. & OMI, E. M. (1974). Persistence of insect viruses in field populations of alfalfa insects. *Journal of Invertebrate Pathology*, **23**, 360–5.

THOMAS, E. D., REICHELDERFER, C. F. & HEIMPEL, A. M. (1973). The effect of soil pH on the persistence of cabbage looper nuclear polyhedrosis virus in soil. *Journal of Invertebrate Pathology*, **21**, 21–5.

THOMPSON, C. G. & SCOTT, D. W. (1979). Production and persistence of the nuclear polyhedrosis virus of the douglas-fir tussock moth, *Orgyia pseudotsugata* (Lepidoptera:Lymantriidae), in the forest ecosystem. *Journal of Invertebrate Pathology*, **33**, 57–65.

TINSLEY, T. W. (1979). The potential of insect pathogenic viruses as pesticidal agents. *Annual Review of Entomology*, **24**, 63–87.

TINSLEY, T. W. & HARRAP, K. A. (1978). Viruses of invertebrates. In *Comprehensive Virology*, vol. 12, chapter 1, ed. H. Fraenkel-Conrat & R. R. Wagner, pp. 1–101. New York: Plenum Press.

VAGO, C., FOSSET, J. & BERGOIN, M. (1966). Dissémination des virus de polyédries par les Éphippigères prédateurs d'insectes. *Entomophaga*, **11**, 177–82.

VAIL, P. V. & JAY, D. L. (1973). Pathology of a nuclear polyhedrosis virus of the alfalfa looper in alternate hosts. *Journal of Invertebrate Pathology*, **21**, 198–204.

VINSON, S. B. & IWANTSCH, G. F. (1980). Host suitability for insect parasitoids. *Annual Review of Entomology*, **25**, 397–419.

WAGNER, G. W., PASCHKE, J. D., CAMPBELL, W. R. & WEBB, S. R. (1974). Proteins of two strains of mosquito iridescent virus. *Intervirology*, **3**, 97–105.

WELLINGTON, E. F. (1954). The amino-acid composition of some insect viruses and their characteristic inclusion-body proteins. *Biochemical Journal*, **57**, 334–8.

WIGLEY, P. J. (1976). The epizootiology of a nuclear polyhedrosis virus disease of the winter moth, *Operophtera brumata* L. at Wistman's Wood, Dartmoor. D. Phil. thesis, University of Oxford.

WILLIAMS, R. C. & SMITH, K. M. (1957). A crystallizable insect virus. *Nature (London)*, **179**, 119–20.

WITT, D. J. & HINK, W. F. (1979). Selection of *Autographa california* nuclear polyhedrosis virus for resistance to inactivation by near ultraviolet, far ultraviolet, and thermal radiation. *Journal of Invertebrate Pathology*, **33**, 222–32.

WITT, D. J. & STAIRS, G. R. (1975). The effects of ultraviolet irradiation on a baculovirus infecting *Galleria mellonella*. *Journal of Invertebrate Pathology*, **26**, 321–7.

WRIGLEY, N. G. (1969). An electron microscope study of the structure of *Sericesthis* iridescent virus. *Journal of General Virology*, **5**, 123–34.

WRIGLEY, N. G. (1970). An electron microscope study of the structure of *Tipula* iridescent virus. *Journal of General Virology*, **6**, 169–73.

YADAVA, R. L. (1971). On the chemical stressors of nuclear-polyhedrosis virus of gypsy moth, *Lymantria dispar* L. *Zeitschrift für Angewandte Entomologie*, **69**, 303–11.

YOUNG, E. C. (1974). The epizootiology of two pathogens of the coconut palm rhinoceros beetle. *Journal of Invertebrate Pathology*, **24**, 82–92.

YOUNG, S. Y. & YEARIAN, W. C. (1979). Soil application of *Pseudoplusia* NPV:

persistence and incidence of infection in soybean looper caged on soybean. *Environmental Entomology*, **8,** 860–4.

YOUNG, S. Y., YEARIAN, W. C. & KIM, K. S. (1977). Effect of dew from cotton and soybean foliage on activity of *Heliothis* nuclear polyhedrosis virus. *Journal of Invertebrate Pathology*, **30,** 237–41.

ZELAZNY, B. (1977). *Oryctes rhinoceros* populations and behaviour influenced by a baculovirus. *Journal of Invertebrate Pathology*, **29,** 210–15.

MORBILLIVIRUS PERSISTENT INFECTIONS IN ANIMALS AND MAN

VOLKER TER MEULEN AND MICHAEL J. CARTER

Institut für Virologie und Immunbiologie, Versbacher Straße 7, D-8700 Würzburg, West Germany

INTRODUCTION

The morbilliviruses, which belong to the family of paramyxoviridae, include measles, canine distemper and rinderpest viruses (Kingsbury *et al.*, 1978). These agents are naturally occurring pathogens which cause acute diseases of clinical importance. The severity of these diseases and the complications which can develop during the infections have excited great interest in the biology of these agents as well as in the pathogenic mechanisms underlying the different disease processes. The progress made in virology in the past has led to the development of vaccines which proved to be effective in controlling morbillivirus infections, and in the case of rinderpest has confined this disease to enzootic areas. However, despite these achievements none of these infections has been eradicated and a morbillivirus infection is still a threat to the natural host. Moreover, measles and canine distemper viruses have been associated with subacute and chronic diseases of the central nervous system, which are considered late complications of the acute infections. From the group of morbilliviruses, only measles and canine distemper virus have been shown to establish persistence in a host, which may give rise to chronic progressing disorders after the long incubation periods on the basis of very complex virus–cell and virus–host interactions.

DISEASES ASSOCIATED WITH PERSISTENT MORBILLIVIRUS INFECTIONS

Infection of a susceptible host with a morbillivirus is followed normally by an acute disease with a clinical symptomatology typical of each virus. The pattern of disease induced by these agents varies greatly, and in the case of canine distemper (CD) or rinderpest this has been attributed to strain differences. In the course of these acute infections complications may develop which influence the outcome of the disease, often leading to a mortality rate of up to 50–80% for CD or rinderpest infection. Recovery from these diseases is associated with life-long immunity to reinfection (Appel *et al.*, 1981). However, in both men and dogs disorders of the central nervous system (CNS) have been observed, developing months or years after the onset of the acute disease, which are associated with a persistent measles or a persistent canine distemper virus infection. The occurrence of such a viral disease in a host which has overcome the initial acute infection underlines the importance of virus persistence in relation to disease processes.

Subacute sclerosing panencephalitis

Subacute sclerosing panencephalitis (SSPE) is a rare and fatal, slowly progressing inflammatory CNS disease which occurs primarily in children and young adults. It belongs to the group of human slow virus diseases associated with persistent measles virus infection, and exhibits epidemiological, clinical and neuropathological features quite different from those seen in acute measles and measles encephalitis (ter Meulen & Hall, 1978).

Epidemiology

Although the majority of SSPE cases have been observed in North America and Europe, this CNS disorder has been found in all ethnic groups examined (McDonald *et al.*, 1974; Okuno *et al.*, 1978; Broor, 1975; Chin-mei & Szu-chih, 1977) and the disease pattern appears to be similar in all countries. Certain racial predispositions have been observed in Southern Africa (McDonald *et al.*, 1974), Israel (Soffer *et al.*, 1975) and in the United States (Jabbour *et al.*, 1972). However, no definite HLA linkage pattern has been recognized for this disease (Kurent *et al.*, 1975; Kreth *et al.*, 1975; Agnarsdottir, 1977). Moreover, there are several reports of the occurrence of SSPE in only one of identical twins which argues that genetic factors may not be important in this disorder (Chao, 1962; Whitaker *et al.*, 1972; D'Onghia *et al.*, 1974; Houff *et al.*, 1979). In addition to these epidemiological findings there are other observations to be considered in the evaluation of this disease. Firstly, it has been found that SSPE has a higher incidence in rural areas than in urban communities (Brody & Detels, 1970; Detels *et al.*, 1973; Halsey *et al.*, 1980). Secondly, it could be shown in several case controlled studies, as well as from surveillance of national SSPE registrees, that approximately 50% of all SSPE patients contracted acute measles before the age of two years. This is a remarkably high figure considering the broader spread of age at contraction of acute measles in the general population (Brody & Detels, 1970; Jabbour *et al.*, 1972; Detels *et al.*, 1973; Bellman & Dick, 1980; Halsey *et al.*, 1980). Thirdly, the disease is 2–3 times more common in boys than in girls and is found significantly more often in lower socio-economic groups and in younger children of large families (Halsey *et al.*, 1980).

The age at which clinical manifestations of SSPE appear ranges between 2–35 years with an average interval between acute measles and SSPE of 6–8 years (Dayan *et al.*, 1967; Cape *et al.*, 1973). The frequency of disease is in the order of one case per million acute measles virus infections; however, in the USA where measles vaccination programmes have been carried out, it appears that the number of SSPE cases is declining, suggesting that measles vaccination may well prevent SSPE by eliminating natural measles virus infection (Halsey *et al.*, 1978). So far, no transmission of SSPE, either horizontally or vertically, has been observed. This suggests the disease is not infectious (Nelson, Dennery, Montpeptit & Furesz, 1972; Glasner & Kirsch, 1975; Gaines & Jabbour, 1979).

Disease pattern
The clinical features of SSPE are quite variable, reflecting the widespread involvement of different CNS tissues in the disease process. In general, three clinical stages can be observed (Kalm, 1952). The first signs of the disease are intellectual deterioration or psychological disturbances which are insidious in their onset lasting for weeks or months. They are followed by neurological and motor dysfunctions which eventually lead to a progressive cerebral degeneration accompanied by symptoms and signs of decortication. The disease process may last from months to years and is highly variable. Remissions are common and may last for weeks to years and be repeated several times. There are a few cases in which remission has lasted more than ten years. It is not known whether the measles infection of the central nervous system is halted or eliminated in these cases (Donner *et al.*, 1972; Risk *et al.*, 1978).

Besides the clinical symptomatology there are immunological and virological findings which are pathognomonic for this disease. CSF specimens of SSPE patients reveal a prominent gammaglobulin elevation which is related to an increase in immunoglobulin G whereas all other proteins are normal. Moreover, the IgG exhibits restricted banding on electrophoresis indicating an oligoclonal population of IgG (Lowenthal, 1964; Vandvik, 1973). Virological–serological studies demonstrated that SSPE patients have exceptionally high antibody titres to measles virus in both CSF and serum specimens. These have not been observed to such an extent in any other disease associated with measles virus. This elevation of measles antibody activity can be measured in all assays available for measles virus serology, whereas antibody titres against the variety of non-viral and viral antigens other than measles virus are unremarkable (Sever & Zeman, 1968; Gerson & Haslam, 1971; Joncas et al., 1974; Blaese & Hofstrand, 1975). The amount of measles virus-specific immunoglobulin in serum has been estimated to be 10–20% of total IgG (Mehta et al., 1977); minor proportions of specific antibodies have also been found to belong to the IgM, IgA, IgD and IgE classes of immunoglobulins (Vandvik et al., 1977; Thomson et al., 1975; Kiessling et al., 1977). The finding of specific IgM antibodies has been a controversial issue as both negative (Najera et al., 1972) and positive results (Connolly et al., 1971; Thomson et al., 1975; Kiessling et al., 1977; Luster et al., 1978) have been reported. It has been claimed that some of the positive results might be due to IgM rheumatoid factor which is found in SSPE sera.

The presence of measles antibodies in the CSF of SSPE patients is the most interesting observation. The finding of low serum/CSF ratios for measles antibodies as compared to antibodies against other unrelated viral antigens (Connolly, 1968; Salmi et al., 1972) and the experimental observations of an intact blood/brain barrier by Cutler et al. (1968) provide strong evidence for a local production of measles antibodies in SSPE. Moreover, measles antibodies are associated with the electrophoretically restricted IgG bands, as demonstrated by absorption experiments using highly concentrated and purified measles virus preparations (Vandvik et al., 1976). Under optimal conditions almost all electrophoretically restricted IgG bands can be removed from SSPE CSF which is in contrast to other CNS diseases (Norrby, 1978; Vartdal et al., 1980). The occurrence of IgG elevation with measles-specific oligoclonal IgG in the CSF of SSPE patients reflects a state of CNS hyperimmunization against this agent and indicates the local production of antiviral antibodies by sensitized lymphocytes which have invaded this compartment.

Neuropathology

Neuropathological examination of SSPE brain reveals a diffuse encephalitis of varying severity both in the grey and white matter. This is characterized by leptomeningeal lymphocytic infiltration, perivascular cuffings consisting of lymphocytes and plasma cells, and a diffuse lymphocytic infiltration. Glial cell proliferation and an increase in fibrous astrocytes can be seen in addition to demyelination in chronic cases. The most striking and characteristic morphological changes are the intranuclear inclusion bodies of Cowdry (1934) which are found in neurones, oligodendro-glial cells and astrocytes. These inclusion bodies contain paramyxovirus nucleocapsid structures which have been identified by immunofluorescent staining as being measles virus specific (Freeman *et al.*, 1967; Connolly *et al.*, 1967; ter Meulen *et al.*, 1967, 1969). Within intranuclear inclusions nucleocapsids have a predominantly 'smooth' appearance whereas 'rough' nucleocapsids are detectable in the cytoplasm of brain cells (Iwasaki & Koprowski, 1974). Moreover, nucleocapsids were also detected in the neuropile of various cellular processes, demonstrating the transport of nucleocapsids by the axonal flow. However, the typical feature of measles infection, the development of giant cell formation, has never been seen in SSPE brain section. As a member of the paramyxoviridae, measles virus is known to acquire its viral envelope from the membrane of the infected cell during release from the host cell by a budding process. Morphological changes of the infected host cell such as an increased cell membrane thickness and an alignment of nucleocapsid under the cell membrane are related to the process of virus budding which can easily be demonstrated in infected tissue cultures. The absence of these changes in CNS cells suggests a defect in measles virus replication in this tissue and is compatible with the failure to isolate infectious virus by conventional techniques.

Canine distemper encephalomyelitis

An infection of the CNS by canine distemper virus (CDV) is a relatively rare complication in the course of acute canine distemper. However, despite the importance of this CNS disease little information is available on the epidemiology, disease pattern or mechanisms producing this disorder.

Epidemiology

Canine distemper is an enzootic disease, present throughout the world, affecting dogs and other members of the canine family. Many wild-life species are susceptible to this agent and probably constitute a constant source of infection for dogs. Transmission of infection occurs either by contact or aerosol since infected animals shed virus in all body secretions (Laidlaw & Dunkin, 1928). Transplacental transmission may also be possible (Krakowka *et al.*, 1974, 1977). Puppies with maternal antibodies are protected, but, once the protection effect wanes, animals are susceptible to infection regardless of their age. A seasonal prevalence of canine distemper has been observed in countries with temperate climates with the highest incidence of clinical disease in the winter (Rockborn, 1958). However, in other studies no such seasonal variations were recorded (Erno & Möller, 1961). This has been taken as an indication that infection can occur throughout the year and can often be subclinical in nature. No epidemiological data are available on the development of CNS complications in the course of canine distemper. So far no genetic or other risk factors have been described which might promote development of this disorder.

Clinical pattern and neuropathological characteristics

Distemper in dogs develops after a short incubation period as an acute or subacute disease clinically characterized by pyrexia, anorexia, nasal discharge and conjunctivitis as well as by gastrointestinal and respiratory symptoms. The severity of this disease and the duration of the clinical symptoms appear to depend on the resistance of the host and on the virulence of the virus strain. Some dogs develop a complicating CNS disease which manifests itself as an acute encephalitis, subacute encephalitis, chronic disseminated demyelinating meningoencephalitis or old dog encephalitis (ODE) (Innes & Saunders, 1962; Cordy, 1942). ODE usually develops in aged, immunized dogs and is therefore of especial interest in the context of persistent morbillivirus infection. The neurological symptoms of these diseases are variable, consisting mainly of disturbances in gait, uncoordination, convulsive seizures, myoclonus and paralysis (McGrath, 1960; Innes & Saunders, 1962). Optic neuritis and retinal lesions have also been observed (Fischer, 1965). Symptoms in the case of subacute or chronic encephalitis may appear weeks or months after recovery from distemper. On the other hand, encephalitis may also develop in dogs without any preceding acute clinical illness, indicating that virus can reach the CNS in the course of a subclinical infection. Dogs with nervous signs usually die but may sometimes recover. In such cases, neurological signs often persist as a result of CNS damage. Dogs with a subacute or chronic encephalomyelitis possess neutralizing antibodies against canine distemper virus in CSF and IgG levels are increased (Appel, 1969, 1970; Cutler & Averill, 1969; Imagawa *et al.*, 1980). Serum titres are also high and some antiviral IgM is present (Appel, 1969). However, so far, there are no reports of oligoclonal immunoglobin bands in CSF specimens as observed in SSPE.

Neuropathological studies have shown that many areas of the CNS are infected by canine distemper virus. The histopathological lesions consist of inflammation, necrosis and demyelination, affecting both white and grey matter in the brain and spinal cord. The cellular infiltrates consist mainly of lymphocytes, plasma cells and undifferentiated mononuclear cells. Structures such as the meninges, cerebral cortex, optic tracts, midbrain, pons and medulla are consistently involved in the disease process and can be considered as predilective sites of virus replication (Cordy, 1942; Gibson *et al.*, 1965; Innes & Saunders, 1962). As in SSPE, Cowdry type A and B inclusion bodies can be detected in glial cells and neurones, but unlike in measles virus infection these structures can be observed in the meningeal and ependymal cells. Electron microscopy and immuno-fluorescent studies have demonstrated that inclusion bodies contain canine distemper virus nucleocapsids (Appel & Gillespie, 1972). The demyelinated areas are characterized by loss of myelin with relative sparing of axons and the presence of gitter cells. In severe cases, however, breakdown of axis cylinders can be seen with the development of dense glial cell hyperplasia. Perivascular cuffing, consisting of lymphocytes and macrophages, may be observed in the neighbourhood of demyelinating plaques. This is typical of post-infectious encephalomyelitides in both animal and man.

MORBILLIVIRUS PERSISTENCE *IN VITRO*

In view of the pathogenic consequences of morbillivirus persistence in the organism, much effort has been invested in the study of this phenomenon *in vitro*. In general, during a persistent infection, cellular growth continues despite the presence of viral genetic information. The extent to which this information is expressed may vary greatly, from infections in which progeny virus is continually shed, to those in which no virus is released and only a small percentage of the cells are found to express viral antigens. In view of this variety, it is possible that there may be no common mechanism for the establishment and/or maintenance of the persistently infected state.

Establishment of morbillivirus persistence

Morbilliviruses readily establish persistent infections in a wide variety of mammalian cell lines. Following an acute phase of virus infection, surviving colonies of cells are seen to grow up (Rustigian, 1962). This so-called 'cell-survivor' technique has been successfully applied in the establishment of a number of cell lines persistently infected with measles and SSPE viruses (Norrby, 1967; Minagawa, 1971a; Knight et al., 1972; Burnstein et al. 1974; Menna et al., 1975; Gould & Linton, 1975; Joseph et al., 1975; Wild & Dugre, 1978; Ju et al., 1978). This suggests either that a small proportion of cells in the population are resistant to the challenge or that the events necessary to establish persistence are relatively rare. When such lines have been cured of virus, or recloned to isolate uninfected cells, they are no more resistant to infection than the original cells (Minagawa, 1971a; Minagawa & Kuwajima, 1975). This supports the latter possibility or implies that such cellular resistance is a transient phenomenon. Other workers have stressed the importance of undiluted passage virus in the establishment of persistence by both measles (Rima et al., 1977; Chiarini et al., 1976) and canine distemper viruses (ter Meulen & Martin, 1976). Such passaging could increase the proportion of defective interfering (DI) viruses in the inoculum and thus favour cell survival during the initial infection.

Antibody has also been used to help establish (Gould & Linton, 1975) or alter the characteristics of a persistent infection (Rustigian, 1966). Antibody can directly block virus-induced cell fusion which is mainly responsible for cell death (Graves et al., 1978). Furthermore antibody acting at the plasma membrane may modulate the actual expression of viral antigens both intracellularly and at the cell surface (Joseph & Oldstone, 1975; Gould & Almeida, 1977; Fujinami & Oldstone, 1979, 1980). This could bring about a more long-term effect on virus-induced cytopathic effect (CPE) and promotion of cell survival. In some cases interferon may be involved in the establishment of persistence as shown for vesicular stomatitis virus (VSV) (Nishiyama, 1977). Measles virus is able to infect both B and T lymphoid cell lines persistently (Joseph et al., 1975; Minagawa et al., 1976; Minagawa & Sakumura, 1977; Ju et al., 1978). These lines are known to be continuously producing low levels of interferon (Adams et al., 1975). A role for this substance cannot therefore be excluded in these cases.

All of these factors could act to aid cell survival of the initial infection. Events following this stage are not clear. Crises of rapidly increasing CPE followed by recovery are common before a more stable cell population is formed (Rustigian, 1966; Minagawa, 1971a; Burnstein et al. 1974; Menna et al., 1975; ter Meulen & Martin, 1976; Ju et al. 1978; Wild & Dugre, 1978). It is possible that during this period virus and cellular replication may become linked. Some cell lines display increased virus expression if cellular metabolism or growth rate are perturbed artificially (Menna et al., 1975; Flanagan & Menna, 1976; May & Menna, 1979), or spontaneously (Minagawa, 1971a). Changes in virus expression with cell cycle phase have also been reported. Cells in stationary phase express viral antigen to a lesser extent than those actively dividing (Ehrnst, 1979). Such a synchronization could involve direct action on the cell nucleus since measles virus is known to induce chromosome breakage in lytic infection (Csonka et al. 1975), rearrangement during persistence (Heneen, 1976) and nucleolar enlargement (Macintyre & Armstrong, 1976; Heneen, 1978). Once established, the persistent infection may be maintained by one or more of these factors acting in concert. Furthermore the actual processes involved may be continuously changing.

Mechanism of persistence

The persistent state may be maintained by several factors. Mechanisms to be considered are integration of viral genetic information into that of the host, virus-mediated interference (DI particles), host-mediated interference (interferon) or selection of temperature-sensitive (ts) virus mutants. All of these processes have been implicated in virus persistence.

Integration of virus information is known to occur during retrovirus replication and has been observed during visna virus multiplication (Haase & Varmus, 1973). There exists one report of measles virus-specific DNA in the host genome (Zhdanov, 1975). However, this has not been confirmed, and so does not seem to be involved in morbillivirus persistence. Unlike the situation in VSV persistent infections, where DI particles have been detected, data on this aspect of persistence is lacking for the morbilliviruses. Morbillivirus DI particles are themselves poorly characterized. There is a similar lack of data concerning the role of interferon in this process, especially since much work has been carried out in cells which do not produce this compound. However, a role for both of these phenomena cannot be excluded and more information is required on both of these points.

During persistent infection by other paramyxoviruses, e.g. Newcastle disease virus (NDV), there is an apparent selection for ts mutants (Preble & Youngner, 1972, 1973) and mutants isolated did not complement each other. A similar situation has also been reported for Sindbis virus (Shenk et al., 1974). The ts nature itself may be unimportant since ts(+) revertants maintain their small-plaque morphology and ability to re-establish persistence (Haspel et al., 1973; Shenk et al., 1974). Furthermore Holland and co-workers have shown that during a persistent infection of L cells by VSV, mutations rapidly accumulate in the virus genome. These were manifested as a variety of small-plaque ts mutants (Holland et al., 1979). Multiple mutations could result in slower virus growth and diminished virulence. Such a situation may well also apply to morbilliviruses. It is frequently observed that virus recovered from persistent infections displays a small-plaque phenotype, is often temperature sensitive for growth, and may establish further persistent infections more readily than the parent virus (Haspel et al., 1973; Gould & Linton, 1975). Virus protein expression (Fisher & Rapp, 1979) or release of infectious virus (Haspel et al., 1973) may often be increased by lowering the temperature of cultivation. Ju et al. (1978) demonstrated a highly heterogeneous group of ts mutants associated with measles virus persistence in two B cell lines. The measles virus released from persistent infections is altered with time (Rustigian, 1966; Wechsler et al., 1979b; Wild et al., 1981). Furthermore Rozenblatt et al. (1979a) identified electrophoretic migration alterations in virus-specific proteins immunoprecipitated from lytic and persistent infections. More recently, studies in our laboratory have utilized monoclonal antibodies directed against the H polypeptide to demonstrate antigenic differences produced in this molecule during persistence (ter Meulen et al., 1981).

The possibility of a host factor in maintenance of the persistent state has been raised by several authors. Ter Meulen & Martin (1976) observed an interferon-like, non-sedimentable, non-UV labile factor in medium from vero cells persistently infected with CDV. These cells do not produce interferon, and the factor concerned was further differentiated from interferon by its sensitivity to neutralization by anti-CDV serum. Winston & Rustigian (1979) made a similar observation using Hela cells persistently infected with Edmonston measles virus. This factor was distinguished from interferon by its specific inhibition of measles virus growth.

Examination of the regulated state within the persistently infected cell has so far been inconclusive. Flanagan & Menna (1976) found that synthesis of protein but not of RNA was required for the expression of haemadsorption following enucleation of the BGM/MV cell line. This could be interpreted as suggesting that mRNA was present but not translated, or that any H protein produced was not inserted into the plasma membrane in the presence of the nucleus. H protein was not however detected in cell lysates. More recently Stephenson et al. (1981) have obtained further evidence for some form of translational control in hamster brain cells persistently infected with the SSPE virus LEC. Membrane protein (M) was readily produced by translation in vitro of extracted mRNA but could not be detected in cell extracts. A lack of M protein in non-productive SSPE cell lines has also been reported by Hall & Choppin (1979) and Lin & Thormar (1980). This finding is of potentially great significance in the understanding of SSPE since M protein might be a trigger in the final assembly of enveloped viruses (McSharry et al., 1971). Lack of this protein could therefore bring about a non-productive infection. Examination of some persistent infections does in fact suggest that virus replication is halted at a late stage and neither RNA nor protein synthesis was required for virus induction by cocultivation techniques (Knight et al., 1972). This may not however always be the case since examination of those cells used by Flanagan & Menna (1976) showed similarities to an early stage in the replication cycle.

The possibility of replication inhibition at more than one stage is also raised by the work of Rustigian et al. (1979), using a Hela cell line persistently infected with Edmonston virus. This line (KllA-HGl) released no infectious virus and had poor membrane antigen expression. However, cocultivation with vero cells susceptible to measles virus infection led to the formation of syncytia containing much virus antigen and nucleocapsid. Replication could still not be completed, even in these cells. The virus was therefore able to replicate further and express F protein activity, but was inhibited at a later stage. Whether this was due to virus dysfunction or cellular regulation is not known.

All morbillivirus persistently infected cells examined contain internal nucleocapsids. These are, as in the lytic infection, of 'smooth' appearance in the nucleus and 'rough' in the cytoplasm. In some cases they are known to be infectious (Rozenblatt *et al.*, 1979b). The infection could thus be transmitted vertically at division and so maintain the persistent state in the absence of virus release. Nucleocapsids have been observed to be excluded from the mitotic apparatus of dividing cells suggesting their distribution would be random. Some persistent infections consist of a mixture of infected and uninfected cells which are inseparable by cloning. This could be explained by dilution of virus structures if cell division proceeds more rapidly than virus replication. Alternatively, non-equal partition may occur (Ehrnst, 1979).

In conclusion, morbillivirus persistence at the cellular level is an extremely complex process. Many factors may be involved in establishment and maintenance of this state. In general, any factor which promotes cell survival of the initial infection may help to establish the persistent stage. There is certainly a virus-specific (Minagawa *et al.*, 1976) and a cell-specific (Cremier *et al.*, 1979) function in establishment of persistence. In this respect persistence is similar to DI virus induction (Kang & Allen, 1978; Kang & Tischfield, 1978). No detailed information is available concerning the roles of DI viruses or interferon in maintenance of persistence, but ts mutants have often been demonstrated. Further investigations along these lines are required.

PATHOGENIC ASPECTS OF MORBILLIVIRUS PERSISTENCE *IN VIVO*

The association of measles and canine distemper viruses with a chronic inflammatory CNS disorder has led to intensive studies of the pathogenic mechanisms leading to these diseases. In dogs and man the development of the CNS disease is preceded by an acute infection, months to years before neurological signs become apparent. This suggests that in both hosts the defence mechanisms which control the acute infection and eliminate infectious virus, may occasionally fail and permit the agents to establish residence in host cells. Obviously, it must be decided during the acute infection whether a CNS complication may develop at a later stage. Therefore, attempts to unravel the pathogenesis of SSPE or canine distemper encephalomyelitis have to investigate the virus–cell and virus–host interactions in subacute or chronic diseases and also during acute infection.

Virus–host interaction in acute morbillivirus infections

Acute measles

The exposure of a seronegative host to measles virus leads to primary infection of the upper respiratory epithelial cells and associated lymphoid tissue. Thereafter, dissemination occurs into the respiratory, intestinal and renal tracts as well as lymphoid tissues in many parts of the body (reviewed by Fraser & Martin, 1978). During viraemia virus is carried by lymphocytes and transported widely, suggesting measles virus has a special preference for these cells. This association has a direct effect and this is illustrated by the development of multinucleated, giant cells in lymphoid tissue and chromosome breakage in peripheral lymphocytes. These changes have been linked to virus replication in these cells (Nichols et al., 1962; Tanzer et al., 1963; Mori, 1969) and moreover measles virus is known to grow well in both B and T lymphocyte tissue cultures (Joseph et al., 1975).

The involvement of the lymphoid cells in measles infection can lead to a functional impairment of the immune system as originally observed by von Pirquet (1908) and confirmed by others (Starr & Berkovich, 1964; Fireman et al., 1969). During or shortly after acute measles, tuberculous children develop a partial or total skin anergy against tuberculin or other antigens, and activation of a tuberculous process or a secondary infection may occur. In most instances, this infection of the lymphoid system is not harmful to healthy children but it is conceivable that occasionally an impairment of the specific immune functions against measles virus itself may occur through measles virus infection. The possibility that the immune system fails to control and eliminate measles virus has been suggested by Burnet (1968) as one possible pathogenic mechanism in SSPE. These observations have stimulated detailed studies to understand the defence mechanisms during measles. It was found that a functional humoral immune response is not essential in overcoming the acute disease and in controlling the virus infection (Good & Zak, 1959). Children with agammaglobulinaemia recover from acute measles without major difficulties in contrast to those with a cellular immune defect. Patients with congenital, acquired or iatrogenic T cell deficiency develop severe measles often accompanied by fatal complications (Gatti & Good, 1970). Moreover, in cases of leukaemia or lymphosarcoma (with or without immunosuppressive treatment) measles virus infection can quite often lead to a CNS complication in the form of an acute or subacute encephalitis developing weeks or months later (Breitfeld *et al.*, 1973; Drysdale *et al.*, 1976; Mellor, 1976; Pullan *et al.*, 1976; Sluga *et al.*, 1975; Smyth *et al.*, 1976; Wolinsky *et al.*, 1977; Haltia *et al.*, 1978).

These clinical studies underline the important role of the cell-mediated immune defence mechanism in overcoming acute measles. Obviously, in cases of a CNS complication, measles virus enters the CNS during viraemia to induce an acute encephalitis (ter Meulen *et al.*, 1972) or enters into a cellular association which later develops into SSPE. Measles virus can only be isolated during the prodromal stage and early at the beginning of the acute measles rash since later with the appearance of neutralizing antibodies all extracellular virus disappears. Therefore, invasion of the CNS at a later stage, when neutralizing antibodies are present, is only conceivable if virus is carried in infected cells (perhaps lymphocytes or macrophages). This hypothesis has not been proven experimentally for measles virus but in general the blood/brain barrier is not impervious to viruses or lymphocytes (Johnson & Mims, 1968; Prineas & Wright, 1978).

Acute canine distemper
The course of infection in acute canine distemper has been well studied experimentally by several groups (Coffin & Liu, 1957; Mickwitz & Schröder, 1968; Appel, 1969, 1970; Vandevelde & Kristensen, 1977). Two days after exposure to virus, CDV antigen was detected by immunofluorescence in mononuclear cells in the tonsils and bronchial lymph nodes. From these areas virus is readily spread by mononuclear blood cell elements to the spleen, thymus, bone marrow and lymph nodes, especially those associated with the intestine. In the experiments of Appel (1969, 1970) approximately 50% of infected animals rapidly developed antibody which elimin-

ated the virus. Viral antigens could no longer be detected in organ material 2–3 weeks after infection. The other 50% of animals, which failed to produce measurable neutralizing antibody titres, developed severe clinical symptoms. Moreover viral antigens were continuously demonstrable in all the above-mentioned tissues as well as the surface epithelia of the alimentary, respiratory and urogenital tracts and in exocrine and endocrine glands. In addition, canine distemper virus could be detected in brain tissue, firstly in the meningeal macrophages, then in cells in the perivascular areas, ependymal region and later in glial cells and neurones. Animals showing a CNS involvement developed neurological signs and convulsions, leading to death 3–4 weeks after infection. However, some dogs developed a CNS disease 40–60 days after infection and the distribution of viral antigens in CNS tissue was similar to cases of late distemper encephalitis (Appel, 1970). A positive fluorescence was detectable in brain, white matter, neurones and glial cells as well as in the pituitary gland. Outside the brain, virus antigens were observed in the epithelial cells of the alimentary, respiratory and urogenital tracts as well as foot-pad gland and epidermal cells.

These earlier investigations were recently extended by Appel and co-workers to characterize the pathways by which canine distemper virus enters the CNS (Summers et al., 1978, 1979). It could be shown that lymphocytes infected with the virus were present in the CNS tissue of all animals, 8–10 days after virus exposure. These cells were widely distributed throughout the CNS, occurring in both perivascular cuffs and adjacent tissue with no apparent predilection for certain sites. These initial inflammatory changes were followed 14–16 days later by demyelination which occurred in well-defined subependymal foci. Moreover, in these areas giant cells containing nucleocapsid structures were identified, indicating virus replication in these cellular elements. At this stage signs of inflammation in brain tissue were hardly detectable, which has been linked to the pronounced lymphopenia existing in these animals (Summers et al., 1979). Lymphopenia is consistently observed in acute canine distemper as a result of lymphatic tissue infection by the virus. Virus can be found in lymphocytes and in lymph nodes which themselves may reveal areas of necrosis, indicating a strong tropism of the virus to this cell type (McCullough et al., 1974). Quite often, this infection has an immunosuppressive effect as detected by laboratory assays (Krakowka et al., 1975). Normally, dogs that recover are able to clear virus from lymphatic tissue by a potent immune response.

However, in those cases where the immune response is impaired, CDV spreads throughout the body causing a generalized infection involving many different organs (Appel & Gillespie, 1972).

The early detection of CDV-infected lymphocytes in the brain tissue of every dog examined is important in relation to the pathogenesis of morbillivirus CNS infection. Obviously, CDV is disseminated within the body during viraemia by lymphocytes. In these cells viral antigens and nucleocapsid structures are recognizable suggesting that partial replication at least is possible. Whether infectious virus is produced and the lymphocytes are destroyed as a result of viral replication is unknown. However, it is conceivable that such a virus infection may lead to an impairment of the immune system. Alteration of immunological functions by virus infection has been shown for a variety of agents including members of the morbilliviruses (Woodruff & Woodruff, 1975a). In general, either enhancement or suppression of immune reactivity may occur which can influence the outcome of the virus infection. Measles and canine distemper infection induce a general suppression of cell-mediated immunity in the host. However, no detailed or definite information is available about the immune reaction against these two agents in those cases which later develop a CNS complication.

The occurrence of a pronounced lymphopenia during the acute disease as observed in acute measles and canine distemper may be significant and such drastic decreases in lymphocyte counts may be the result of lymphocyte destruction by the virus or alteration of lymphocyte traffic. It has been observed that virus infection can affect distribution of lymphocytes in the body as well as the lymphocyte recirculation between blood and lymphoid tissue (Woodruff & Woodruff, 1975b). Alteration of recirculation through lymphoid tissue could interfere with the induction of immunity, since interaction between immunologically competent cells and antigen occurs there. Studies with Friend leukaemia virus or Newcastle disease virus in mice reveal that these agents interfere with the homing of lymphocytes to lymph nodes or spleen (Bainbridge & Bendinelli, 1972; Woodruff & Woodruff, 1972). The mechanism by which such disturbances in lymphocyte distribution are induced by viruses is unknown. However, it has been suggested that the integrity of the lymphocyte surface is essential for these cells to recirculate normally (Woodruff & Gesner, 1969). Cell membrane changes occur in the course of measles and CDV replication and absorption of virus-specific immunoglobulins may take place. This could perhaps lead to abnormal distribution of lymphocytes in the body. Whether this phenomenon is connected with the processes by which lymphocytes reach the central nervous system in acute canine distemper is at present unknown.

Virus–host interactions in persistent morbillivirus infections

The interaction of infectious virus with susceptible cells may lead to cell destruction, persistent infection or cell transformation. In acute measles or canine distemper, lytic infection with accompanying cell death is the predominant form of infection. Clinical signs are mainly the result of cell destruction and inflammatory reactions towards virus infection. A symbiosis between virus and host cells is established in many persistent infections of tissue culture systems and gross morphological changes are not necessarily produced. However, little is known about the influence of virus persistence on cell function. Studies of neural cells in cultures (Oldstone *et al.*, 1977; Halbach & Koschel, 1979; Koschel & Halbach, 1979) have shown that a significant effect on luxury functions such as production of acetylcholine, acetyltransferase or acetylcholineesterase is observed in persistent lymphocytic choriomeningitis (LCM), measles or rabies virus infection, without interference with cell morphology, growth rate or protein synthesis. Such changes occurring in cells of the central nervous system could have a profound effect on brain function leading eventually to a disease with a defined clinical symptomatology. The virus–cell interaction at a molecular level may be expected to play an important role in the extent to which a persistent infection may interfere with brain function.

Molecular biological aspects of morbillivirus persistence in CNS tissue

CNS infection by measles or canine distemper virus does not necessarily lead to virus persistence. It is usual for an acute disease process to develop when these agents enter the CNS compartment (ter Meulen *et al.*, 1972; Appel & Gillespie, 1972). Normally, virus spread throughout certain brain areas is followed by destruction of brain cells and signs of inflammation. Only occasionally does infection of brain cells not result in destructive lytic infection. In these instances it is possible that a persistent infection is established and rapid spread and dissemination of virus in CNS tissue is prevented by unknown factors. Such events may be a prerequisite for the later onset of subacute or chronic encephalomyelitis.

No information on the molecular basis of morbillivirus persistence in CNS tissue in available. The main theoretical models for RNA virus persistence implicate virus mutation, defective interfering particles, interferon, selection of susceptible host cells and integration of virus genetic material into that of the host cell by means of reverse transcription (reviewed by Rima & Martin, 1977; Morgan & Rapp, 1977). Some evidence for all of these phenomena has been found in tissue cultures persistently infected with measles or canine distemper viruses although the precise role for any one of them is not clear.

In SSPE these mechanisms of persistence have been discussed and proposed, but so far no experimental evidence is available which can explain this phenomenon unequivocally. The occasional isolation of measles virus from SSPE brain tissue, and its aetiological connection with this rare disease, led to the suggestion that these isolates were a mutant or a variant of measles virus. This was supported by the observation that SSPE isolates vary in host range, growth rate, plaque size, type of haemagglutination, neuropathogenicity as well as electrophoretic mobility of viral proteins, mRNAs and complexity of viral genome RNAs (reviewed by Fraser & Martin, 1978; ter Meulen et al., 1981). In addition, differences in antigenicity of certain structural proteins and a reduced sensitivity to neutralization by immune sera in comparison to measles virus, have been described. However, none of these findings represents a single, stable property of SSPE viruses which is solely a characteristic of measles virus isolates from CNS tissue, since these changes have also been occasionally observed between 'regular' measles viruses. Moreover, there is no epidemiological evidence for typical clusters or spread of infection to suggest the existence of an independent SSPE strain within the measles virus population. Therefore, at the present time it seems to be a valid assumption that the virus which has already caused acute measles may also be responsible for the development of SSPE in each patient. It is conceivable that during persistent infection of the brain by measles virus, mutational changes occur as have been observed during vesicular stomatitis virus persistent infections (Holland et al., 1979; Rowlands et al., 1980). This could explain the observed biological and biochemical differences between measles and SSPE viruses.

The mode of measles virus infection of CNS tissue, and of persistence in SSPE, are the main pathogenic questions in this disease. Direct information on the molecular biological events allowing such virus–cell interaction in brain tissue is not available, but recently some indirect evidence was obtained by analysing the response of patients against individual virus structural polypeptides using immunoprecipitation techniques. It could be shown that in general, SSPE patients lack an antibody response to measles virus M protein, a major structural protein of the virus, whereas the other polypeptides were immunoprecipitated (Hall *et al.*, 1979b; Stephenson & ter Meulen, 1979; Wechsler *et al.*, 1979a). This finding in SSPE is very surprising in view of the humoral hyperimmune reaction of these patients to the other structural proteins of measles virus. This may imply that in SSPE, measles virus M protein is not synthesized in infected brain cells or perhaps is so strongly bound to intracellular structures that exposure to the immune system does not take place. Alternatively, it is conceivable that an immune defect towards M protein could exist, possibly as the consequence of lymphocyte infection by the virus during acute infection. None of these different hypotheses has been proven for SSPE patients. Yet some experimental evidence from tissue culture studies would suggest a deficiency in M protein synthesis in non-productive cell lines persistently infected with measles virus (Hall & Choppin, 1979; Lin & Thormar, 1980; Stephenson *et al.*, 1981a, b). Moreover, attempts to detect measles virus M protein in SSPE brain material directly by a modified immunoprecipitation technique failed to show the presence of this viral protein (Hall & Choppin, 1981). Certainly, a more definite proof of these interpretations will come from studies of fresh SSPE brain material using monoclonal antibodies to measles virus M protein since the methods hitherto used for detection of protein synthesis in persistent infection *in vitro* or organ material are not sensitive enough to prove the absence of certain proteins unequivocally. However, the apparent lack of measles virus M protein in SSPE patients would suggest that the expression of this particular viral protein is blocked either at the level of transcription or at the level of translation. This process may be host cell controlled since it is possible to overcome it by cocultivation experiments as demonstrated by successful isolation of infectious SSPE virus from brain tissue.

In contrast to these observations, evidence from chronic canine distemper encephalomyelitis or old dog encephalitis suggest that CDV persistence may involve a different mechanism. Recently, Imagawa and co-workers (1980) were successful in isolating infectious canine distemper virus directly from infected brain tissue without application of cocultivation techniques. Moreover, frozen brain from some animals could be titrated and assayed directly in ferrets which are very susceptible to canine distemper virus. These results are surprising since both diseases have certain similarities to SSPE where identical experiments have always failed. Obviously, canine distemper virus replicates fully in brain cells, yielding infectious virus, yet rapid viral spread within the CNS as in acute infection, does not take place. The fact that a full replication cycle occurs, with synthesis of all structural proteins, is reflected by a complete immune response against the viral polypeptides. Immunoprecipitation of radio-actively labelled canine distemper virus polypeptides by serum and CSF specimens from dogs with chronic encephalomyelitis or old dog encephalitis revealed the presence of antibodies directed against all of the virus polypeptides including M (Hall *et al.*, 1979a).

The virological data obtained in SSPE and canine distemper virus-induced CNS diseases suggest that morbillivirus persistence may operate by different molecular biological mechanisms. However, these virological differences do not lead to diverse diseases and the CNS disorders induced by the two agents have many clinical features and neuropathological changes in common.

Immune response and disease process
One of the immunological hallmarks of SSPE is an oligoclonal hyperimmune response to measles virus antigens. Such high antibody activities can be detected in serum and CSF specimens at any time during the course of this CNS process. Since the first stage of the disease is usually diagnosed retrospectively, no information is available about the measles antibody titres at the early period or within the time period after acute measles and onset of SSPE. Recently, we had the opportunity to test a serum specimen from an SSPE case, which was taken for another medical reason, one year before the development of the neurological symptomatology. This specimen contained high measles antibody titres indicating that hyperimmunization against measles virus was already present before, or at the beginning of, this disease. The pronounced oligoclonal hyperimmune response to measles virus is very unusual for a virus infection since a polyclonal antibody response is normally found. Such an unusual reaction could be explained if SSPE patients were genetically high responders to measles antigens. However, such a situation seems unlikely since SSPE siblings have normal measles

antibody levels (Jabbour & Sever, 1968; Brody *et al.*, 1972). It is conceivable that in SSPE, suppressor cells which normally control antibody synthesis to measles virus have been affected at the time of acute measles. If such a situation were to occur a hyperimmune reaction could develop as found in SSPE.

The presence of oligoclonal antibodies in SSPE has led to the hypothesis that such a response could favour the occurrence of virus mutants, since antibodies directed against only certain determinants of the parental virus would increase the probability of a mutant escaping neutralization. The development of mutants in the presence of antiviral antibodies has been seen in influenza as well as visna virus infections. Influenza mutants have been selected under suitable experimental conditions *in vivo* and *in vitro* (Laver & Webster, 1968). A recurrent infection of influenza virus in the same host could never be linked to a mutant virus. However, such a phenomenon does occur in visna (Narayan *et al.*, 1977) as well as in equine infectious anaemia (Kono *et al.*, 1973). In these diseases mutants develop in the presence of host antibodies, and give rise to recurrent infections. These are responsible for the progressive nature of the visna disease process (Gudnadottir, 1974; Narayan *et al.*, 1977). and for the recurring relapses in equine infectious anaemia. Such phenomena have not been detected in SSPE patients. Moreover, evidence for an immune pathological disease process as suggested for visna (Nathanson *et al.*, 1976), has not been obtained for SSPE. So far, the reported specific and non-specific cellular immunity blocking factors associated with the plasma of SSPE patients (Allen *et al.*, 1973; Swick *et al.*, 1976; Ahmed *et al.*, 1974) obviously do not operate *in vivo*. Treatment of patients with immunosuppressive drugs, or removal of such factors from the CNS by drainage of cerebrospinal fluid, did not lead to any observable clinical changes. In addition, the interpretation of a pathogenic involvement of antibodies or antigen–antibody complexes in the disease is complicated by reports that SSPE has been observed in patients with hypogammaglobulinaemia and immunodeficiency syndrome (Hanissian *et al.*, 1972; Allison, 1972). Therefore, the available data do not support the notion that the antimeasles virus immune response in SSPE contributes to the pathological processes in the CNS at the stage when the disease is clinically recognizable. However, it is still conceivable that the immune system plays an important pathogenic role at the time when measles virus first enters the CNS compartment.

In contrast to SSPE, the immune response to CDV persistent infection of brain tissue may be directed against some antigens of the CNS itself (Koestner & Krakowka, 1977). It was found that sera from dogs with distemper-associated demyelinating encephalomyelitis possessed a demyelinating activity when tested in explant cultures (Koestner *et al.*, 1974). The activity was associated with immunoglobulin M and believed to represent an antibody activity against basic myelin protein (Krakowka *et al.*, 1973). However, serum specimens from some control animals also contained such activity which suggests that this *in vitro* demyelinating effect may not play an important role in the pathogenesis of CDV induced encephalomyelitis. On the other hand, Krakowka and co-workers (1978) have recently carried out some interesting experiments which indicate that an antiviral immune response may modulate a CNS infection. Neonatal gnotobiotic dogs which had passively received antibodies against CDV, were infected with the virus. Both these animals and control dogs developed similar CNS diseases yet examination revealed distinct neuropathological and immunohistological differences between the groups. In the control group, small foci of subtle neuropathological changes consisting of neuronal degeneration accompanied by microgliosis were restricted to the grey matter of the frontal lobe. In contrast, brains from animals which had received antibodies, showed extensive lesions in the spinal cord as well as in the grey matter of the frontal lobe. Moreover, viral antigen distribution resembled that observed in ODE. These studies indicated clearly that the extent of immune protection can influence the course of infection and subsequent development of the disease process. Similar considerations may be important in SSPE since epidemiological studies indicate that a relatively high percentage of SSPE patients do acquire acute measles within the first two years of life (Agnarsdottir, 1977). Perhaps maternal antibodies are able to exert a similar effect on the course of measles virus infection in humans.

CONCLUSIONS AND FUTURE ASPECTS

It is surprising that a morbillivirus persistent infection associated with a disease process is only detectable in the central nervous system. Obviously, this does not exclude the possibility that these agents also establish persistence in other tissues, but functional impairment of the organ, leading ultimately to clinical symptomatology and morphological changes, has not been observed outside the CNS. It is conceivable that the anatomical and physiological peculiarities of the brain play an important role in the extent to which a virus infection can cause dysfunction. In this context, it has to be emphasized that the nervous system contains a highly differentiated cell population with complex functionally integrated cell-to-cell connections and highly specialized cytoplasmic membranes. Moreover, CNS tissue is unique in its high metabolic rate and relative lack of regenerative capacity. Consequently a persistent virus infection might interfere with normal functions whereas in other organs, with lower energy requirements and higher rates of regeneration, persistent infections could perhaps be tolerated.

The development of a life-long immunity with the presence of measurable antibody titres after acute measles or canine distemper strongly suggests persistence of these agents in the host and this is supported by the observation that multiple reinfection with these agents is not required for this type of immunity. More than 100 years ago Panum (1847) observed on the Faeroe Islands an outbreak of measles after this infectious disease had disappeared for 60 years. Almost everyone of the inhabitants acquired measles except those who were older than 60 years and had contacted measles in early childhood. These persons were protected by an effective immunity in the absence of re-exposure to measles virus. Perhaps measles virus resides in lymphoid tissue itself, and thus produces a continuous immune stimulation as suggested by the isolation of measles virus from lymph node material in a case of SSPE (Horta-Barbosa, 1971). Certainly this aspect has to be further studied and evaluated in relationship to immune persistence.

The ubiquitous occurrence of measles and canine distemper viruses and their association with chronic disease processes has led to speculation that these agents may also be responsible for other diseases of the CNS. One of the prime candidates is multiple sclerosis (MS) in which both viruses have been incriminated (reviewed by ter Meulen & Stephenson, 1981). Measles virus infection has been linked to MS on the basis of serological studies. It could be shown by many groups (reviewed by Norrby, 1978) that the majority of MS patients revealed a statistically significant, higher antibody titre than control groups. In addition, this increase in measles antibody titres was also demonstrable in CSF associated with an oligoclonal immune reaction. However, in contrast to SSPE where the majority of oligoclonal CSF IgG reacts with measles virus, in MS only a very small fraction of oligoclonal IgG carries antimeasles activities. Moreover, other antiviral antibodies were found in both serum and CSF of a number of patients. Those activities were directed against a variety of viruses such as rubella, mumps, herpes simplex and parainfluenza viruses (Salmi *et al.*, 1972; Norrby *et al.*, 1974; Nordal *et al.*, 1979). In CSF these antibodies were also of oligoclonal nature, present in small quantities and occasionally occurring simultaneously in the same patient. The fact that the specificity of most oligoclonal IgG in MS patients is not identifiable and not associated with any antiviral activity to known viruses, has led to the interpretation that the detectable viral antibodies may represent an immunological epiphenomenon which is not aetiologically related to MS. In general, the oligoclonal antibody response with synthesis of antiviral antibodies in the CNS tissue by invading lymphocytes, may represent a regular local immune response to virus replication in CNS tissue. Such a phenomenon has been shown in cases of mumps meningitis (Vandvik *et al.*, 1978a) and chronic progressive rubella panencephalitis (Vandvik *et al.*, 1978b). In these virus infections, virus has been demonstrated in brain tissue and it is possible that the presence of these viral antigens lead to invasion of sensitized lymphocytes. So far, all attempts to isolate a virus from MS brain tissue, which could be aetiologically linked to this disease, have failed. However, recently measles virus RNA was detected in MS brain tissue by a nucleic acid hybridization technique (Haase *et al.*, 1981). By applying *in situ* hybridization using a labelled randomly transcribed cDNA from measles virus virion RNA, a positive reaction was obtained in two foci in one of four cases of MS. This observation is quite interesting and may lead to a resurrection of theories suggesting measles virus as a possible aetiological factor for this disease; however, at present more information about the biochemical probe used and the specificity of the reaction is required before presence of measles virus in MS brain tissue can be accepted.

The possible association of canine distemper virus with multiple sclerosis has been derived from epidemiological studies (Cook & Dowling, 1977; Cook *et al.*, 1978; Cook *et al.*, 1979a, b). It was found that MS patients did have more contact with dogs exposed to CDV than a control population. Moreover, in one serological study it was claimed that MS patients exhibit canine distemper-specific antibodies, since these patients revealed a slightly higher activity to the virus antigens than matched controls (Cook *et al.*, 1979). However, other epidemiological evaluations did not confirm the possible association of canine distemper virus with MS. (Poskanzer *et al.*, 1974; Bauer & Wikstrom 1977; Kurtzke & Priester, 1979; Nathanson *et al.*, 1978). Moreover, in a detailed immunological study, in which the immune response of MS patients to measles and canine distemper virus was analysed by an immunoprecipitation technique, it could be shown that MS patients do not have CDV-specific antibodies (Stephenson *et al.*, 1980a). MS patients immunoprecipitated all measles virus polypeptides except the M protein, but only the nucleocapsid and F proteins of canine distemper virus. In contrast, sera from dogs infected with canine distemper virus immunprecipitated all canine distemper virus polypeptides. If MS patients had been exposed to canine distemper and infected by this virus, a full immune response should be expected. Therefore, the observed immune reactions to canine distemper virus in MS serum specimens can be interpreted merely as cross-reaction between antigenically related proteins.

Despite the fact that at present morbillivirus persistent infections could only be identified as aetiological agents for few CNS diseases, there is circumstantial evidence that these agents may be associated with other disorders as well. Advancement of our understanding of these processes will greatly depend on the progress made in analysing the different interactions of morbillivirus with cells and with the host. Increasing knowledge of the molecular basis of morbillivirus persistence will aid the search for viral genomic information in suspected diseases and also lead to a better understanding of host reactions to persistent infection.

This work is supported by the Deutsche Forschungsgemeinschaft. We thank Helga Kriesinger for typing the manuscript.

REFERENCES

ADAMS, A., STRANDER, H. & CANTELL, K. (1975). Sensitivity of the Epstein-Barr virus transformed human lymphoid cell lines to interferon. *Journal of General Virology*, **28**, 207–17.

AGNARSDOTTIR, G. (1977). Subacute sclerosing panencephalitis. In *Recent Advances in Clinical Virology*, ed. Waterson, pp. 21–49. Edinburgh: Churchill Livingstone.

AHMED, A. A., STRONG, D. M., SELL, K. W., THURMAN, G. B., KNUDSEN, R. C., WISTAR, R. JR & GRACE, W. R. (1974). Demonstration of a blocking factor in the plasma and spinal fluid of patients with subacute sclerosing panencephalitis. I. Partial characterization. *Journal of Experimental Medicine*, **139**, 902–24.

ALLEN, J., OPPENHEIM, J., BRODY, J. A. & MILLER, J. (1973). Labile inhibitor of lymphocyte transformation in plasma from a patient with subacute sclerosing panencephalitis. *Infection and Immunity*, **8**, 80–2.

ALLISON, A. C. (1972). Immunity and immunopathology in virus infections. *Annals of the Institute of Pasteur*, **123**, 585–608.

APPEL, M. J. (1969). Pathogenesis of canine distemper. *American Journal of Veterinary Research*, **30**, 1167.

APPEL, M. J. (1970). Distemper pathogenesis in dogs. *Journal of the American Veterinary Medical Association*, **156**, 1681.

APPEL, M. J. & GILLESPIE, J. H. (1972). Canine distemper virus. *Virology Monographs*, **11**, 1–96.

APPEL, M. J., GIBBS, E. P., MARTIN, S. J., TER MEULEN, V., RIMA, B. K., STEPHENSON, J. R. & TAYLOR, W. P. (1981). Morbillivirus diseases of animals and man. In *Comparative Diagnosis of Viral Diseases*, ed. E. Kurstak, London & New York: Academic Press.

BAINBRIDGE, D. R. & BENDINELLI, M. (1972). Circulation of lymphoid cells in mice infected with Friend leukemia virus. *Journal of the National Cancer Institute*, **49**, 773–81.

BAUER, H. J. & WIKSTROM, J. (1977). Multiple sclerosis and house pets. *Lancet*, 1029.

BELLMAN, M. H. & DICK, G. (1980). Surveillance of subacute sclerosing panencephalitis. *British Medical Journal*, **281**, 393–4.

BLAESE, R. M. & HOFSTRAND, H. (1975). Immunocompetence in patients with SSPE. *Archives of Neurology*, **32**, 494–5.

BREITFELD, V., HASHIDA, Y., SHERMAN, F. E., ODAGIRI, K. & YUNI, E. J. (1973). Fatal measles infection in children with leukemia. *Laboratory Investigations*, **28**, 279.

BRODY, J. A. & DETELS, R. (1970). Subacute sclerosing panencephalitis: a zoonosis following aberrant measles. *Lancet*, **ii**, 500–1.

BRODY, J. A., DETELS, R. & SEVER, J. L. (1972). Measles-antibody titres in siblings of patients with subacute sclerosing panencephalitis and controls. Lancet, **i**, 177–8.

BROOR, S. (1975). Virological and pathological study of SSPE. *Indian Journal of Medical Research*, **63**, 671.

BURNET, F. M. (1968). Measles as an index of immunological function. *Lancet*, **ii**, 610–13.

BURNSTEIN, T., JACOBSEN, L. B., ZEMAN, W. & CHEN, T. T. (1974). Persistent infection of BSC-1 cells by defective measles virus derived from subacute sclerosing panencephalitis. *Infection and Immunity*, **10**, 1378–82.

CAPE, C. A., MARTINEZ, A. J., ROBERTSON, J. T., HAMILTON, R. & JABBOUR, J. T. (1973). Adult onset of subacute sclerosing panencephalitis. *Archives of Neurology*, **28**, 124–7.

CHAO, D. (1962). Subacute inclusion body encephalitis. Report of three cases. *Pediatrics*, **61**, 501–10.

CHIARINI, A., SINATRA, A., AMMATUNA, P. & DI STEFANO R. (1976). Studies on a measles virus variant inducing persistent infections in cultured cells. I. Isolation and characterization of plaque purified virus clones. *Archives of Virology*, **52**, 47–58.

CHIN-MEI, Y. & SZU-CHIH, W. (1977). Subacute sclerosing panencephalitis. Clinical and immunological investigations of two cases. *Chinese Medical Journal*, **6**, 419–22.

COFFIN, D. L. & LIU, C. (1957). Studies on canine distemper infection by means of fluorescein-labeled antibody. II. The pathology and diagnosis of the naturally occurring disease in dogs and the antigenic nature of the inclusion body. *Virology*, **3**, 132–45.

CONNOLLY, J. H. (1968). Additional data on measles virus antibody and antigen in subacute sclerosing panencephalitis. *Neurology*, **18**, 87–9.

CONNOLLY, J. H., ALLEN, I. V., HURWITZ, L. J. & MILLER, J. H. D. (1967). Measles-virus antibody and antigen in subacute sclerosing panencephalitis. *Lancet*, **i**, 542–4.

CONNOLLY, J. H., HAIRE, M. & HADDEN, D. S. M. (1971). Measles immunoglobulins in subacute sclerosing panencephalitis. *British Medical Journal*, **1**, 23–5.

COOK, S. D. & DOWLING, P. C. (1977). A possible association between house pets and multiple sclerosis. *Lancet*, **i**, 980–2.

COOK, S. D., DOWLING, P. C. & RUSSELL, W. C. (1979a). Neutralizing antibodies to canine distemper and measles virus in multiple sclerosis. *Journal of Neurological Science*, **41**, 61–70.

COOK, S. D., DOWLING, P. C., NORMAN, J. & JABLON, S. (1979b). Multiple sclerosis and canine distemper in Iceland. *Lancet*, 380–1.

COOK, S. D., NATELSON, B. H., LEVIN, B. E., CHAVIS, P. S. & DOWLING, P. C. (1978). Further evidence of a possible association between house dogs and multiple sclerosis. *Annals of Neurology*, **3**, 141–3.

CORDY, D. R. (1942). Canine encephalomyelitis. *Cornell Veterinary*, **32**, 11–28.

COWDRY, E. B. (1934). The problem of intranuclear inclusions in virus diseases. *Archives of Pathology*, **18**, 527–42.

CREMIER, N. E., OSHIRO, L. S. & HAGENS, S. J. (1979). Cyclic expression of antigen and infectious virus in a BHK cell line (0–853) persistently infected with an SSPE strain of measles virus. *Journal of General Virology*, **42**, 637–40.

CSONKA, E., RUZICKSA, P. & KUCH, A. S. (1975). Chromosome aberrations in cells infected with various strains of measles virus. *Acta Microbiologica Academica Scienciarum Hungarica*, **22**, 41–4.

CUTLER, R. W. P. & AVERILL, D. R. (1969). Cerebrospinal fluid gamma globulins in canine distemper encephalitis. *Neurology*, **19**, 1111–14.

CUTLER, R. W. P., MERLER, E. & HAMMERSTAD, J. P. (1968). Production of antibody by the central nervous system in subacute sclerosing panencephalitis. *Neurology*, **18**, 129–32.

DAYAN, A. D., GOSTLING, J. V. T., GREAVES, J. L., STEVENS, D. W. & WOODHOUSE, M. A. (1967). Evidence of a pseudomyxovirus in the brain in subacute leucoencephalitis. *Lancet*, **i**, 980–1.

DETELS, R., BRODY, J. A., McNEW, J. & EDGAR, A. H. (1973). Further epidemiological studies of subacute sclerosing panencephalitis. *Lancet*, **ii**, 11–14.

DONNER, M., WALTIMO, O., PORRAS, J., FORSIUS, H. & SAUKKONEN, A. K. (1972). Subacute sclerosing panencephalitis as a cause of chronic dementia and relapsing brain disorder. *Journal of Neurosurgery and Psychiatry*, **35**, 180–5.

DRYSDALE, H. C., JONES, L. F., OPPENHEIMER, D. R. & TOMLINSON, A. H. (1976). Measles inclusion-body encephalitis in a child with treated acute lymphoblastic leukaemia. *Journal of Clinical Pathology*, **29**, 865.

EHRNST, A. (1979). Growth phase related loss of measles virus surface-associated antigens and cytotoxic susceptibility in persistently infected cells. *Journal of General Virology*, **45**, 547–56.

ERNO, H. & MÖLLER, T. (1961). Epidemiological studies on dog distemper. *Nordic Veterinary Medicine*, **16**, 654–74.

FIREMAN, P., FRIDAY, G. & KUMATE, J. (1969). Effect of measles vaccine on immunologic responsiveness. *Pediatrics*, **43**, 264–72.

FISCHER, K. (1965). Einschlußkörperchen bei Hunden mit Staupeenzephalitis und anderen Erkrankungen des Zentralnervensystems. *Pathological Veterinary*, **2**, 380–410.

FISHER, L. E. & RAPP, F. (1979). Temperature-dependent expression of measles virus structural proteins in persistently infected cells. *Virology*, **94**, 55–60.

FLANAGAN, T. D. & MENNA, J. H. (1976). Induction of measles virus hemagglutinin in a persistently infected, non-virogenic line of cells (BGM/MV). *Journal of Virology*, **17**, 1052–5.

FRASER, K. B. & MARTIN, S. J. (1978). *Measles Virus and its Biology*. London & New York: Academic Press.

FREEMAN, J. M., MAGOFFIN, R. L., LENNETTE, E. H. & HERNDON, R. M. (1967). Additional evidence of the relations between subacute inclusion-body encephalitis and measles virus. *Lancet*, **ii**, 129–31.

FUJINAMI, R. S. & OLDSTONE, M. B. A. (1979). Antiviral antibody reacting on the plasma membrane alters measles virus expression inside the cell. *Nature*, **279**, 529–30.

FUJINAMI, R. S. & OLDSTONE, M. B. A. (1980). Alterations in expression of measles virus polypeptides by antibody: molecular events in antibody-induced antigenic modulation. *Journal of Immunology*, **125**, 78–85.

GAINES, K. J. & JABBOUR, J. T. (1979). Subacute sclerosing panencephalitis during pregnancy. *Archives of Neurology*, **46**, 314–16.

GATTI, J. M. & GOOD, R A. (1970). The immunological deficiency diseases. *Medical Clinic of North America*, **54**, 281–307.

GERSON, K. L. & HASLAM, R. H. A. (1971). Subtle immunologic abnormalities in four boys with subacute sclerosing panencephalits. *New England Journal of Medicine*, **285**, 78–81.

GIBSON, J. P., GRIESEMER, R. A. & KOESTNER, A. (1965). Experimental distemper in the gnotobiotic dog. *Journal of Pathological Veterinary*, **2**, 1–19.

GLASNER, H. & KIRSCH, W. (1975). Ätiopathogenetische und therapeutische Aspekte der subakuten sclerosierenden Encephalitis. *Archiv für die Psychiatrie und Nervenkrankheiten*, **221**, 29–38.

GOOD, R. A. & ZAK, S. J. (1959). Disturbances in gamma-globulin synthesis as experiments of nature. *Pediatrics*, **18**, 109–49.

GOULD, E. A. & LINTON, P. E. (1975). The production of a temperature-sensitive persistent measles virus infection. *Journal of General Virology*, **28**, 21–8.

GOULD, J. J. & ALMEIDA, J. D. (1977). Antibody modification of measles in vitro infection. *Journal of Medical Virology*, **1**, 111–18.

GRAVES, M. C., SILVER, S. M. & CHOPPIN, P. W. (1978). Measles virus polypeptide synthesis in infected cells. *Virology*, **86**, 254–63.

GUDNADOTTIR, M. (1974). Visna-maedi in sheep. *Progress in Medical Virology*, **18**, 336–49.

HAASE, A. T. & VARMUS, H. E. (1973). Demonstration of a DNA provirus in the lytic growth of visna virus. *Nature*, **254**, 237–9.

HAASE, A. T., VENTURA, P., GIBBS, C. J. JR & TOURTELOTTE, W. W. (1981). Measles virus nucleotide sequences: detection by hybridization in situ. *Science*, **212**, 672–5.

HALBACH, M. & KOSCHEL, K. (1979). Impairment of hormone dependent signal transfer by chronic SSPE virus infection. *Journal of General Virology*, **42**, 615–19.

HALL, W. W. & CHOPPIN, P. W. (1979). Evidence for the lack of synthesis of the M polypeptide of measles virus in brain cells from SSPE. *Virology*, **99**, 443–7.

HALL, W. W. & CHOPPIN, P. W. (1981). Measles virus proteins in the brain tissue of patients with subacute sclerosing panencephalitis. Absence of the M protein. *New England Journal of Medicine*, **304**, 1152–5.

HALL, W. W., IMAGAWA, D. T. & CHOPPIN, P. W. (1979a). Immunological evidence for the synthesis of all canine distemper virus polypeptides in chronic neurological diseases in dogs. Chronic distemper and old dog encephalitis differ from SSPE in man. *Virology*, **98**, 283–7.

HALL, W. W., LAMB, R. A. & CHOPPIN, P. W. (1979b). Measles and subacute sclerosing panencephalitis virus proteins: lack of antibodies to the M protein in patients with subacute sclerosing panencephalitis. *Proceedings of the National Academy of Sciences, USA*, **76**, 2047–51.

HALSEY, N. A., MODLIN, J. F. & JABBOUR, J. T. (1978). Subacute sclerosing panencephalitis (SSPE): an epidemiologic review. In *Persistent Viruses*, ed. J. G. Stevens, G. J. Todaro & C. F. Fox, pp. 101–14. New York & London: Academic Press.

HALSEY, N. A., MODLIN, J. F., JABBOUR, J. T., DUBEY, L., EDDINS, D. L. & LUDWIG, D. A. (1980). Risk factors in subacute sclerosing panencephalitis: a case-control study. *American Journal of Epidemiology*, **111**, 415–24.

HALTIA, M., PAETAU, A., VAHERI, A. & SALMI, A. (1978). Measles encephalopathy during immunosuppression (MEI). *Scandinavian Journal of Infectious Diseases*, **10**, 159.

HANISSIAN, A. S., JABBOUR, J. T., DE LAMPERENS, S., GARCIA, J. H. & HORTA-BARBOSA, L. (1972). Subacute encephalitis and hypogamma-globulinemia. *American Journal of Diseases in Children*, **123**, 151–5.

HASPEL, M. V., KNIGHT, P. R., DUFF, R. G. & RAPP, F. (1973). Activation of a latent measles virus infection in hamster cells. *Journal of Virology*, **12**, 690–5.

HENEEN, W. K. (1976). The chromosome complement of a measles-carrier human cell line in comparison to the cell line of origin. *Hereditas*, **83**, 91–104.

HENEEN, W. K. (1978). Silver staining and nucleolar patterns in human heteroploid and measles-carrier cells. *Hereditas*, **88**, 213–27.

HOLLAND, J. J., GRABAU, E. A., JONES, C. L. & SEMLER, B. L. (1979). Evolution of multiple genome mutations during long-term persistent infection by vesicular stomatitis virus. *Cell*, **16**, 495–504.

HORTA-BARBOSA, L., HAMILTON, R., WITTIG, B., FUCCILLO, D. A., SEVER, J. L. & VERNON, M. L. (1971). Subacute sclerosing panencephalitis: isolation of suppressed measles virus from lymph node biopsy. *Science*, **173**, 840–1.

HOUFF, S. A., MADDEN, D. L. & SEVER, J. L. (1979). Subacute sclerosing panencephalitis in only one of identical twins. *Archives of Neurology*, **36**, 854–6.

IMAGAWA, D. T., HOWARD, E. B., VAN PELT, L. F., RYAN, C. P., BUI, H. D. & SHAPSHAK, P. (1980). Isolation of canine distemper virus from dogs with chronic neurological diseases. *Proceedings of the Society of Experimental Biology and Medicine*, **164**, 355–62.

INNES, J. R. M. & SAUNDERS, L. Z. (1962). *Comparative Neuropathology*. New York & London: Academic Press.

IWASAKI, Y. & KOPROWSKI, H. (1974). Cell to cell transmission of virus in the central nervous system. *Laboratory Investigations*, **31**, 187–96.

JABBOUR, J. T., DUENAS, D. A., SEVER, J. L., KREBS, H. M. & HORTA-BARBOSA, L. (1972). Epidemiology of subacute sclerosing panencephalitis (SSPE). *Journal of the American Medical Association*, **220**, 959–62.

JABBOUR, J. T. & SEVER, J. L. (1968). Serum measles antibody titres in patients with subacute sclerosing panencephalitis compared with parents and siblings. *Journal of Pediatrics*, **73**, 905–7.

126 VOLKER TER MEULEN AND MICHAEL J. CARTER

JOHNSON, R. T. & MIMS, C. A. (1968). Pathogenesis of viral infections of the nervous system. *New England Journal of Medicine*, **278**, 84–92.

JONCAS, J., GEOFFROY, G., MCLAUGHLIN, B., ALBERT, G., LAPOINTE, N., DAVID, P., LAFONTAINE, R. & GRANGER-JULIEN, M. (1974). Subacute sclerosing panencephalitis. Elevated Epstein-Barr virus antibody titres and failure of amantadine therapy. *Journal of Neurological Sciences*, **21**, 381–90.

JOSEPH, B. S., LAMPERT, P. W. & OLDSTONE, M. B. A. (1975). Replication and persistence of measles virus in defined subpopulations of human leukocytes. *Journal of Virology*, **16**, 1638–49.

JOSEPH, B. S. & OLDSTONE, M. B. A. (1975). Immunologic injury in measles virus infection. II. Suppression of immune injury through antigenic modulation. *Journal of Experimental Medicine*, **142**, 864–76.

JU, G., UDEM, S., RAZER-ZISMAN, B. & BLOOM, B. (1978). Isolation of a heterogenous population of temperature sensitive mutants of measles virus from persistently infected human lymphoblastoid cell lines. *Journal of Experimental Medicine*, **147**, 1637–52.

KALM, H. (1952). Über die Stellung der Panencephalitis (Pette-Döring) zur Leucoencéphalite sclérosante subaigue (van Bogaert). *Deutsche Zeitschrift für Nervenheilkunde*, **168**, 322.

KANG, C. Y. & ALLEN, R. (1978). Host function dependent induction of defective interfering particles of vesicular stomatitis virus. *Journal of Virology*, **25**, 202–6.

KANG, C. Y. & TISCHFIELD, J. A. (1978). Host gene control in generation of vesicular stomatitis defective interfering virus particles. In *International Virology IV*, p. 240. Centre for Agricultural Publishing and Documentation, Wageningen.

KIESSLING, W. R., HALL, W. W., YUNG, L. L. L. & TER MEULEN, V. (1977). Measles-virus-specific immunoglobulin M response in subacute sclerosing panencephalitis. *Lancet*, **i**, 324–7.

KINGSBURY, D. W., BRATT, M. A., CHOPPIN, P. W., HANSON, R. P., HOSAK, Y., TER MEULEN V., NORRBY, E., PLOWRIGHT, W., ROTT, R. & WUNNER, W. H. (1978). Paramyxoviridae. *Intervirology*, **10**, 137–52.

KNIGHT, P., DUFF, R. & RAPP, F. (1972). Latency of human measles virus in hamster cells. *Journal of Virology*, **10**, 995–1001.

KOESTNER, A. & KRAKOWKA, S. (1977). A concept of virus-induced demyelinating encephalomyelitis relative to an animal model. In *Slow Virus Infections of the Central Nervous System*, ed. V. ter Meulen & M. Katz. Berlin: Springer-Verlag.

KOESTNER, A., MCCULLOUGH, B., KRAKOWKA, G. S., LONG, J. F. & OLSEN, R. G. (1974). Canine distemper. A virus-induced demyelinating encephalomyelitis. In *Slow Virus Diseases*, ed. W. Zeman & E. H. Lennette, pp. 86–101. Baltimore: Williams & Wilkins.

KONO, Y., KOBAYASHI, K. & FUKUNAGA, Y. (1973). Antigenic drift of equine infectious anaemia virus in chronically infected horses. *Archiv für die gesamte Virusforschung*, **41**, 1–10.

KOSCHEL, K. & HALBACH, M. (1979). Rabies virus infection selectively impairs membrane receptor functions in neuronal model cells. *Journal of General Virology*, **42**, 627–32.

KRAKOWKA, S., COCKERELL, G. L. & KOESTNER, A. (1975). Effects of canine distemper virus on lymphoid function in vitro and in vivo. *Infection and Immunity*, **11**, 1069–78.

KRAKOWKA, S., CONFER, A. & KOESTNER, A. (1974). Evidence for transplacental transmission of canine distemper virus. *American Journal of Veterinary Research*, **35**, 1251–3.

KRAKOWKA, S., HOOVER, E. A., KOESTNER, A. & KETERING, K. (1977). Experimental naturally occurring transplacental transmission of canine distemper virus. *American Journal of Veterinary Research*, **38**, 919–22.

KRAKOWKA, S., McCULLOUGH, B., KOESTNER, A. & OLSEN, R. G. (1973). Myelin-specific autoantibodies associated with central nervous system demyelination in canine distemper virus infection. *Infection and Immunity*, **8**, 819–27.

KRAKOWKA, S., MADOR, R. A. & KOESTNER, A. (1978). Canine distemper virus-associated encephalitis: modification by passive antibody administration. *Acta Neuropathologica (Berlin)*, **43**, 235–41.

KRETH, H. W., TER MEULEN V. & ECKERT, G. (1975). HL-A and subacute sclerosing panencephalitis. *Lancet*, **ii**, 415–16.

KURENT, J. E., SEVER, J. L. & TERASAKI, P. I. (1975). HA-AW29 and subacute sclerosing panencephalitis. *Lancet*, **i**, 927–8.

KURTZKE, J. F. & PRIESTER, W. A. (1979). Dogs, distemper, and multiple sclerosis in the United States. *Acta Neurologica Scandinavia*, **60**, 312–19.

LAIDLAW, P. P. & DUNKIN, G. W. (1928). A report upon the cause and prevention of dog distemper. *British Veterinary Journal*, **84**, 596–637.

LAVER, W. G. & WEBSTER, R. C. (1968). Selection of antigens mutants of influenza viruses. Isolation and peptide mapping of their hemagglutinating protein. *Virology*, **34**, 193–202.

LIN, F. H. & THORMAR, H. (1980). Absence of M protein in a cell-associated subacute sclerosing panencephalitis virus. *Nature*, **285**, 490–2.

LOWENTHAL, A. (1964). *Ager Gel Electrophoresis in Neurology*. Amsterdam: Elsevier.

LUSTER, M. I., ARMEN, R. C., HALLUM, J. V., AHMED, A. & LESLIE, G. A. (1978). Immunoglobulin class distribution of measles virus antibodies in serum and spinal fluids of patients with subacute sclerosing panencephalitis. *International Archives of Allergy and Applied Immunology*, **56**, 488–92.

McCULLOUGH, B., KRAKOWKA, S. & KOESTNER, A. (1974). Experimental canine distemper virus-induced demyelination. *Laboratory Investigations*, **31**, 216–22.

McDONALD, R., KIPPS, A. & LEARY, P. M. (1974). Subacute sclerosing panencephalitis in the Cape province. *South African Medical Journal*, **48**, 7–9.

McGRATH, J. T. (1960). Distemper complex. In *Neurologic Examination of the Dog*, pp. 127–36. Philadelphia, Pennsylvania: Lea & Febiger.

MACINTYRE, E. H. & ARMSTRONG, J. A. (1976). Fine structural changes in human astrocyte carrier lines for measles virus. *Nature*, **263**, 232–4.

McSHARRY, J. J., COMPANS, R. W. & CHOPPIN, P. W. (1971). Proteins of vesicular stomatitis virus and of phenotypically mixed vesicular stomatitis virus-simian virus 5 virions. *Journal of Virology*, **8**, 722–9.

MAY, J. D. & MENNA, J. H. (1979). Changes in the virus-host-cell relationship in a stable non-virogenic cell line persistently infected with measles virus (BGM/MV). *Journal of General Virology*, **45**, 185–94.

MEHTA, P. D., KANE, A. & THORMAR, H. (1977). Quantitation of measles virus-specific immunoglobulins in serum, CSF, and brain extract from patients with subacute sclerosing panencephalitis. *Journal of Immunology*, **118**, 2254–61.

MELLOR, D. (1976). Encephalitis and encephalopathy in childhood leukaemia. *Developmental Medicine and Child Neurology*, **18**, 90.

MENNA, J. H., COLLINS, A. R. & FLANAGAN, T. D. (1975). Characterization of an *in vitro* persistent-state measles virus infection and virological characterization of the BGM/MV cell line. *Infection and Immunity*, **11**, 152–8.

TER MEULEN, V., ENDERS–RUCKLE, G. MÜLLER, D. & JOPPICH, G. (1969). Immunhistological, microscopical and neurochemical studies on encephalitides.

III. Subacute progressive panencephalitis. Virological and immunohistological studies. *Acta Neuropathological (Berlin)*, **12**, 244–59.

TER MEULIN, V. & HALL, W. W. (1978). Slow virus infections of the nervous system: virological, immunological and pathogenetic considerations. *Journal of General Virology*, **41**, 1–25.

TER MEULEN, V., LÖFFLER, S., CARTER, M. J. & STEPHENSON, J. R. (1981). Antigenic characterization of measles and SSPE virus haemagglutinin by monoclonal antibodies. *Journal of General Virology*, (in press).

TER MEULEN, V. & MARTIN, S. J. (1976). Genesis and maintenance of a persistent infection by canine distemper virus. *Journal of General Virology*, **32**, 431–40.

TER MEULEN, V., MÜLLER, D. & JOPPICH, C. (1967). Fluorescence microscopy studies of brain tissue from a case of subacute progressive panencephalitis. *German Medical Monthly*, **12**, 438–41.

TER MEULEN, V., MÜLLER, D., KÄCKELL, Y., KATZ, M. & MEYERMANN, R. (1972). Isolation of infectious measles virus in measles encephalitis. *Lancet*, **ii**, 1172–5.

TER MEULEN, V. & STEPHENSON, J. R. (1981). The possible role of viral infections in multiple sclerosis and other related demyelinating diseases. In *Multiple Sclerosis*, ed. J. Hallpike, C. Adams & W. Tourtellotte. London: Chapman & Hall (in press.)

TER MEULEN, V., STEPHENSON, J. R. & KRETH, H. W. (1981). Subacute sclerosing panencephalitis. In *Comprehensive Virology*, vol. 18, ed. H. Fraenkel-Conrat & R. R. Wagner. New York: Plenum Press. (In press.)

MICKWITZ, C. U. V. & SCHRÖDER, H. D. (1968). Histologische, fluoreszenzserologische und histochemische Untersuchungen zum Verhalten der Einschlußkörperchen bei der Staupe des Hundes. *Zentralblatt der Veterinärmedizin*, **15**, 453.

MINAGAWA, T. (1971a). Studies on the persistent infection with measles virus in Hela cells. I. Clonal analysis of cells of carrier cultures. *Japanese Journal of Microbiology*, **15**, 325–31.

MINAGAWA, T. (1971b). Studies on the persistent infection with measles virus in Hela cells. II. The properties of carried virus. *Japanese Journal of Microbiology*, **15**, 333–40.

MINAGAWA, T. & KUWAJIMA, S. (1975). Studies on the persistent infection of measles virus in Hela (Hela/MV) cells. VI. Electron microscopic observation of Hela/MV cells. *Japanese Journal of Microbiology*, **20**, 347–50.

MINAGAWA, T., SAKUMURA, T., KUWAJIMA, S., YAMAMOTO, T. K. & IIDA, H. (1976). Characterization of measles viruses in establishment of persistent infections in human lymphoid cell line. *Journal of General Virology*, **33**, 361–79.

MINAGAWA, T. & SAKUMURA, T. (1977). Growth of measles virus in various human lymphoid cell lines. *Microbiology and Immunology*, **21**, 23–31.

MORGAN, E. M. & RAPP, R. (1977). Measles virus and its associated diseases. *Bacteriological Reviews*, **41**, 636–66.

MORI, M. (1969). Some aspects of chromosome aberrations in leucocyte cultures from measles patients. *La Kromosomo*, **77–8**, 2510–16.

NAJERA, R., SAIZ, G., HERRERA, I. & VALENCIANO, L. (1972). Serological and tissue culture observations from cases of subacute sclerosing panencephalitis. *Annals of the Institute of Pasteur*, **123**, 565–70.

NARAYAN, O., GRIFFIN, D. E. & CHASE, J. (1977). Antigenic shift of visna virus persistently infected sheep. *Science*, **197**, 376–8.

NATHANSON, N., MILLER, A., MARTIN, J. & FISCHMANN, H. (1978). Multiple sclerosis, canine distemper and measles immunization. *Lancet*, 1204.

NATHANSON, N., PANITCH, H., PALSSON, P. A., PETURSSON, G. & GEORGSSON, G.

(1976). Pathogenesis of visna. II. Effect of immunosuppression upon early central nervous system lesions. *Laboratory Investigations*, **35**, 444–51.

NELSON, R. F., DENNERY, J. M., MONTPEPTIT, V. & FURESZ, J. (1972). SSPE and pregnancy. *Lancet*, **i**, 1289.

NICHOLS, W. W., LEVEN, A., HALL, B. & ÖSTERGREN, G. (1962). Measles-associated chromosome breakage. *Hereditas*, **48**, 367–70.

NISHIYAMA, Y. (1977). Studies of L cells persistently infected with VSV: factors involved in the regulation of persistent infections. *Journal of General Virology*, **35**, 265–79.

NORDAL, H. J., VANDVIK, B. & NORRBY, E. (1979). Oligoclonal virus antibodies in healthy and neurological patients. In *Humoral Immunity in Neurological Diseases*, ed. D. Karcher, A. Lowenthal & A. D. Strosberg pp. 249–302. New York: Plenum Press.

NORRBY, E. (1967). A carrier cell line of measles virus in Lu 106 cells. *Archiv für die gesamte Virusforschung*, **20**, 215–24.

NORRBY, E. (1978). Viral antibodies in multiple sclerosis. In *Progress in Medical Virology*, vol. 24, ed. J. L. Melnick, pp. 1–39. Basel: S. Karger.

NORRBY, E., LINK, H. & OLSSON, J. E. (1974). Comparison of measles virus antibody titers in cerebrospinal fluid and serum from patients with multiple sclerosis and from controls. *Archives of Neurology*, **30**, 285–92.

OKUNO, Y., NAKAO, T., ISHIDA, N., KONNO, T., MIZUTANI, H., SATO, T., ISOMUR, S., UEDA, S., KITAMURA, I. & KAJI, M. (1978). An epidemiological study of subacute sclerosing panencephalitis in Japan 1976. *Biken Journal*, **21**, 9–14.

OLDSTONE, M. B. A., HOLMSTOEN, J. & WEESE, R. M. (1977). Alterations of acetylcholine enzymes in neuroblastoma cells persistently infected with lymphocytic chloriomeningitis virus. *Journal of Cell Physiology*, **91**, 459–72.

D'ONGHIA, C. A., LEFÈVRE, A. O., CANELSA, H. M., GROSSMANN, R. M., SALLES-GOMES, L. F. & SALDANHA, P. H. (1974). Subacute sclerosing panencephalitis in only one member of monozygotic twin pair. *Journal of Neurological Science*, **21**, 323–33.

PANUM, P. L. (1847). Iagttageiser anstillede under maeslingeapidemien paa faeröerne i aaret 1846. *Bibliothek for Laeger*, **1**, 270–344.

PIRQUET, C. VON (1908). Das Verhalten der kutanen Tuberculinreaktion während der Masern. *Deutsche Medizinische Wochenschrift*, **34**, 1297–300.

POSKANZER, D. C., PRENNEY, L. B. & SHERIDAN, J. L. (1977). House pets and multiple sclerosis. *Lancet*, 1204.

PREBLE, O. T. & YOUNGNER, J. S. (1972). Temperature sensitive mutants isolated from L Cells persistently infected with Newcastle disease virus. *Journal of Virology*, **9**, 200–6.

PREBLE, O. T. & YOUNGNER, J. S. (1973). Selection of temperature sensitive mutants during persistent infection: role in maintenance of persistent Newcastle disease virus infections of L cells. *Journal of Virology*, **12**, 481–91.

PRINEAS, J. W. & WRIGHT, R. G. (1978). Macrophages, lymphocytes, and plasma cells in the perivascular compartment in chronic multiple sclerosis. *Laboratory Investigations*, **38**, 409–21.

PULLAN, C. R., NOBLE, T. C., SCOTT, D. J., WISNIEWKI, K. & GARDNER, P. S. (1976). Atypical measles infections in leukaemic children on immunosuppressive treatment. *British Medical Journal*, **1**, 1562.

RIMA, B. K., DAVIDSON, W. B. & MARTIN, S. J. (1977). The role of defective interfering particles in persistent infection of vero cells by measles virus. *Journal of General Virology*, **35**, 89–97.

RIMA, B. & MARTIN, S. J. (1977). Persistent infection of tissue culture cells by RNA viruses. *Medical Microbiology and Immunology*, **162**, 89–118.

RISK, W. S., HADDAD, F. S. & CHEMALI, R. (1978). Substantial spontaneous long-term improvement in subacute sclerosing panencephalitis. *Archives of Neurology*, **35**, 494–502.

ROCKBORN, G. (1958). A study of serological immunity against distemper in an urban dog population. *Archiv für die gesamte Virusforschung*, **8**, 493–9.

ROWLANDS, D., GRABAU, E., SPINDLER, K., JONES, C., SEMLER, B. & HOLLAND, J. (1980). Virus protein changes and RNA termini alterations evolving during persistent infection. *Cell*, **19**, 871–80.

ROZENBLATT, S., GORECKI, M., SHURE, H. & PRIVES, L. L. (1979a). Characterization of measles virus specific proteins synthesised in vitro and in vivo from acutely and persistently infected cells. *Journal of Virology*, **29**, 1099–106.

ROZENBLATT, S., KOCH, T., PINHASI, O. & BRATASIN, S. (1979b). Infective substructures of measles virus from acutely and persistently infected cells. *Journal of Virology*, **32**, 329–33.

RUSTIGIAN, R. (1962). A carrier state in Hela cells with measles virus (Edmonston strain) apparently associated with non-infectious virus. *Virology*, **16**, 101–4.

RUSTIGIAN, R. (1966). Persistent infection of cells in culture by measles virus. I. Development and characteristics of Hela sublines persistently infected with complete virus. *Journal of Bacteriology*, **92**, 1792–804.

RUSTIGIAN, R., WINSTON, S. H. & DARLINGTON, R. W. (1979). Variable infection of vero cells and homologous interferences after cocultivation with Hela cells with persistent defective infection by Edmonston measles virus. *Infection and Immunity*, **23**, 775–86.

SALMI, A. A., NORRBY, E. & PANELIUS, M. (1972). Identification of different measles virus-specific antibodies in the serum and cerebrospinal fluid from patients with subacute sclerosing panencephalitis and multiple sclerosis. *Infection and Immunity*, **6**, 248–54.

SEVER, J. L. & ZEMAN, W. (1968). Serological studies of measles and subacute sclerosing panencephalitis. *Neurology*, **18**, 95–7.

SHENK, T. E., KOSHELNYK, K. A. & STOLLAR, V. (1974). Temperature sensitive virus from Aedes albopictus cells chronically infected with sindbis virus. *Journal of Virology*, **13**, 439–47.

SLUGA, E., BUDKA, H., JELLINGER, K. & PICHLER, E. (1975). SSPE-like inclusion body disorder in treated childhood leukemia. *Acta Neuropathologica (Berlin)*, **6**, 267.

SMYTH, D., TRIPP, J. H., BRETT, E. M., MARSHALL, W. W., ALMEIDA, J., DAYAN, A. D., COLEMAN, J. C. & DAYTON, R. (1976). Atypical measles encephalitis in leukaemic children in remission. *Lancet*, **ii**, 574.

SOFFER, D., RANNON, L., ALTER, M., KAHANA, E. & FELDMAN, S. (1975). Subacute sclerosing panencephalitis: incidence among ethnic groups in Israel. *Israel Journal of Medical Sciences*, **11**, 1–4.

STARR, S. & BERKOVICH, S. (1964). Effects of measles, gammaglobulin modified measles and vaccine measles on the tuberculin test. *New England Journal of Medicine*, **270**, 386–91.

STEPHENSON, J. R. & TER MEULEN, V., (1979). Antigenic relationships between measles and canine distemper virus. Comparison of immune response in animals and humans to individual virus-specific polypeptides. *Proceedings of the National Academy of Sciences, USA*, **76**, 6601–5.

STEPHENSON, J. R., TER MEULEN, V. & KIESSLING, W. (1980). Search for canine-distemper-virus antibodies in multiple sclerosis. *Lancet*, 772–5.

STEPHENSON, J. R., SIDDELL, S. G. & TER MEULEN, V. (1981a). Persistent and lytic infections with SSPE virus; a comparison of the synthesis of virus-specific polypeptides. *Journal of General Virology*, (in press).

STEPHENSON, J. R., SIDDELL, S. G. & TER MEULEN, V. (1981b). Comparison of lytic and persistent measles virus infections by analysis of the synthesis, structure and antigenicity of intracellular virus-specific polypeptides. In *Replication of Negative Strand Viruses*, ed. D. H. L. Bishop & R. W. Compans, pp. 573–7. Elsevier/ North Holland, New York, Amsterdam, Oxford.

SUMMERS, B. A., GREISEN, H. A. & APPEL, M. J. G. (1978). Possible initiation of viral encephalitis in dogs by migrating lymphocytes infected with distemper virus. *Lancet*, **ii**, 187–90.

SUMMERS, B. A., GREISEN, H. & APPEL, M. J. G. (1979). Early events in canine distemper demyelinating encephalomyelitis. *Acta Neuropathologica (Berlin)*, **46**, 1–10.

SWICK, H. M., BROOKS, W. H., ROSZMAN, T. L. & CALDWELL, D. (1976). A heat-stable blocking factor in the plasma of patients with subacute sclerosing panencephalitis. *Neurology*, **26**, 84–8.

TANZER, J., STOITCHKOV, Y., HAREL, P. & BOIRON, M. (1963). Chromosomal abnormalities in measles. *Lancet*, **ii**, 1070–1.

THOMSON, D., CONNOLLY, J. H., UNDERWOOD, B. O. & BROWN, F. (1975). A study of immunoglobulin M antibody to measles, canine distemper and rinderpest viruses in sera of patients with subacute sclerosing panencephalitis. *Journal of Clinical Pathology*, **28**, 543–6.

VANDEVELDE, M. & KRISTENSEN, B. (1977). Observations on the distribution of canine distemper virus in the central nervous system of dogs with demyelinating encephalitis. *Acta Neuropathologica (Berlin)*, **40**, 233–6.

VANDVIK, B. (1973). Immunopathological aspects in the pathogenesis of subacute sclerosing panencephalitis, with special reference to the significance of the immune response in the central nervous system. *Annales of Clinical Research*, **5**, 308–15.

VANDVIK, B., NATVIG, J. B. & NORRBY, E. (1977). Subclass restriction of oligoclonal measles virus-specific IgG antibodies in patients with subacute sclerosing panencephalitis and in a patient with multiple sclerosis. *Scandinavian Journal of Immunology*, **6**, 651–7.

VANDVIK, B., NORRBY, E., NORDAL, H. J. & DEGRÉ, M. (1976). Oligoclonal measles virus-specific IgG antibodies isolated from cerebro-spinal fluids, brain extracts, and sera from patients with subacute sclerosing panencephalitis and multiple sclerosis. *Scandinavian Journal of Immunology*, **5**, 980–92.

VANDVIK, B., NORRBY, E., STEEN-JOHNSON, J. & STENSVOLD, K. (1978A). Mumps meningitis. Prolonged pleocytosis and mumps virus-specific oligoclonal IgG antibodies in the cerebrospinal fluid. *European Neurology*, **17**, 13–22.

VANDVIK, B., WEIL, M. L., GRANDIEN, M. & NORRBY, E. (1978b). Progressive rubella virus panencephalitis: synthesis of oligoclonal virus-specific IgG antibodies and homogenous free light chains in the central nervous system. *Acta Neurologica Scandinavia*, **57**, 53–64.

VARTDAL, F., VANDVIK, B. & NORRBY, E. (1980). Viral and bacterial antibody responses in multiple sclerosis. *Annals of Neurology*, **8**, 248–55.

WECHSLER, S. L., RUSTIGIAN, R., STALLCUP, K. C., BYERS, K. B., WINSTON, S. H. & FIELDS, B. N. (1979b). Measles virus-specific polypeptide synthesis in 2 persistently infected Hela cell lines. *Journal of Virology*, **31**, 677–84.

WECHSLER, S. L., WEINER, H. L. & FIELDS, B. N. (1979a). Immune response in subacute sclerosing panencephalitis: reduced antibody response to the matrix protein of measles virus. *Journal of Immunology*, **123**, 884–9.

WHITAKER, J. N., SEVER, J. L. & ENGEL, W. K. (1972). Subacute sclerosing panencephalitis in only one of identical twins. *New England Journal of Medicine*, **287**, 364–6.

WILD, T. F., BERNARD, A. & GREELAND, T. (1981). Measles virus: evolution of a persistent infection in BGM cells. *Archives of Virology*, **67**, 297–308.

WILD, T. F. & DUGRE, R. (1978). Establishment and characterisation of a subacute sclerosing panencephalitis. (Measles) virus persistent infection in BGM-cells. *Journal of General Virology*, **39**, 113–24.

WINSTON, S. H. & RUSTIGIAN, R. (1979). Enhancement and suppression by actinomycin D of a vero cell non-transmissible measles infection. *Infection and Immunity*, **24**, 967–70.

WOLINSKY, J. S., SWOVELAND, P., JOHNSON, K. P. & BARINGER, J. R. (1977). Subacute measles encephalitis complicating Hodgkin's disease in an adult. *Annals of Neurology*, **1**, 452.

WOODRUFF, J. F. & WOODRUFF, J. J. (1975a). The effect of viral infections on the function of the immune system. In *Viral Immunology and Immunopathology*, ed. A. L. Notkins. pp. 393–418. New York & London: Academic Press.

WOODRUFF, J. F. & WOODRUFF, J. J. (1975b). T. lymphocyte interaction with viruses and virus-infected tissues. In *Progress in Medical Virology*, vol. 19, ed. J. L. Melnick, pp. 120–60. Basel: S. Karger.

WOODRUFF, J. J. & GESNER, B. M. (1969). The effect of neuraminidase on the fate of transfused lymphocytes. *Journal of Experimental Medicine*, **129**, 551–67.

WOODRUFF, J. J. & WOODRUFF, J. F. (1972). Virus-induced alterations of lymphoid tissues. III. Fate of radiolabelled thoracic duct lymphocytes in rats inoculated with Newcastle disease virus. *Cellular Immunology*, **5**, 307–17.

ZHDANOV, V. (1975). Integration of viral genomes. *Nature*, **256**, 471–3.

CLASSICAL HERPES LATENCY REVISITED

P. WILDY, H. J. FIELD AND A. A. NASH

Department of Pathology, Tennis Court Road, Cambridge, UK

INTRODUCTION

In most virus infections of vertebrates the virus quickly establishes itself in susceptible cells, it is replicated and after a few days it and its progeny are eliminated by host-mediated influences. The host is then immune. Some viruses have evolved strategies by which they regularly avoid elimination and despite the development of immunity they persist in the host for long periods. The most elegant stratagem which has been recognized for more than half a century is the classical latency and reactivation exhibited by varicella-zoster and some other neurotropic herpesviruses. The enormous survival advantage gained by that virus has been pointed out by Hope-Simpson (1965); whereas a virus such as measles requires a sizeable host population e.g. 3×10^5 to survive varicella can make do with about 10^3.

The phenomenon is of course well known and the classical hypothesis, which we examine here, will be familiar. However, it is important to state the hypothesis if only to draw attention to the terminology we use. (*a*) There is an acute infection which may remain localized or become generalized – this is the *primary* infection. (*b*) there is apparent recovery but nonetheless some virus remains dormant in nervous tissue – in particular certain sensory ganglion cells; this is *latency*. (*c*) Virus may be reawakened spontaneously or as a result of external stimuli – so that infective virus may once again be isolated – this is *reactivation*. (*d*) The reactivated virus may on occasion initiate a peripheral lesion in the dermatome relating to the sensory ganglion – this is a *recrudescent lesion* and the phenomenon is *recrudescence*. (*e*) The whole phenomenon requires translation of virus from the periphery to the sensory ganglion and back again – that is believed to be via the axoplasm and will be called *axonal transport*. (*f*) Instances are described where virus evidently reactivates and passes to a peripheral site but fails to cause a noticeable lesion, although it probably multiplies and can be isolated; this is *recurrence*.

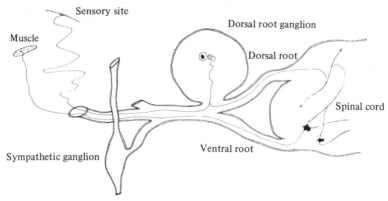

Fig. 1. Diagram showing the relationship of those parts of the nervous system involved in classical latency.

THE NEURODERMATOME

The primitive plan upon which all vertebrates are built is bilaterally symmetrical and segmental. Though this pattern is perturbed during development it is retained to a remarkable degree by the nervous system. Each segment of the cerebrospinal axis originates two dorsal sensory roots and two ventral motor roots; these combine to form a nerve trunk on each side comprising both sensory and motor nerve fibres. The cell bodies of the sensory neurones are situated in the dorsal root ganglia and the axons of these cells pass centripetally to the dorsal columns of the spinal cord and centrifugally to the skin. Each dorsal root ganglion innervates a patch of skin (Fig. 1), called a dermatome. The preferred area of operation by neurotropic herpesviruses and that ordinarily involved in classical latency is confined to the dermatome, the sensory fibres and the dorsal root ganglion. It is convenient to call this unit of the peripheral nervous system the *neurodermatome*.

This operational unit the neurodermatome is not anatomically isolated from the rest of the nervous system. Indeed there are nervous connexions with the central nervous system and via the *rami communicantes* with the sympathetic nervous system – viruses can and do pass by these connexions.

WHICH HERPESVIRUSES NATURALLY SHOW CLASSICAL LATENCY?

A large number of herpesviruses have been delineated. These are recognized by their characteristic morphology and all have large

DNA genomes. The viruses tend to persist in their natural hosts (Klein, 1976); some of them show classical latency, others do not – for example most of the herpesviruses that persist in the kidney.

The viruses that interest us here are necessarily neurodermatropic. They have a predilection for tissues of ectodermal origin (skin, pharyngeal mucosa, nervous tissue and adrenal medulla) but are not necessarily confined to them. They are very much host orientated – several examples are listed by McCarthy (1969) where infection of unnatural host species may result in florid disease (e.g. B virus of monkeys infecting man). It is therefore very important in compiling a list of viruses showing classical latency (Table 1) to restrict our consideration to the behaviour of viruses in their natural hosts for if we are defining a category at all it is a category of virus–host interaction – not of virus alone.

One criterion that has to be satisfied before a virus is acceptable in the list is evidence that it can remain latent in a dorsal root ganglion and be reactivated. At present evidence for this is provided by techniques of explant cultures and cocultivation (Stevens & Cook, 1971) but in the future other means of detecting intact genomes might equally well serve the purpose. Where data are lacking there could be a number of less satisfactory reasons for including a virus in our list; for instance, a virus for which recrudescence in a dermatome can be provoked by a specific stimulus would merit inclusion, for example bovine mammillitis virus (Probert & Povey, 1975).

Herpesviruses are notoriously unamenable material for the taxonomist; no satisfactory subdivision of the so called family has been devised. It is scarcely surprising therefore that the viruses that cause classical latency do not fall into any one established taxon. However, where acceptable viruses are listed (Table 1) several points are worth noting as matters of fact: (a) The viruses at the top of the table belong with Melnick's (1971) group A, i.e. they have wide host range in tissue culture and are not greatly cell associated. These viruses cause well-defined primary lesions in their natural hosts, seldom produce generalized rashes, are prone to spread to the nervous system and may be reactivated from dorsal root ganglia of immune animals by explant culture. There is no general pattern in the structure of their DNA, for example herpes simplex 1 and 2 and bovine mammillitis viruses have two inverted internal repeats of terminal sequences (as do Marek's disease virus and turkey herpes virus; P. Sheldrick, personal communication). On the other hand

Table 1. *Herpes viruses as candidates for classical latency*

Virus	Melnick's group[13]	%G+C[14]	Typical primary infection[a] Local or widespread	Typical primary infection[a] Pulmonary involvement	Natural host	Other experimental hosts	Recurrence or recrudescence	Reactivation from nervous tissue
B virus[1] (*Herpesvirus simiae*)	A	75	L	–	Rhesus	(Man), rabbit guinea-pigs, other simians	+ usually lethal	+
Pseudorabies virus[2] (Aujeszky's disease virus)	A	73	L-W	+	Swine	Rabbits, calves, mice and many other species	+ usually lethal	+
Spider monkey herpes virus[3]	A	71	L-W	+	Ateles	Guinea-pigs marmosets, rabbits	+? usually lethal	
Infectious bovine rhinotracheitis virus[4]	A	71	W[b]	+	Cattle	Rabbits, cotton tail rabbits	+	–?
Squirrel monkey herpes virus[d5] (*Herpesvirus tamarinus*)	A	67	L	?	Squirrel monkey	Mice, rabbits	+? +	+
Herpes simplex virus 2[6]	A	68	L	–	Man	Mice, rabbits, guinea-pigs	+	+
Herpes simplex virus 1[6]	A	66	L	–	Man		+	+
Bovine mammillitis virus[7]	A	64	L-W	–	Cattle	Mice, rabbits	+	–?

Virus		No.	L–W[b]		Natural host	Experimental host		
Equine abortion virus[8]	A	56		+	Horses	Hamsters, mice	+	–?
SA8[9]	A		L?	?	Vervet	Rabbit	+?	+
Varicella-zoster virus[10]	B	46	W	+	Man	None[e]	+	+
Feline rhinotracheitis virus[11]	B	46	W	+	Cats		+	+
Canine herpesvirus[12]	B	33	L	+	Dogs		+	–

[a] In the natural host, immunocompetent, non-neonate; [b] Genital infection is markedly localized; [c] Not to be confused with *Herpesvirus saimiri*; [d] Not to be confused with *Herpesvirus ateles*; [e] D-herpes virus of patas monkey (serologically related to varicella-zoster) may be a useful model. (Felsenfeld & Schmidt (1975). Immunological relationship between delta herpesvirus of patas monkeys and varicella-zoster virus of humans. *Infection and Immunity*, **12**, 261–6).

+?, probably; –?, tried but not yet successful; L, localized; W, widespread; 1–14 References (see following list):

1 Barahona et al., 1974; Gay & Holden, 1933; Hull & Nash, 1960; Melnick & Banker, 1954; Sabin & Wright, 1934; Vizoso, 1975a, b.
2 Baskerville, 1973; Baskerville et al., 1971; Becker, 1968; Field & Hill, 1974; Fraser & Ramachandran, 1969; Hurst, 1933; McCracken et al., 1973; McFerran & Dow, 1965; Sabó & Rajčáni, 1976; Toneva, 1976.
3 Barahona et al., 1974; Hull, 1972.
4 Andrewes et al., 1977; Davies & Duncan, 1974; Gibbs et al., 1975; Lupton & Reed, 1979; Narita et al., 1978; Sheffy et al., 1972.
5 Barahona et al., 1974; Holmes et al., 1964, 1966; McCarthy & Tosolini, 1975.
6 Baringer, 1975; Donnenberg et al., 1980b; Hill et al., 1980; Openshaw et al., 1979; Scriba, 1975, 1977; Spruance et al., 1977; Stevens & Cook, 1971; Stevens et al., 1972.
7 Martin et al., 1969; Probert & Povey, 1975.
8 R. Burrows (personal communication); Jackson et al., 1977; Plummer et al., 1970; Pursell et al., 1979.
9 Barahona et al., 1974; Malherbe & Harwin, 1958.
10 Bastian et al., 1974; Head & Campbell, 1900.
11 Gaskell & Povey, 1979; Love, 1971; Plummer et al., 1970.
12 Cornwell & Wright, 1969; Percy et al., 1971.
13 Melnick, 1971.
14 Andrewes et al., 1977; Honess & Watson. 1977.

pseudorabies has only one such internal repeat and resembles equine abortion virus (P. Sheldrick, personal communication) which has so far defied determined attempts to demonstrate reactivation by explantation (R. Burrows, personal communication). It is striking that the G + C molar proportions in the DNA of these viruses occupy the top end of the spectrum exhibited by the gamut of herpes viruses – i.e. 64% to 75% (cf. Honess & Watson, 1977). Probably several of these viruses are related since there is, for example, considerable sequence homology between herpes simplex type 1 and 2 and bovine mammillitis virus and strong antigenic crossing between these viruses and SA8 and B virus. (See Honess & Watson, 1977). (b) The viruses at the bottom of the table belong to Melnick's group B – they have narrow host range and tend to remain cell associated in tissue culture. These viruses tend to produce generalized rashes as part of the primary infection and respiratory involvement is prominent and probably necessary in the life cycle of the virus. The examples listed happen to have very low G + C molar proportions.

The distribution of examples suggests that there are two distinct subgroups of viruses that produce classical latency in their natural hosts. We do not suggest that this has taxonomic importance but it is interesting nosologically.

Finally, the two prototypes showing classical latency are herpes simplex and varicella-zoster viruses. Recurrent herpes simplex has been recognized for centuries:

> O'er ladies' lips, who straight on kisses dream,
> Which oft the angry Mab with blisters plagues
> *Romeo and Juliet*, Act I, Scene iv.

The primary infection was unrecognized until just before the war (Dodd, Buddingh & Johnson, 1939; Burnet & Williams, 1939). Before that herpes stood out as an infectious virus which was recognized only in patients who had neutralizing antibodies to it (Andrewes & Carmichael, 1930). Indeed Doerr (1938) regarded herpes as an entity that did not rely for its survival by infection between hosts and he visualized it as an endogenously generated phenomenon.

The recurrent disease has been described by many (e.g. Wheeler, 1975). Possibly the best account is that of Spruance *et al.* (1977) who studied 80 patients. Twenty-nine of these were used to provide a

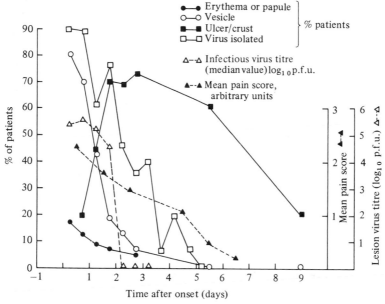

Fig. 2. Summary of recurrent herpes in man – after Spruance *et al.* (1977).

clear account of clinical findings. The remainder were used for virological studies. The general pattern is displayed in Fig. 2. Two-thirds of the patients had prodromal symptoms which generally lasted a few hours only. Pain was experienced early, gradually reducing. The progression erythema→ papule→ vesicle→ ulcer→ healing, was usual. No infectious virus was isolated during the prodromal or erythematous stage though in 3/6 attempts virus was retrieved by cocultivation. Infectious virus was isolated from 6/10 patients with papules and all patients with vesicles. As the vesicles ruptured, the success of isolation diminished quite rapidly and after two days virus could be isolated from fewer than half the patients though variation was such that 1/14 patients yielded virus as late as five days. Daniels, Le Goff & Notkins (1975) reported similar findings on fewer patients but included two other important facts. First, in two immunocompromised patients recurrent lesions persisted and virus was isolated on several occasions up to three weeks. Secondly, they demonstrated that infectious virus was frequently in the form of immune complexes.

The primary infection – varicella – has on the other hand been recognized for many years though it was not until the last century that it was separated from smallpox. Its relationship with shingles was not finally demonstrated until Kundratitz (1925) demonstrated that the virus present in the lesions of zoster produced varicella when inoculated into a non-immune host. But curiously the relationship of zoster with the sensory ganglia has been recognized for a longer time (von Bärensprung, 1863; Head & Campbell, 1900).

EXPERIMENTAL MODELS: HOW VALID ARE THEY?

Early examples of reactivation occurring in various laboratory animals began with spontaneous reactivation of herpes in rabbits (Perdrau, 1938). Good & Campbell (1948) precipitated recurrence in rabbit brain by anaphylaxis and Schmidt & Rasmussen (1960) by adrenalin; a finding confirmed in rabbit eye by Laibson & Kibrick (1966). Plummer, Hollingsworth, Phuangsab & Bowling (1970) detected 'latent' infection in spinal cord tissue of rabbits by cocultivation techniques which were exploited to excellent effect in rabbits and mice (Stevens & Cook, 1971; Stevens, Nesburn & Cook, 1972).

The predilection of herpes virus for nervous tissue and its movement in the nervous system had been shown much earlier. Friedenwald (1923) observed histological lesions in the trigeminal ganglia of rabbits infected on the cornea; a finding extended by Goodpasture & Teague (1923), Goodpasture (1925, 1929), Marinesco & Draganesco (1923) and Marinesco (1932), all using the rabbit.

Some aspects of herpes pathogenesis, latency and reactivation are reproducible in animal models using herpes simplex virus. However, few of the models provide a complete reproduction of the disease in man. The main points to note are that models employing herpes simplex virus in mice, rabbits and guinea-pigs resemble the human disease in many respects though recurrence and recrudescence have only been observed in mice inoculated on the ear flap (Hill, Field & Blyth, 1975) and in guinea-pigs inoculated intravaginally or in the foot pad (Scriba, 1975; Donnenberg, Chaikof & Aurelian, 1980). Many other models for herpes simplex have been described in which recurrence of infectious virus was observed *in vivo* but usually in the *absence* of obvious clinical signs. For example, following ocular inoculation of rabbits, virus may be recovered several weeks later from conjunctival swabs (Anderson, Margruder & Kilbourne, 1961; Nesburn, Elliot & Leibowitz, 1967).

In the model described by Hill *et al.* (1975) using mice sub-cutaneously inoculated in the ear pinna under standardized conditions spontaneous recrudescence characterized by redness and occasionally vesicles occurred at intervals. Some mice had more than one recrudescence and in certain instances infective virus was recovered from their red ears (Hill *et al.*, 1975; Harbour, Hill & Blyth, 1981). The frequency of these recurrences was increased by such treatments as ultraviolet light (Blyth, Hill, Field & Harbour, 1976) and epidermal stripping with sellotape (Hill, Blyth & Harbour, 1978). For some time this group of workers alone were able to reproduce these effects though others using identical techniques and even the same virus strain were unable to do so. Further investigation showed the importance of the mouse strain; in general inbred lines, particularly BALB/c, gave recurrent lesions less readily than outbred mice (H. J. Field, unpublished observations) and infectious virus could rarely be reisolated from the skin of inbred lines following a reactivation stimulus (Harbour *et al.*, 1981).

Using cocultivation techniques Hill, Harbour & Blyth (1980) were occasionally able to isolate virus from the skin of latently infected mice in the absence of recrudescence. It seems reasonable to view this as subclinical recurrence following reactivation in the ganglion; the finding is certainly similar to that of Spruance *et al.* (1977) who isolated virus from patients in the prodromal phase of recrudescence by similar techniques. However, a very different pattern of behaviour is found in the guinea-pig. Both Scriba (1975, 1977) and Donnenberg *et al.* (1980b) using different strains of animal were regularly able to isolate virus from extraneuronal sites by cocultivation techniques and less regularly from sensory ganglia. It has been suggested that this may mean that virus becomes latent in these tissues. However, in these animals recrudescence is frequent (45% to 90% of animals recrudesce). In Scriba's (1975) series the duration of clinical lesions varied between 2 days and 4 weeks and the median recrudescence was about once every 6–7 weeks; so at any point in time it is not surprising that Donnenberg *et al.*, (1980b) found a high proportion of animals yielding virus, viz. 5/18 by direct isolation from the footpad and 26/48 by cocultivation.

At the time of writing we are uncertain whether the guinea-pig model is qualitatively or merely quantitatively different from the mouse and rabbit models and the natural infection in man and we must await further information. What does seem clear is that the mouse and rabbit models are particularly suitable for the study of latency and reactivation while the guinea-pig model should be valuable for this study of factors underlying recurrence and recrudescence.

AN EVALUATION OF THE CLASSICAL HYPOTHESIS

The site of latent virus

The dorsal root ganglion has been favoured as the probable site harbouring latent virus in natural infections because of the striking correspondence between the distribution of recrudescences and histological changes noted in the related sensory ganglia. This was supported by clinical observation (Cushing, 1904; Carton & Kilbourne, 1952); herpes eruptions regularly followed rhizotomy. Similar findings were made by others (reviewed by Baringer, 1975). These findings did not exclude the skin as the site of latency; indeed some championed this view supposing the dormant virus to be triggered into activity by nervous stimulation.

The compelling evidence that the usual site of latency is the sensory ganglion comes from the important finding that in man and animals classically latent herpes viruses can be regularly reactivated by explanting ganglionic tissue in culture (for references see Table 1). In all instances except herpes simplex in the guinea-pig (cf. previous section) such *in vitro* reactivation has succeeded regularly only in tissues containing nerve cells. Indeed Cook & Stevens (1976) who rigorously tested many tissues of mice that had been intravenously injected with herpes simplex virus were only successful in reactivating virus from sensory ganglia, from brain tissue and from the adrenal gland. Nesburn, Cook & Stevens (1972) made similar observations in rabbits. Similarly explant cultures of patches of skin taken from the sites at which herpes recurred in man failed to reactivate virus (Rustigian *et al.*, 1966).

Attempts to detect herpes specific nucleotide sequences in tissues have been bedevilled by problems of insensitivity of the hybridization techniques, nevertheless Puga, Rosenthal, Openshaw & Notkins (1978) have achieved this using sensory ganglia from large numbers of latently infected mice. This is sound evidence, provided that the sequences detected derive from resident genomes. In addition to the ganglionic site, there has been speculation for some time that other neurological sites could also be a haven for latent virus. First, latent virus has been reactivated from the autonomic nervous system of man (Warren *et al.*, 1978) and of mice (Price, Katz & Notkins, 1975; Price & Schmitz, 1978). At the moment the practical significance of these observations is unknown. The second possible alternative site for latency is the cells of the central nervous system. Cook & Stevens (1976) and Knotts, Cook & Stevens (1973) demonstrated that in mice, following intravenous injection, and in rabbits latent virus could be detected by culture of explanted central nervous tissue. More recently Cabrera *et al.* (1980) showed that herpes genomes could be detected in the brains of 90% of mice after ocular inoculation although in this instance virus was not readily reactivated using standard culture techniques. Furthermore there are at least two reports of the detection of herpes simplex virus genomes in the central nervous system of man. This was reported by Sequera *et al.* (1979) using DNA-hybridization and by Brown *et al.* (1979) by means of the novel technique of rescue of putative defective viruses from cell cultures by superinfection with ts-mutants. Both these reports involve the application of techniques which are still early in their development and which detect incomplete as well as entire genomes. It will be of interest to see whether these findings can usefully be extended by others. The presence of herpes virus genomes in the central nervous system of man has been proposed as a possible precursor of encephalitis. Alternatively, such virus might provide a provocative stimulus for demyelination; indeed Kristensson *et al.* (1979) provides some experimental evidence for this. The possibility of more subtle disturbances of the nervous system has recently been discussed by Sequera *et al.* (1979). Further speculation in this area must await more definite information about the frequency of the phenomenon of herpes simplex latency in the central nervous system.

In general we conclude that the sensory ganglion is the most important site in man which harbours herpes simplex virus and like viruses in classical latency.

Axonal transport

For more than half a century the axon has been the preferred conduit by which virus travels from the periphery to the sensory ganglion, to the central nervous system and within the central nervous system. From time to time controversies have erupted, sometimes fierce, when other hypotheses have been advanced. Five routes have been considered: (1) virus travels within axons (Good-pasture, 1929); (2) virus travels within the periaxonal spaces (Pay-ling Wright, 1953); (3) virus travels by multiplying in Schwann cells stepping-stone fashion within nerves (Johnson, 1964); (4) virus travels in lymphatic channels in the epineurium; and (5) virus traffic is haematogenous but localization occurs because nerve impulses render ganglionic cells unduly susceptible (Field, 1952).

Evidence countering hypotheses 2–5 has been discussed previously (Cook & Stevens, 1973; Kristensson, Lycke & Sjöstrand, 1974). In fact experiments using rabbits and mice overwhelmingly support transport within the axon cylinder as being the most important route by which virus spreads from the initial infection site to the neurones of the peripheral nervous system. While it must be accepted that the other routes mentioned above, particularly the sequential infection of glial cells, may have a secondary role, a number of facts point to the major importance of axonal transport. (a) Rate of translocation from the skin to the ganglion. This has been measured for herpes simplex by Kristensson et al. (1974), Cook & Stevens (1973) and for pseudorabies by Field & Hill (1974) and McCracken, McFerran & Dow (1973). In all cases the rates obtained were in the range 2–10 mm/h which is consistent with the rates of movement of macromolecules and organelles by retrograde axonal transport (reviewed by Kristensson, 1978). The rates observed for the trans-location of herpes viruses are also found to apply to other neurotro-pic viruses, for example polio (Bodian & Howe, 1941) and rabies (Dean, Evans & McClure, 1963); the latter has long been thought to translocate by a similar mode. (b) Interruption of axonal flow. Complete severing of the peripheral nerve prevents the passage of virus to the ganglion (Wildy, 1967) but particularly elegant experi-ments were carried out by Kristensson et al., (1974) in which minimal trauma to the nerve produced by freezing or the disruption of the microtubular integrity by the injection of colchicine was also found to prevent the passage of herpes simplex virus. (c) The preferential adsorption of HSV to synaptosomes rather than neuronal perikarya. Evidence for the latter was obtained in organ-ized cultures of mammalian dorsal root ganglia (Kristensson, Shep-pard & Bornstein, 1974) and later by fractionation methods (Vahlne et al., 1978), (d) Direct morphological observation of intact virions within the axoplasm of peripheral nerves (Hill, Field & Roome, 1972; Cook & Stevens, 1973; Baringer & Swoveland, 1974; Kris-tensson, Ghetti & Wisniewski, 1974). However, the electron microscopical evidence must be qualified since we cannot be sure that the particles seen are the effective virions. Thus, though axonal transport can be said to be the important method of virus translocation we cannot tell in what form the virus travels.

Which cell harbours the virus?

The neuronal body has been presumed to harbour latent herpes virus. This was strongly supported by the evidence in mice and rabbits that virus is regularly reactivable from ganglia (which contain neuronal bodies) and never from nerve roots which do not. (Baringer & Swoveland, 1973; Cook, Bastone & Stevens, 1974). The latter authors have followed the course of reactivation of virus in ganglia implanted in Millipore chambers and found that virus antigen (immunofluorescence), virus particles (electron microscopy) and thymidine incorporation into virus DNA (autoradiography) all commence in neurones before they do in satellite cells thus supporting the concept that virus remains latent within neurones. Similar evidence was found by Narita, Inui, Namba & Shimizu (1978) who followed *in vivo* reactivation of infectious bovine rhinotracheitis in calf ganglia.

More recently the ingenious experiments of McLennan & Darby (1980) have made the probability nearly certain. Latency was induced in mice with a ts-mutant of virus. Some weeks later, virus was reactivated by explantation at either the permissive or the non-permissive temperature. At the permissive temperature reactivation regularly occurred but at the non-permissive temperature the only finding was the development of virus antigen in single neurones. Confirmatory evidence was provided by reactivation *in vivo* after nerve section.

The conclusion is inescapable that in these examples it is the neurone that harbours latent virus and we expect the same to be true in man.

The nature of latency and reactivation

Static, dynamic or a combination of both?
Virologists have been intrigued to know whether latency is static, like lysogeny or a dynamic grumbling infection such as can be produced in tissue cultures under neutralizing antibody (cf. Wheeler, 1960) and now we must add the possibility that a combination of both is necessary (Klein, 1976; Schwartz, Whetsell & Elizan, 1978).

The question could perhaps be solved by studying disaggregated ganglionic cells. Walz, Yamamoto & Notkins (1976) found that 0·1% cells caused infective centres, i.e. about 1% of neurones (calculated as 8×10^5 neurones per ganglion; Puga *et al.*, 1978) but this approach has yielded disappointing results in our laboratory (J. L. McLennan, personal communication). It might be solved if DNA extracted from ganglia were shown to contain herpes-specific sequences covalently linked with host sequences, but attempts to demonstrate this have so far failed for technical reasons. It might be solved if examination of latently infected ganglia could be shown to contain absolutely no cells expressing herpes genomes. There are in fact a number of accounts showing that such ganglia rarely have positive cells by immunofluorescence. McLennan (1981) found no immunofluorescent cells when examining serial sections of entire ganglia from 50 latently infected mice. McDougall & Galloway (1978) using *in situ* hybridization found that many neurones contain virus mRNA, but here there is some doubt whether these cells were in the process of reactivation. Baringer & Swoveland (1974) examined serial sections through entire ganglia and found occasional single abnormal cells containing virus particles. The number of such cells was low – on average less than one per ganglion. The overall approach whereby whole ganglia are found to contain no infectious virus unless they are reactivated *in vivo* by nerve section or by explant *in vitro* certainly favours a static form of latency but it may be questioned whether isolation techniques are sensitive enough to exclude the presence of small numbers of virions. Indeed Schwartz *et al.* (1978) reported isolation of infective virus from ganglia of latent mice using very sensitive organ cultures; free reactivated virus is to be expected from time to time but a special feature of the study was that the infective centres dwindled with time. But because latency may be established with ts-mutants (Lofgren, Stevens, Marsden & Subak-Sharpe, 1977; McLennan & Darby, 1980) we favour a static process. In spite of the possibility of leakiness, the mutants used held up well in tissue culture cells and *in vivo*.

The antiviral drugs acyclovir and bromovinyldeoxyuridine were found to be effective against acute herpes infection in mice and to prevent the establishment of latency in their ganglia. However, they had no effect in ablating established latency (Field *et al.*, 1979; Blyth, Harbour & Hill, 1980; Klein, Friedman-Kien & De Stefano, 1979; Field & De Clercq, 1981). This argues strongly against a dynamic process at least in the short term (one month).

Thus the likelihood of a static mechanism being important is on balance high. However, the following consideration suggests that dynamic processes may also play a part in the long term. Current dogma has it that during productive infection with herpes viruses the cells are killed. It might be that neurones which are exceptional in so many ways behave differently in this respect. However, McLennan & Darby (1980) made the interesting observation that neurones in which virus had been reactivated *in vivo* by nerve section subsequently reactivated at a greatly reduced rate when the ganglia were explanted, strongly suggesting that a neurone may only reactivate once. This is consistent with the behaviour of zoster where a period of latency commonly lasts for 70 or 80 years before recrudescence but less easily explains clinical herpes simplex which often occurs very many times. It might be expected that the subject would run out of latently infected neurones after a time. Though there is a tendency for recrudescence to become less frequent with age there are other explanations for this.

If we admit that both static and dynamic processes are at work in latency, how do they interact? *A priori*, virus spread between neuronal bodies seems improbable since they are deficient in virus binding sites (Vahlne *et al.*, 1978) even though direct infection *in vivo* has been achieved by intrasciatic injection (Field *et al.*, 1979) and we see occasional foci containing more than one antigen-bearing neurone. Satellite cells are unlikely to play a part since they appear to undergo abortive infection leading to the formation of naked or hollow capsids (Becker, 1968; Cook & Stevens, 1973; Hill & Field, 1973; Dillard, Cheatham & Moses, 1972; McCracken & Dow, 1973). A mechanism is provided by the 'round trip' hypothesis of Klein (1976) whereby reactivated virus is supposed to translocate intraaxonally to infect epidermal cells where it is amplified and then to infect nerve termini of neighbouring axons thereby inducing latency in fresh neuronal bodies.

We therefore prefer the combined hypothesis in which virus remains static within neuronal bodies and only occasionally reactivates with a fair chance of it establishing latency in new neurones by the 'round trip' mechanism.

The initiation of latency and state of the genome
Latency has been established in mice using ts-mutants (Lofgren *et al.* 1977; McLennan & Darby, 1980). Since the temperature of the

mouse ganglion is non-permissive for these viruses it seems that latent infections can be initiated without the need for a round of multiplication in the neurone. The suggestion must be qualified because these mutants are always 'leaky' albeit at a low rate in this instance and because the neurone may behave differently *in vivo* from tissue culture cells. However, the case is supported by the finding (Sekizawa, Openshaw, Wohlenberg & Notkins, 1980) that in mice given neutralizing antibody, latency was established within one or two days; in fewer than half was infective virus detected in the sensory ganglia.

If latency is static we wish to know in what form it exists. It is aesthetically attractive to imagine that the virus genome is integrated into the host DNA as in lysogeny. If this is so the whole genome must be represented. But as Fenner *et al.* (1974) have pointed out, integration is unnecessary since neurones do not divide.

Precipitation of reactivation and control of latency

Much of the information on the precipitation of reactivation comes from the more complex phenomena underlying the precipitation of recurrence or recrudescence (see p. 149). Now we adhere strictly to reactivation within the ganglion.

Early work showed that infective virus reactivated spontaneously in rabbit brain (Perdrau, 1938) or by anaphylaxis (Good & Campbell, 1948) or adrenalin (Schmidt & Rasmussen, 1960; Laibson & Kilbrick, 1966). Stevens & Cook (1973) failed to precipitate reactivation by cortisone or artificial fever as had Schmidt & Rasmussen (1960) before them. They also failed to reactivate virus with adrenalin, cyclophosphamide, irradiation, the graft virus host reaction or implantation of tumour cells. They did have success by superinfecting with pneumococci. Several workers have since achieved reliable reactivation *in vivo* by nerve section (Walz, Price & Notkins, 1974; Price & Schmitz, 1978; McLennan & Darby, 1980) and it may be that the technique of explantation acts in the same way. Otherwise reactivation has been precipitated by cyclophosphamide and X-irradiation (Sekizawa *et al.*, 1980), hair plucking (Hurd & Robinson, 1976), burning with solid CO_2 (Openshaw, Puga & Notkins, 1979), UV-treatment (Blyth *et al.*, 1976) and sellotape stripping (Hill *et al.*, 1978) – procedures which also induce recrudescent lesions.

Stevens & Cook (1974) suggested that virus-specific antibody might hold virus in a latent state. Their experiments depended on the finding that latently infected ganglia were less likely to reactivate when planted in Millipore chambers with immune mice than when implanted in non-immune mice. Indeed they showed also that passively administered herpes simplex IgG prevented reactivation. It seemed possible that the antibody interacted with some surface antigen expressed at the cell surface and that this in turn held down the virus. However, no such antigens were ever detected. Recently, this suggestion has been questioned by Sekizawa *et al.* (1980), who established latency by infecting mice and giving them rabbit anti-herpes antiserum. After a few months the antibody had cleared yet the virus remained latent in the sensory ganglia. Mice did not develop antibody unless reactivation was precipitated by a solid CO_2 burn. It looks as though neutralizing antibody does not after all control latency but as Nash (1981) points out non-neutralizing antibodies have not been excluded.

At present there seems no need to invoke any arguments about factors which hold virus in a static latent state. Perhaps the virus just likes to be that way unless positively provoked; what the stimulus might be is anyone's guess at present since it is difficult to determine a final common path for the stimuli listed above.

RECURRENCE AND RECRUDESCENCE

From what we have so far concluded it seems that recurrence must occur after reactivation and axonal translocation have occurred. It by no means follows that recurrence is a necessary consequence. Similarly recrudescence must follow recurrence though it is again not a necessary consequence.

The stimuli that provoke recrudescence in man have been known for a long time but have been subject to special investigation in few instances. For example pyrexia (Warren, Carpenter & Boak, 1940), sunburn and UV light (Blyth *et al.*, 1976; Wheeler, 1975) and skin trauma (Hill *et al.*, 1978; Hurd & Robinson, 1976) have all been examined. There are also accounts of similar stimuli providing recurrence without recrudescence and there is no reason to doubt that recurrence is a necessary prelude to recrudescence. However, the sequence of events leading up to it is so complex that except for immunological factors (see p. 151) and mediators of inflammation –

in particular the prostaglandins – little direct use can yet be made of the findings.

What is required is an understanding of how the stimuli may alter the physiology within the neurodermatome and so provoke the processes. Probably such information will not be long coming – for instance Enerbäck, Kristensson & Olsson (1980) have evidence that quantitative changes which appear to accelerate axonal transport follow various stimuli.

Harbour, Blyth & Hill (1978) recorded that recrudescence of herpes in mice was promoted by prostaglandin E2 as well as by UV-irradiation which increases prostaglandin levels (they also found that buffered saline could do the same). *In vitro* studies have shown that prostaglandins increase herpes virus yields and that inhibitors of prostaglandin synthesis reduce them (Inglot, 1969; Newton, 1979). Hill & Blyth (1976) have interpreted the findings by assuming that spontaneous reactivation is a relatively common event so that virus is frequently arriving at the periphery but usually in insufficient quantities to infect more than a few epidermal cells and not enough to cause a lesion. Thus it should be possible, using sensitive techniques, to detect infective virus in the skin resulting from such minimal recurrences. Indeed this has been reported (Hill *et al.*, 1980).

The conclusion that may be entertained is that latency and reactivation in the spinal ganglion represent a phenomenon quite distinct from recurrence of infection or recrudescence at the periphery and that the latter are triggered by stimuli at the periphery – the so called skin trigger hypothesis of Hill & Blyth (1976).

FACTORS AFFECTING LATENCY, REACTIVATION, RECURRENCE AND RECRUDESCENCE

Host factors

There is a positive association between HLA antigen A1 and the incidence of recrudescent herpes and suggestive associations with A29 and B8. A negative association of less strength has been noted with A10 (Russell & Schlaut, 1977). No explanations for this are available and it is not even clear whether such individuals are especially susceptible to infection with the virus.

Similarly, genetical associations have been noted in the ability of different strains of mice to give recrudescent lesions (Blyth *et al.*, 1980). It is unknown whether this correlates with genetically determined resistance levels to herpes simplex type 1 (Lopez, 1980) which appear to be mediated by NK cells, or resistance to type 2 virus, which is sex linked and which is expressed by macrophage function (Mogensen, 1980).

Immunocompromised hosts, not surprisingly, tend to produce recrudescences (reviewed by Merigan & Stevens, 1971). We have already discussed the strong predilection of the guinea-pig to recrudesce.

Virus attributes

Viruses are of course well known to vary in virulence for particular host species. It is interesting that thymidine kinase deficient mutants tend to be avirulent and multiply poorly *in vivo* (Field & Wildy, 1978; Field & Darby, 1980). These certainly tend to induce latency with more difficulty than wild-type viruses and multiplication is often restricted in nervous tissue. In some instances these differences are large (Tenser, Miller & Rapp, 1979; Tenser & Dunstan, 1979). Hill (personal communication), finds virus-strain differences in connection with the frequency of recrudescence.

THE IMMUNOLOGY OF HERPES INFECTION IN RELATION TO LATENCY

A variety of immunological mechanisms exists which counters infection with herpes simplex virus (see Table 2). It is important in this section to consider which of these mechanisms, if any, are involved in latency, recurrences and recrudescences. As already discussed in a previous section, antiherpes antibodies appear to be of limited importance in maintaining latency. However, certain immunosuppressive regimes have been successful in precipitating virus recurrences, particularly those concerned with the destruction of lymphoid cells. It is now well established that cell-mediated immunity (CMI) is important for the protection of the host against herpes (Lodmell, Niwa, Hayashi & Notkins, 1973; Rager-Zisman & Allison, 1976; Oakes, 1975), and defects in CMI have been implicated in herpes recurrences. Wilton, Ivanyi & Lehner (1972) and Shillitoe, Wilton & Lehner (1977) demonstrated that patients

Table 2. *Immunological mechanisms involved in immunity to herpes*

Cytotoxic T cells	Sethi & Brandis (1977)
	Pfizenmaier *et al.* (1977)
	Nash *et al.* (1980b)
NK cells	Ching & Lopez (1979)
	Rager-Zisman & Allison (1980)
ADCC	Shore, Cromeans & Romano (1976)
Macrophages	Zisman, Hirsch & Allison (1970)
	Morahan & Morse (1979)
DTH	Nash *et al.* (1980a)
	Nash *et al.* (1981b)
Suppressor T cells	Nash, Phelan, Gell & Wildy (1981)
Suppressor B cells	Nash & Gell (1980)
Interferon	Gresser, Tovey, Maury & Bandu (1976)
	Sonnenfeld & Merigan (1979)
Antibody and complement	Burns, Billups & Notkins (1975)
	Oldstone & Lampert (1979)

undergoing recrudescence have impaired cell-mediated immunity. Though competent with respect to the lymphoproliferative response to herpes antigens, their T lymphocytes nevertheless failed to make macrophage migration inhibition factor (MIF) in response to herpes while this response was unimpaired with respect to other antigens. Similar lymphokine deficiency has been demonstrated using the leucocyte inhibition factor (LIF) test (O'Reilly, Chibbaro, Anger & Lopez, 1977) and has been confirmed by Donnenberg *et al.* (1980a, b) in guinea-pigs infected with type 2 herpes virus. These findings probably have no relation to the control of latency and reactivation but they are extremely relevant to recurrence and recrudescence in man and guinea-pig. The results may be interpreted in several ways. For example, they may reflect a reduced proportion of lymphokine-producing cells in the T cell population or a preponderance of suppressor cells. Both would result in diminished clearance of virus and would most likely act at the stages of recurrence or recrudescence.

Work with mice has progressed swiftly in the past three years. Using the mouse ear model Nash and colleagues have provided the following background information (Figs. 3 and 4). In BALB/c mice growth of virus in the primary infection follows a course similar to previous studies. Virus titres in the inoculated pinna become maximal on the third day and gradually dwindle to become unde-

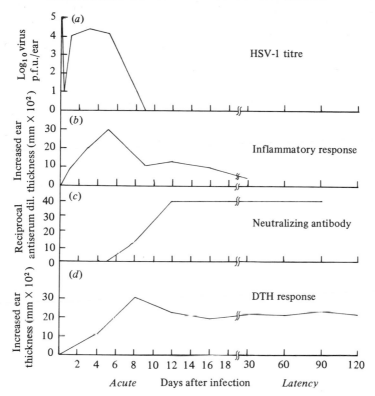

Fig. 3. General immunological and virological properties of the mouse ear model following infections with 10^5 p.f.u. herpes simplex virus type 1 (clone SC_{16}). Details of figure (a) to (d) can be found in Nash, Field & Quartey-Papafio (1980a) and Nash, Quartey-Papafio & Wildy (1980b).

tectable after a week. Primary ear swelling (a combination of acute inflammation and developing CMI) follows a similar course initially, but remains elevated for longer. Delayed type hypersensitivity (DTH) (produced by challenging the opposite ear with herpes) is first demonstrated on day 4 and remains inducible for the life of the animal. Neutralizing antibody becomes detectable between the fifth and eighth day and reaches a stable level by the twelfth day. The draining lymph node swells to five or ten times its normal size and remains larger than normal for a long period. Lymphocytes taken from the draining lymph node show cytotoxic activity for a short period reaching a maximum at about six or seven days and becoming undetectable beyond the twelfth day. However, lymphocytes exhibiting a lymphoproliferative response appear by the eighth day and remain active for very long periods – probably for the lifetime of the mouse. When the cells from the draining node are

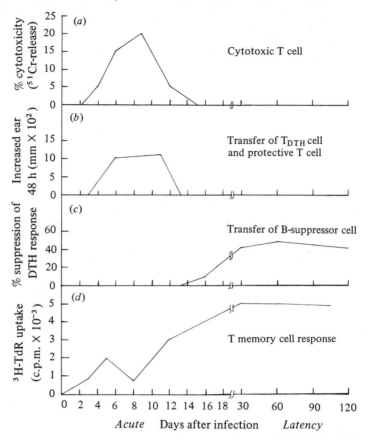

Fig. 4. Properties of cells from the lymph node draining the ear of BALB/c mice infected with 10^5 p.f.u. herpes simplex virus type 1 (clone SC_{16}). Details of figures (a) to (d) can be found in Nash, Field & Quartey-Papafio (1980a): Nash & Gell (1980) and Nash, Quartey-Papfio & Wildy (1980b).

examined by transfer to syngeneic recipients which have never encountered the virus, the period during which DTH can be transferred is restricted to the first ten days and is accompanied by T cells which exhibit a strong antiherpes activity. The failure to transfer DTH beyond ten days could be in part explained by the appearance in the draining lymph node of cells which suppress the expression of DTH, when transferred to immune mice. This suppressive effect is mediated by Ig +ve, thy 1.2-ve lymphocytes (presumably B cells) and hence might be thought of as an antibody-mediated effect were it not for the fact that antibody against herpes does not suppress DTH (Nash & Gell, 1980). A particular point of interest is that this suppressor activity is confined to the draining lymph nodes and is not found in contralateral lymph nodes. We supposed this asymmetry to be associated with latency since it did not occur in vaccinia-infected animals (Nash, unpublished) but whether it is associated causally is not clear. This suppression can readily be demonstrated when animals are thymectomized as adults and in this instance the suppressor activity is found in contralateral lymph nodes, spleen, as well as the draining nodes.

The obvious hypothesis was that protection was the product of cytotoxic T cells and T cells mediating DTH since both populations were transferred simultaneously. This hypothesis was reinforced by the finding that the major histocompatibility restriction for T cells involved in the rapid elimination of infectious herpes is H-2 K/I_A (Nash, Phelan & Wildy, 1981a). Since it is known that cytotoxic T cells are restricted by the K and D region (Pfizenmaier et al., 1977) and DTH T cells by the H-2I_A subregion (Nash et al., 1981a), then this suggests that a combination of both (or another I_A restricted population) is required for protection. The possible importance of DTH was further highlighted in experiments in which the induction of DTH was tolerized (Nash, Gell & Wildy, 1981b) and consequently the response was not transferable which resulted in an impairment of the clearance of infectious virus, despite the presence of cytotoxic cells in the transferred cell mixture (Nash, unpublished). However, it is also clear that DTH is not absolutely essential for immunity to the virus since mice tolerized for DTH are protected against subsequent infectious virus challenges (Nash et al., 1981b). This paradoxical situation can best be explained if one argues that DTH-type responses are important early in the defence against virus infections and once other arms of immunity have become established such as neutralizing antibody and cytotoxic cell responses then DTH may become superfluous. It is nevertheless clear that defects in CMI as measured by lymphokine production (generally considered products of a DTH reaction) correlate with the high incidence of recrudescences in humans and guinea-pigs. Consequently it is important to study the effect of a paralysed DTH system in controlling recurrences and recrudescences in latently infected mice as well as in guinea-pigs.

What is most striking is that while tolerance to DTH and suppressor cell activity have been found against both induction and expression of DTH, no such effect has yet been demonstrated for cytotoxicity in vitro or from protection in vivo. Indeed cell-mediated protection is so strong in herpes infected mice that one is tempted to

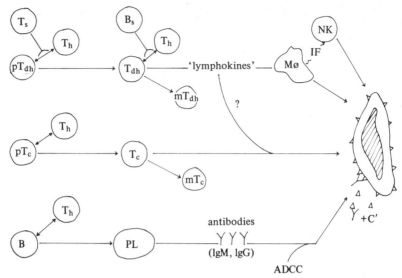

Fig. 5. A schematic representation of the proposed T cell mechanisms involved in the induction of protective immune responses, as defined using the mouse ear model. The virus infected target cell is representative of either a neurone or epidermal skin cell during the primary infection or following recurrences and recrudescences. The term 'lymphokines' is used in its broadest sense and includes interferon. Abbreviations used in the figure are as follows: T_s, suppressor T cell; T_h, helper T cell; T_{dh}, delayed hypersensitivity T cell; T_c, cytotoxic T cell; pT, precursor T cell; mT, Memory T cell; B_s, suppressor B cell; B, activated B cell; PL, plasma cell; Mø, activated macrophage; NK, natural killer cell; ADCC, antibody-dependent cell cytotoxicity; IF, interferon; C', complement; $\triangle$, herpes virus antigens; —(, suppression of immune cell function.

suggest that this might be responsible for the difficulties in detecting recurrence and recrudescence in these animals.

To summarize we envisage that three protective T cell mechanisms are initiated following infection with herpes virus (Fig. 5). We believe that all three are active during the primary infection and also during recurrences and recrudescences. One of these, namely DTH, can be subject to a particularly strong immunological control by suppressor cells; the result of this is failure to rapidly localize antiviral cells at the infection site. Probably the induction of suppressor cells to DTH results in impairment of normal surveillance mechanisms thus allowing the break through of reactivating virus to produce clinical recrudescence. At present less is known about the control of the other two arms of immunity in herpes infections, although suppressor cells active against influenza-specific cytotoxic T cells have been described (Leung, Ashman, Ertl & Ada, 1980). A key cell in our scheme is the T helper which we believe to be essential for the induction of all three responses, and in particular for DTH (Nash & Gell 1981) and neutralizing antibody production (Burns, Billups & Notkins, 1975). In the end though whether antibody or lymphocytes play a definitive role in maintaining the latent state is still an open question.

GENERAL CONCLUSIONS

We have delineated five aspects of the phenomenon of classical latency with herpes viruses in their natural hosts: (1) latency in the neurone, (2) reactivation of infective virus in the neurone, (3) axonal transport, (4) recurrence of virus at the periphery, and (5) recrudescent lesions. We think it likely that in mice and rabbits the last two phenomena are difficult to reproduce with herpes simplex virus but they appear to follow virtually automatically in the guinea-pig resulting in apparent discrepancy.

It follows that mice are particularly useful for studying latency and reactivation, and the guinea-pig for recurrence and recrudescence. When the evidence from all sources is amassed the following composite picture results. It appears that a primary herpes infection results in a round of local multiplication in the epidermal cells of the neurodermatome and some progeny virus is taken up by the nerve terminals and translocated centrifugally by axonal transport. Virus becomes latent in neurones especially those of the sensory ganglion and requires no round of multiplication at this stage.

It persists in those neurones in static form; otherwise nothing else can be concluded definitely about the stage of latency. Though various stimuli are recognized that lead to reactivation nothing definite can be concluded about the mechanism of reactivation nor of factors maintaining latency. It seems likely that a neurone can only reactivate once thus a dynamic process such as the 'round trip' hypothesis must be invoked to maintain latency.

Recurrence and recrudescence represent a quite separate phenomenon and are probably modulated immunologically. There is evidence that defects in lymphokine production may be responsible for recrudescences in certain individuals but of course other mechanisms are quite likely to be discovered.

Recrudescence must be moderated by inflammatory factors but in addition there is limited evidence that prostaglandins may actually enhance virus replication in such lesions.

Scrutiny of the literature seems to reveal two groups of virus naturally showing classical latency. As further information accumulates it will be interesting to see whether these are true or spurious groups.

EPILOGUE

In this chapter we have explored the classical theory of herpes latency. In one respect we are astonished at the large amount of work in the past half century which has resulted in so little shift of opinion about the nature of herpes latency. Perhaps we are all too intent upon fitting our findings with established dogma and too prone to overlook or neglect findings that do not fit in.

On the other hand we do espy novel approaches which will certainly answer a good number of questions if those with the expertise can be made interested in the problem. First, molecular biologists are exhorted to turn their attention to this important practical problem. Once the difficulty of sensitivity is overcome their task should be simple. For example, to judge whether herpes sequences are integrated or not, the new DNA technology should yield probes capable of high resolution and sensitivity. Secondly, immunologists are exhorted to address their attention to this academically satisfying yet practical problem. Thirdly, pathologists and pharmacologists are exhorted to turn their minds to the factors that cause damage in herpes and finally virologists of all kinds are exhorted to forget their narrow interests in various odd areas of the subject and to address themselves to the problem latency – reactivation – recurrence and recrudescence which causes much accumulated misery to about 2×10^9 human beings and a number of other animals each year.

REFERENCES

ANDERSON, W. A., MARGRUDER, B. & KILBOURNE, E. D. (1961). Induced reactivation of herpes simplex virus in healed rabbit corneal lesions. *Proceedings of the Society for Experimental Biology*, **107**, 608–32.

ANDREWES, C. H. & CARMICHAEL, E. A. (1930). A note on the presence of antibodies to herpes simplex virus in post-encephalitic and other human sera. *Lancet*, **i**, 857–8.

ANDREWES, C. H., PEREIRA, H. G. & WILDY, P. (1977). *Viruses of Vertebrates*. London: Ballière Tindall.

BARAHONA, H., MELENDEZ, L. V. & MELNICK, J. L. (1974). A compendium of herpesviruses isolated from non-human primates. *Intervirology*, **3**, 175–92.

VON BÄRENSPRUNG, F. G. F. (1863). Beiträge zur Kenntniss des Zoster. *Dritte Folge. Charite-Krankenhauses*, **11**, 96–116.

BARINGER, J. R. (1975). Herpes simplex virus infection of nervous tissue in animals and man. *Progress in Medical Virology*, **20**, 1–26.

BARINGER, J. R. & SWOVELAND, P. (1973). Recovery of herpes simplex virus from human trigeminal ganglions. *New England Journal of Medicine*, **288**, 648–50.

BARINGER, J. R. & SWOVELAND, P. (1974). Persistent herpes simplex virus infection in rabbit trigeminal ganglia. *Laboratory Investigation*, **30**, 230–40.

BASTIAN, F. O., RABSON, A. S., YEE, C. L. & TRALKA, T. S. (1974). *Herpesvirus varicellae* isolated from human dorsal root ganglion. *Archives of Pathology*, **97**, 331–2.

BASKERVILLE, A. (1973). The histopathology of experimental pneumonia in pigs produced by Aujeszky's disease virus. *Research in Veterinary Science*, **14**, 223–8.

BASKERVILLE, A., McCRACKEN, R. M. & McFERRAN, J. B. (1971). The histopathology of experimental rhinitis in pigs produced by a strain of Aujeszky's disease virus. *Research in Veterinary Science*, **12**, 323–6.

BECKER, C.-H. (1968). Die Multiplikation des Aujeskyschen Virus in den Spinalganglien des Kanishens. *Archivs für experimentelle Veterinärmedzin, Berlin*, **22**, 363–81.

BLYTH, W. A., HARBOUR, D. A. & HILL, T. J. (1980). Effect of acyclovir on recurrence of herpes simplex skin lesions in mice. *Journal of General Virology*, **48**, 417–19.

BLYTH, W. A., HILL, T. J., FIELD, H. J. & HARBOUR, D. A. (1976). Reactivation of herpes simplex infection by ultraviolet light and possible involvement of prostaglandins. *Journal of General Virology*, **33**, 547–50.

BODIAN, D. & HOWE, H. A. (1941). The rate of progression of poliomyelitis virus in nerves. *Bulletin of the Johns Hopkins Hospital*, **69**, 79–85.

BROWN, S. M., SUBAK-SHARPE, J. H., WARREN, K. G., WROBLEWSKA, Z. & KOPROWSKI, H. (1979). Detection by complementation of defective or uninducible (herpes simplex type 1) virus genomes latent in human ganglia. *Proceedings of the National Academy of Sciences, USA*, **76**, 2364–8.

BURNET, F. M. & WILLIAMS, S. W. (1939). Herpes simplex: a new point of view. *Medical Journal of Australia*, **1**, 637–42.

BURNS, W. H., BILLUPS, L. C. & NOTKINS, A. L. (1975). Thymic dependence of viral antigens. *Nature, London*, **256**, 654–6.

CABRERA, C. V., WOHLENBERG, C., OPENSHAW, H., REY-MENDEZ, M., PUGA, A. & NOTKINS, A. L. (1980). Herpes simplex virus DNA sequences in the CNS of latently infected mice. *Nature, London*, **288**, 288–90.

CARTON, C. A. & KILBOURNE, E. D. (1952). Activation of latent herpes by trigeminal sensory root section. *New England Journal of Medicine*, **246**, 172–6.

CHING, C. & LOPEZ, C. (1979). Natural killing of herpes simplex virus type 1-infected target cells: normal human responses and influence of antiviral antibody. *Infection and Immunity*, **26**, 49–56.

COOK, M. L., BASTONE, V. B. & STEVENS, J. G. (1974). Evidence that neurons harbour latent herpes simplex virus. *Infection and Immunity*, **9**, 946–51.

COOK, M. L. & STEVENS, J. G. (1973). Pathogenesis of herpetic neuritis and ganglionitis in mice: evidence for intra-axonal transport of infection. *Infection and Immunity*, **7**, 272–88.

COOK, M. L. & STEVENS, J. G. (1976). Latent herpetic infections following experimental viraemia. *Journal of General Virology*, **31**, 75–80.

CORNWELL, H. J. C. & WRIGHT, N. G. (1969). Neonatal canine herpes-virus infection: A review of present knowledge. *Veterinary Record*, **84**, 2–6.

CUSHING, H. (1904). Perineal zoster with notes upon cutaneous segmentation post-axial to the lower limb. *American Journal of Medical Science*, **127**, 375–91.

DANIELS, C. A., LE GOFF, S. G. & NOTKINS, A. L. (1975). Shedding infectious virus/antibody complexes from vesicular lesions of patients with recurrent herpes labialis. *Lancet*, **ii**, 524–8.

DAVIES, D. H. & DUNCAN, J. R. (1974). The pathogenesis of recurrent infections with infectious bovine rhinotracheitis virus induced in calves by treatment with corticosteroids. *Cornell Veterinarian*, **64**, 340–66.

DEAN, D. J., EVANS, W. M., McCLURE, R. C. (1963). Pathogenesis of rabies. *Bulletin of the World Health Organization*, **29**, 803–11.

DILLARD, S. H., CHEATHAM, W. J. & MOSES, H. L. (1972). Electron microscopy of zosteriform herpes simplex infection in the mouse. *Laboratory Investigation*, **26**, 391–402.

DODD, K., BUDDINGH, G. J. & JOHNSTON, L. (1939). Herpetic stomatitis. *American Journal of Diseases of Children*, **58**, 907.

DOERR, R. (1938). Herpes febrilis. In *Handbuch der Virusforschung*, ed. R. Doerr & C. Hallauer, vol. 1, pp. 44–5. Wien: Springer.

DONNENBERG, A. D., BELL, R. B. & AURELIAN, L. (1980a). Immunity to herpes simplex virus type 2. Development of virus-specific lymphoproliferative and LIF responses in HSV-2 infected guinea pigs. *Cellular Immunology*, **56**, 526–39.

DONNENBERG, A. D., CHAIKOF, E. & AURELIAN, L. (1980b). Immunity to herpes simplex virus type 2: cell mediated immunity in latently infected guinea pigs. *Infection and Immunity*, **30**, 90–109.

ENERBÄCK, L., KRISTENSSON, K. & OLSSON, T. (1980). Cytophotometric quantification of retrograde axonal transport of a fluorescent tracer (primuline) in mouse facial neurons. *Brain Research*, **186**, 21–32.

FENNER, F., McAUSLAN, B. R., MIMS, C. A., SAMBROOK, J. & WHITE, D. O. (1974). *Biology of Animal Viruses*. New York: Academic Press.

FIELD, E. J. (1952). Pathogenesis of herpetic encephalitis following corneal and masseteric inoculation. *Journal of Pathology and Bacteriology*, **64**, 1–11.

FIELD, H. J., BELL, S. E., ELION, G. B., NASH, A. A. & WILDY, P. (1979). Effect of acycloguanosine treatment on acute and latent herpes simplex infections in mice. *Antimicrobial Agents and Chemotherapy*, **15**, 554–61.

FIELD, H. J. & DARBY, G. (1980). Pathogenicity in mice of strains of herpes simplex virus which are resistant to acyclovir *in vitro* and *in vivo*. *Antimicrobial Agents and Chemotherapy*, **17**, 209–16.

FIELD, H. J. & DE CLERCQ, E. (1981). Effects of oral treatment with acyclovir and bromovinyldeoxyuridine on the establishment and maintenance of latent herpes simplex infection in mice. *Journal of General Virology* (in press).

FIELD, H. J. & HILL, T. J. (1974). The pathogenesis of pseudorabies in mice following peripheral inoculation. *Journal of General Virology*, **23**, 145–57.

FIELD, H. J. & WILDY, P. (1978). The pathogenicity of thymidine kinase-deficient mutants of herpes simplex virus in mice. *Journal of Hygiene (Cambridge)*, **81**, 267–77.

FRASER, G. & RAMACHANDRAN, S. P. (1969). Studies on the virus of Aujeszky's disease. I. Pathogenicity in rats and mice. *Journal of Comparative Pathology*, **79**, 435–44.

FRIEDENWALD, J. S. (1923). Studies on the virus of herpes simplex. *Archives of Ophthalmology*, **52**, 105–31.

GASKELL, R. M. & POVEY, R. C. (1979). Feline rhinotracheitis: sites of virus replication and persistence in acutely and persistently infected cats. *Research in Veterinary Science*, **27**, 107–74.

GAY, F. P. & HOLDEN, M. (1933). The herpes encephalitis problem. *Journal of Infectious Diseases,* **53**, 287–303.

GIBBS, E. P. J., PITZOLIS, G. & LAWMAN, M. J. P. (1975). Use of corticosteroids to isolate IBR from cattle in Cyprus after respiratory disease and ataxia. *Veterinary Record*, **96**, 464–6.

GOOD, R. A. & CAMPBELL, B. (1948). The precipitation of latent herpes simplex encephalitis by anaphylactic shock. *Proceedings of the Society for Experimental Biology and Medicine*, **68**, 82–7.

GOODPASTURE, E. W. (1925). Intranuclear inclusions in experimental herpetic lesions of rabbits. *American Journal of Pathology*, **1**, 1–9.

GOODPASTURE, E. W. (1929). Herpetic infection with especial reference to involvement of the nervous system. *Medicine*, **8**, 223–43.

GOODPASTURE, E. W. & TEAGUE, O. (1923). Experimental production of herpetic lesions in organs and tissues of the rabbit. *Journal of Medical Research*, **44**, 121–33.

GRESSER, I., TOVEY, M. G., MAURY, C. & BANDU, M.-T. (1976). Role of interferon in the pathogenesis of virus diseases in mice as demonstrated by the use of anti-interferon serum. II. Studies with herpes simplex, Moloney sarcoma, vesicular stomatitis, Newcastle disease and influenza viruses. *Journal of Experimental Medicine*, **144**, 1316–23.

HARBOUR, D. A., BLYTH, W. A. & HILL, T. J. (1978). Prostaglandins enhance spread of herpes simplex virus in cell cultures. *Journal of General Virology*, **41**, 87–95.

HARBOUR, D. A., HILL, T. J. & BLYTH, W. A. (1981). Acute and recurrent herpes simplex in several strains of mice. *Journal of General Virology*, **55** (in press).

HEAD, H. & CAMPBELL, A. W. (1900). The pathology of herpes zoster and its bearing on sensory localization. *Brain*, **23**, 353–523.

HILL, T. J. & BLYTH, W. A. (1976). An alternative theory of herpes-simplex recurrence and a possible role for prostaglandins. *Lancet*, **i**, 397–9.

HILL, T. J., BLYTH, W. A. & HARBOUR, D. A. (1978). Trauma to the skin causes recurrence of herpes simplex in the mouse. *Journal of General Virology*, **39**, 21–8.

HILL, T. J., FIELD, H. J. & ROOME, A. P. C. (1972). Intra-axonal location of herpes simplex virus particles. *Journal of General Virology*, **15**, 253–5.

HILL, T. J. & FIELD, H. J. (1973). The interaction of herpes simplex virus with cultures of peripheral nervous tissue: an electron microscopic study. *Journal of General Virology*, **21**, 123–33.

HILL, T. J., FIELD, H. J. & BLYTH, W. A. (1975). Acute and recurrent infection with herpes simplex virus in the mouse: a model for studying latency and recurrent disease. *Journal of General Virology*, **28**, 341–53.

HILL, T. J., HARBOUR, D. A. & BLYTH, W. A. (1980). Isolation of herpes simplex virus from the skin of clinically normal mice during latent infection. *Journal of General Virology*, **47**, 205–7.

HOLMES, A. W., CALDWELL, R. G., DEDMON, R. E. & DEINHARDT, F. (1964). Isolation and characterization of a new herpes virus. *Journal of Immunology*, **92**, 602–10.

HOLMES, A. W., DEVINE, J. A., NOWAKOWSKI, E. & DEINHARDT, F. (1966). The epidemiology of a herpesvirus infection of New World monkeys. *Journal of Immunology*, **90**, 668–71.

HONESS, R. W. & WATSON, D. H. (1977). Unity and diversity in the herpesviruses. *Journal of General Virology*, **37**, 15–17.

HOPE-SIMPSON, R. E. (1965). The nature of herpes zoster: a long term study and a new hypothesis. *Proceedings of the Royal Society of Medicine*, **58**, 9–20.

HULL, R. N., DWYER, A. C., HOLMES, A. W., NOWAKOWSKI, E., DEINHARDT, F., LENETTE, E. H. & EMMONS, R. W. (1972). Recovery and characterization of a new simian herpesvirus from a fatally infected spider monkey. *Journal of the National Cancer Institute*, **49**, 225–32.

HULL, R. & NASH, J. C. (1960). Immunization against B virus infection. I. Preparation of an experimental vaccine. *American Journal of Hygiene*, **71**, 15–28.

HURD, J. & ROBINSON, T. W. E. (1976). Herpes simplex: aspects of reactivation in a mouse model. *Journal of Antimicrobial Chemotherapy*, **3**, 99–106.

HURST, E. W. (1933). Studies on pseudorabies (infectious bulbar paralysis, mad itch). I. Histology of the disease with a note on the symptomatology. *Journal of Experimental Medicine*, **58**, 415–33.

INGLOT, A. D. (1969). Comparison of the antiviral activity *in vitro* of some non-steroidal anti-inflammatory drugs. *Journal of General Virology*, **4**, 203–14.

JACKSON, T. A., OSBURN, B. I., CORDY, D. R. & KENDRICK, J. W. (1977). Equine herpes 1 infection of horses: studies on the experimentally induced neurologic disease. *American Journal of Veterinary Research*, **38**, 709–19.

JOHNSON, R. T. (1964). The pathogenesis of herpes encephalitis. I. Virus pathways to the nervous system of suckling mice demonstrated by fluorescent antibody staining. *Journal of Experimental Medicine*, **119**, 343–56.

KLEIN, R. J. (1976). Pathogenetic mechanisms of recurrent herpes simplex virus infections. *Archives of Virology*, **51**, 1–13.

KLEIN, R. J., FRIEDMAN-KIEN, A. E. & DE STEFANO, E. (1979). Latent herpes simplex virus infections in sensory ganglia of hairless mice prevented by acycloguanosine. *Antimicrobial Agents and Chemotherapy*, **15**, 723–9.

KNOTTS, F. B., COOK, M. L. & STEVENS, J. G. (1973). Latent herpes simplex virus in the central nervous system of rabbits and mice. *Journal of Experimental Medicine*, **138**, 740–4.

KRISTENSSON, K. (1978). Retrograde transport of macromolecules in axons. *Annual Review of Pharmacology and Toxicology*, **18**, 97–110.

KRISTENSSON, K., GHETTI, B. & WISNIEWSKI, H. M. (1974). Study on the propagation of herpes simplex virus (type 2) into the brain after intraocular injection. *Brain Research*, **69**, 189–201.

KRISTENSSON, K., LYCKE, E. & SJÖSTRAND, J. (1974). Spread of herpes simplex virus in peripheral nerves. *Acta Neuropathologica, Berlin*, **17**, 44–53.

KRISTENSSON, K., SHEPPARD, R. D. & BORNSTEIN, M. (1974). Observations on the uptake of herpes simplex virus in organized cultures of mammalian nervous tissue. *Acta Neuropathologica, Berlin*, **28**, 37–44.

KRISTENSSON, K., SUENNERHOLM, B., PERSSON, L., VAHLNE, A. & LYCKE, E. (1979). Latent herpes simplex virus trigeminal ganglionic infection in mice and demyelination in the central nervous system. *Journal of Neurological Science*, **43**, 253–64.

KUNDRATITZ, K. (1925). Experimentelle Übertragungen von Herpes zoster auf Menschen und die Beziehungen von Herpes zoster zu Varicellen. *Zeitschrift für Kinderheilkranken*, **39**, 379–87.

LAIBSON, P. R. & KIBRICK, S. (1966). Reactivation of herpetic keratitis by epinephrine in rabbit. *Archives of Opthalmology*, **75**, 254–60.

LEUNG, K., ASHMAN, R. B., ERTL, H. C. J. & ADA, G. L. (1980). Selective suppression of the cytotoxic T cell response to influenza virus in mice. *European Journal of Immunology*, **10**, 803–10.

LODMELL, D. L., NIWA, A., HAYASHI, K. & NOTKINS, A. L. (1973). Prevention of cell-to-cell spread of herpes simplex virus by leukocytes. *Journal of Experimental Medicine*, **137**, 706–20.

LOFGREN, K. W., STEVENS, J. G., MARSDEN, H. S. & SUBAK-SHARPE, J. H. (1977). Temperature sensitive mutants of herpes simplex virus differ in the capacity to establish latent infections in mice. *Virology*, **76**, 440–3.

LOPEZ, C. (1980). Genetic resistance to herpesvirus infections: role of natural killer cells. In *Genetic Control of Natural Resistance to Infection and Malignancy*, ed. E. Skamene, P. A. L. Kongshaun & M. Landy, pp. 253–65. New York: Academic Press.

LOVE, D. N. (1971). Feline herpesvirus associated with interstitial pneumonia in a kitten. *Veterinary Record*, **89**, 178–81.

LUPTON, H. W. & REED, D. E. (1979). Experimental infection of eastern cotton tail rabbits *(Sylvilagus floridanus)* with infectious bovine rhinotracheitis virus. *American Journal of Veterinary Research*, **40**, 1329–31.

McCARTHY, K. (1969). In *Hazards of Handling Simians*, ed. F. T. Perkins & P. N. Donoghue, *Laboratory Animals Handbook*, vol. 4, p. 121. Laboratory Animals Ltd.

McCARTHY, K. & TOSOLINI, F. A. (1975). Hazards from simian herpes viruses: reactivation of skin lesions with virus shedding. *Lancet*, **i**, 649–50.

McCRACKEN, R. M. & DOW, C. (1973). An electron microscopic study of Aujeszky's disease. *Acta Pathologica (Berlin)*, **25**, 207–19.

McCRACKEN, R. M., McFERRAN, J. B. & DOW, C. (1973). The neural spread of pseudorabies virus in calves. *Journal of General Virology*, **20**, 17–28.

McDOUGALL, J. K. & GALLOWAY, D. A. (1978). Detection of viral nucleic acid sequences using *in situ* hybridization. In *Persistent Viruses. ICN/UCLA Symposium on Molecular and Cellular Biology, XI*, ed. J. G. Stevens, G J. Todaro & C. F. Fox, pp. 181–8. New York: Academic Press.

McFERRAN, J. B. & DOW, C. (1965). The distribution of the virus of Aujeszky's disease (pseudorabies virus) in experimentally infected swine. *American Journal of Veterinary Research*, **26**, 631–5.

McLENNAN, J. L. (1981). Studies on herpes simplex virus latency. Ph.D. Thesis, Cambridge University.

McLENNAN, J. L. & DARBY, G. (1980). Herpes simplex virus latency: the cellular location of virus in dorsal root ganglia and the fate of the infected cell following virus activation. *Journal of General Virology*, **51**, 233–43.

MALHERBE, M. & HARWIN, R. (1958). Neurotropic virus in African monkeys. *Lancet*, **ii**, 530.

MARINESCO, G. (1932). Recherches sur la pathologie de certaines encéphalomyélites à ultravirus. *Revue Neurologique (Paris)*, **1**, 1–37.

MARINESCO, G. & DRAGANESCO, S. (1923). Recherches experimentales sur le neurotropinisme du virus herpétique. *Annales de l'Institut Pasteur (Paris)*, **37**, 753–83.

MARTIN, W. B., JAMES, Z. H., LAUDER, I. M., MURRAY, M. & PIRIE, H. M. (1969). Pathogenesis of bovine mammillitis virus infection in cattle. *American Journal of Veterinary Research*, **30**, 2151–66.

MELNICK, J. L. (1971). Classification of animal viruses. *Progress in Medical Virology*, **13**, 462–84.

MELNICK, J. L. & BANKER, D. D. (1954). Isolation of B virus (herpes group) from the central nervous system of a rhesus monkey. *Journal of Experimental Medicine*, **100**, 181–94.

MERIGAN, T. C. & STEVENS, D. A. (1971). Viral infections in man associated with acquired immunological deficiency states. *Federation Proceedings*, **30**, 1858–64.

MOGENSEN, S. C. (1980). Genetics of macrophage-controlled natural resistance to hepatitis induced by herpes simplex virus type 2 in mice. In *Genetic Control of Natural Resistance to Infection and Malignancy*, ed. E. Skamene, P. A. L. Kongshaum & M. Landy, pp. 291–6. New York: Academic Press.

MORAHAN, P. S. & MORSE, S. S. (1979). Macrophage-virus interactions. In *Virus-Lymphocyte Interactions: Implications for Disease*, ed. M. R. Proffitt, pp. 17–35. Amsterdam: Elsevier/North Holland.

NARITA, M., INUI, S., NAMBA, K. & SHIMIZU, Y. (1978). Neural changes in recurrent infection of infectious bovine rhinotracheitis virus in calves treated with dexamethasone. *American Journal of Veterinary Research*, **39**, 1399–403.

NASH, A. A. (1981). Antibody and latent herpes simplex infections. *Immunology Today*, **2**, 19–20.

NASH, A. A., FIELD, H. J. & QUARTEY-PAPAFIO, R. (1980a). Cell mediated immunity in herpes simplex virus-infected mice: induction, characterization and antiviral effects of delayed type hypersensitivity. *Journal of General Virology*, **48**, 351–7.

NASH, A. A. & GELL, P. G. H. (1980). Cell-mediated immunity in herpes simplex virus infected mice: suppression of delayed hypersensitivity by an antigen specific B lymphocyte. *Journal of General Virology*, **48**, 359–64.

NASH, A. A. & GELL, P. G. H. (1981). The delayed hypersensitivity T cell and its interaction with other T cells: with special reference to virus infections. *Immunology Today*, **2**, 162–5.

NASH, A. A., GELL, P. G. H. & WILDY, P. (1981b). Tolerance and immunity in mice infected with herpes simplex virus: simultaneous induction of protective immunity and tolerance to delayed-type hypersensitivity. *Immunology*, **43**, 153–9.

NASH, A. A., PHELAN, J., GELL, P. G. H. & WILDY, P. (1981). Tolerance and immunity in mice infected with herpes simplex virus: studies on the mechanism of tolerance to delayed type hypersensitivity. *Immunology*, **43**, 363–9.

NASH, A. A., PHELAN, J. & WILDY, P. (1981a). Cell-mediated immunity in herpes simplex virus-infected mice: H-2 mapping of the delayed-type hypersensitivity response and the antiviral T cell response. *Journal of Immunology*, **126**, 1260–2.

NASH, A. A., QUARTEY-PAPAFIO, R. & WILDY, P. (1980b). Cell mediated immunity in herpes simplex virus-infected mice: functional analysis of lymph node cells during periods of acute and latent infection, with reference to cytoxic and memory cells. *Journal of General Virology*, **49**, 309–17.

NESBURN, A. B., COOK, M. L. & STEVENS, J. G. (1972). Latent herpes simplex virus. Isolation from rabbit trigeminal ganglia between episodes of recurrent ocular infection. *Archives of Opthalmology*, **88**, 412–18.

NESBURN, A. B., ELLIOT, J. M. & LEIBOWITZ, H. M. (1967). Spontaneous reactivation of experimental herpes simplex keratitis in rabbits. *Archives of Ophthalmology*, **78**, 523–9.

NEWTON, A. A. (1979). Inhibitors of prostaglandin synthesis as inhibitors of herpes simplex virus replication. *Advances in Opthalmology*, **38**, 58–63.

OAKES, J. E. (1975). Role for cell mediated immunity in the resistance of mice to subcutaneous herpes simplex infection. *Infection and Immunity*, **12**, 166–72.

OLDSTONE, M. B. A. & LAMPERT, P. W. (1979). Antibody mediated complement dependent lysis of virus infected cells. *Springer Seminars in Immunopathology*, **2**, 261–83.

OPENSHAW, H., PUGA, A. & NOTKINS, A L. (1979). Latency and reactivation of herpes simplex virus in sensory ganglia of mice. *Developments in Immunology*, **7**, 301–6.

O'REILLY, R. J., CHIBBARO, A., ANGER, E. & LOPEZ, C. (1977). Cell mediated responses in patients with recurrent herpes simplex infections. II. Infection associated deficiency of lymphokine production in patients with recurrent herpes labialis or genitalis. *Journal of Immunology*, **118**, 1095–102.

PAYLING WRIGHT, G. (1953). Nerve trunks as pathways in infection. *Proceedings of the Royal Society of Medicine*, **46**, 319–30.

PERCY, D. H., CARMICHAEL, L. E., ALBERT, D. M., KING, J. M. & JONAS, A. M. (1971). Lesions in puppies surviving infection with canine herpesvirus. *Veterinary Pathology*, **8**, 37–53.

PERDRAU, J. R. (1938). Persistence of the virus of herpes in rabbits immunized with living virus. *Journal of Pathology*, **47**, 447–55.

PFIZENMAIER, K., STARZINSKI-POWITZ, A., RÖLLINGHOFF, M., FALKE, D. & WAGNER, H. (1977). T cell mediated cytotoxicity against herpes simplex virus infected target cells. *Nature, London*, **265**, 630–2.

PLUMMER, G., HOLLINGSWORTH, D. C., PHUANGSAB, A. & BOWLING, C. P. (1970). Chronic infections by herpes simplex viruses and by the horse and cat herpes viruses. *Infection and Immunity*, **1**, 351–5.

PRICE, R. W., KATZ, B. J. & NOTKINS, A. L. (1975). Latent infection of the peripheral autonomic nervous system with herpes simplex virus. *Nature, London*, **257**, 686–8.

PRICE, R. W. & SCHMITZ, J. (1978). Reactivation of latent herpes simplex virus infection of the autonomic nervous system by postganglionic neurectomy. *Infection and Immunity*, **19**, 523–32.

PROBERT, M. & POVEY, R. C. (1975). Experimental studies concerning the possibility of a latent carrier state in bovine herpes mammillitis (BHM). *Archives of Virology*, **48**, 29–38.

PUGA, A., ROSENTHAL, J. D., OPENSHAW, H. & NOTKINS, A. L. (1978). Herpes simplex virus DNA and mRNA sequences in acutely and chronically infected trigeminal ganglia of mice. *Virology*, **89**, 102–11.

PURSELL, A. R., SANGSTER, L. T., BYARS, T. D., DIVERS, T. J. & COLE, JR, J. R. (1979). Neurologic disease induced by equine herpesvirus I. *Journal of the American Veterinary Medical Association*, **175**, 473–4.

RAGER-ZISMAN, B. & ALLISON, A. C. (1976). Mechanism of immunological resistance to herpes simplex virus 1 (HSV-1) infection. *Journal of Immunology*, **116**, 35–40.

RAGER-ZISMAN, B. & ALLISON, A. C. (1980). The role of natural killer (NK) cells in resistance to herpes simplex virus 1 (HSV-1) infection. *Parasite Immunology*, **2**, 71–84.

RUSSELL, A. S. & SCHLAUT, J. (1977). Association of HLA-A1 antigen and susceptibility to cold sores. *Archives of Dermatology*, **113**, 1721–2.

RUSTIGIAN, R., SMULOW, J. B., TYE, M., GIBSON, W. A. & SHINDELL, E. (1966). Studies on latent infection of skin and oral mucosa in individuals with recurrent herpes simplex. *Journal of Investigative Dermatology*, **47**, 218–21.

SABIN, A. B. & WRIGHT, A. M. (1934). Acute ascending myelitis following a monkey bite with the isolation of a virus capable of producing the disease. *Journal of Experimental Medicine*, **59**, 115–36.

SABÓ, A. & RAJČÁNI, J. (1976). Latent pseudorabies virus infection in pigs. *Acta Virologica, Prague*, **20**, 208–14.

SCHMIDT, J. R. & RASMUSSEN, A. F. (1960). Activation of latent herpes simplex encephalitis by chemical means. *Journal of Infectious Diseases*, **106**, 154–8.

SCHWARTZ, J., WHETSELL, JR, W. O. & ELIZAN, T. S. (1978). Latent herpes simplex virus infection of mice. Infectious virus in homogenates of latently infected dorsal root ganglia. *Journal of Neuropathology and Experimental Neurology*, **37**, 45–55.

SCRIBA, M. (1975). Herpes simplex infection in guinea pigs: an animal model for studying latent and recurrent herpes simplex virus infection. *Infection and Immunity*, **12**, 162–5.

SCRIBA, M. (1977). Extraneural localization of herpes simplex virus in latently infected guinea pigs. *Nature, London*, **267**, 529–31.

SEKIZAWA, T., OPENSHAW, H., WOHLENBERG, C. & NOTKINS, A. L. (1980). Latency of herpes simplex virus in absence of neutralizing antibody: model of reactivation. *Science*, **210**, 1026–8.

SEQUERA, L. W., JENNINGS, L. C., CARRASSO, L. H., LORD, M. A., CURRY, A. & SUTTON, R. N. P. (1979). Detection of herpes simplex viral genome in brain tissue. *Lancet*, **ii**, 609–12.

SETHI, K. K. & BRANDIS, H. (1977). Specifically immune mouse T-cells can destroy H-2 compatible murine target cells infected with herpes simplex virus types 1 and 2. *Zeitschrift für Immunitätsforschung*, **150**, 162–73.

SHEFFY, B. E., DAVIES, D. H. & BAKER, J. A. (1972). Reactivation of a bovine herpesvirus after corticosteroid treatment. *Proceedings of the Society of Experimental Biology and Medicine*, **140**, 974–6.

SHILLITOE, E. J., WILTON, J. M. A. & LEHNER, T. (1977). Sequential changes in cell-mediated immune responses to herpes simplex virus after recurrent herpetic infection in humans. *Infection and Immunity*, **18**, 130–7.

SHORE, S. L., CROMEANS, T. L. & ROMANO, T. J. (1976). Immune destruction of virus infected cells early in the infectious cycle. *Nature, London*, **262**, 695–6.

SONNENFELD, G. & MERIGAN, T. C. (1979). The role of interferon in viral infections. *Springer Seminars in Immunopathology*, **2**, 311–38.

SPRUANCE, S. L., OVERALL, J. C. JR, KERN, E. R., KRUEGER, G. G., PLIAM, V. & MILLER, W. (1977). The natural history of recurrent herpes simplex labialis. *New England Journal of Medicine*, **297**, 69–74.

STEVENS, J. G. & COOK, M. L. (1971). Latent herpes simplex virus in spinal ganglia of mice. *Science*, **173**, 843–5.

STEVENS, J. G. & COOK, M. L. (1973). Latent herpes simplex virus in sensory ganglia. *Perspectives in Virology*, **8**, 171–88.

STEVENS, J. G. & COOK, M. L. (1974). Maintenance of latent herpetic infection. An apparent role for anti-viral IgG. *Journal of Immunology*, **113**, 1685–93.

STEVENS, J. G., NESBURN, A. B. & COOK, M. L. (1972). Latent herpes simplex virus from trigeminal ganglia of rabbits with recurrent eye infection. *Nature, London*, **235**, 216–17.

TENSER, R. B. & DUNSTAN, M. E. (1979). Herpes simplex virus thymidine kinase expression in infection of the trigeminal ganglion. *Virology*, **99**, 417–22.

TENSER, R. B., MILLER, R. L. & RAPP, F. (1979). Trigeminal ganglion infection by thymidine kinase-negative mutants of herpes simplex virus. *Science*, **205**, 915–17.

TONEVA, V. (1976). Le diagnostic dans la maladie d'Aujeszky. *La Clinica Veterinaire*, **99**, 119–38.

VAHLNE, A., NYSTRÖM, B., SANDBERG, M., HAMBERGER, A. & LYCKE, E. (1978). Attachment of herpes simplex virus to neurons and glial cells. *Journal of General Virology*, **40**, 359–71.

VIZOSO, A. D. (1975a). Recovery of herpes simiae (B virus) from both primary and latent infections in rhesus monkeys. *British Journal of Experimental Pathology*, **56**, 485–8.

VIZOSO, A. D. (1975b). Latency of herpes simiae (B virus) in rabbits. *British Journal of Experimental Pathology*, **56**, 489–94.

WALZ, M. A., PRICE, R. W. & NOTKINS, A. L. (1974). Latent ganglionic infection with herpes simplex virus types 1 and 2: viral reactivation *in vivo* after neurectomy. *Science*, **184**, 1185–7.

WALZ, M. A., YAMAMOTO, H. & NOTKINS, A. L. (1976). Immunological response restricts number of cells in sensory ganglia infected with herpes simplex virus. *Nature, London*, **264,** 554–6.

WARREN, K. G., BROWN, S. M., WROBLEWSKA, Z., GILDEN, D., KOPROWSKI, H. & SUBAK-SHARPE, J. (1978). Isolation of latent herpes simplex virus from the superior cervical and vagus ganglions of human beings. *New England Journal of Medicine*, **298,** 1068–9.

WARREN, S. L., CARPENTER, C. M. & BOAK, R. A. (1940). Symptomatic herpes, a sequela of artificially induced fever. *Journal of Experimental Medicine*, **71,** 155–67.

WHEELER, C. E. JR (1960). Further studies on the effect of neutralizing antibody upon the course of herpes simplex infections in tissue culture. *Journal of Immunology*, **84,** 394–403.

WHEELER, C. E. JR (1975). Pathogenesis of recurrent herpes simplex infections. *Journal of Investigative Dermatology*, **65,** 341–6.

WILDY, P. (1967). The progression of herpes simplex virus to the central nervous system of the mouse. *Journal of Hygiene, Cambridge*, **65,** 173–92.

WILTON, J. M. A., IVANYI, L. & LEHNER, T. (1972). Cell-mediated immunity in Herpesvirus hominis infections. *British Medical Journal*, **1,** 723–6.

ZISMAN, B., HIRSCH, M. S. & ALLISON, A. C. (1970). Selective effects of anti-macrophage serum, silica and anti-lymphocyte serum on pathogenesis of herpes virus infection of young adult mice. *Journal of Immunology*, **104,** 1155–9.

PERSISTENCE OF EPSTEIN-BARR VIRUS INFECTION

M. A. EPSTEIN

Department of Pathology, University of Bristol Medical School, University Walk, Bristol BS8 1TD, UK

INTRODUCTION

In virology, the term 'persistence' is used with remarkable looseness to cover a wide variety of situations which could not conceivably be grouped together under any single, properly defined heading. Indeed, this point is well illustrated by the contributions to the present symposium.

As far as Epstein-Barr (EB) virus (Epstein, Achong & Barr, 1964) is concerned, persistence of the agent within the infected individual must be considered in the context of such differing cell–virus interactions as latency, transformation, and productive infection, and must take into account clearly characterized humoral and cell-mediated immunological responses which profoundly affect each of these relationships. It should also be noted that the type of cell serving as target for the virus likewise influences the outcome of infection; indeed, cellular constraints on EB virus gene expression have not, in general, received sufficient attention.

GENERAL BIOLOGY OF EB VIRUS INFECTION

Epidemiology

It is now well established that EB virus is a herpesvirus of man, spread by horizontal infection and ubiquitous in all human populations (W. Henle & G. Henle, 1979). At the present time, only B lymphocytes have been shown to have receptors for the virus (Pattengale, Smith & Gerber, 1973; Jondal & Klein, 1973) and these receptors appear to correspond to, and may indeed be identical with, C3 receptors (Yefenof, Klein, Jondal & Oldstone, 1976; Yefenof & Klein, 1977).

Natural primary infection usually takes place in childhood without disease manifestations and is accompanied by seroconversion with development of specific antibodies directed against virus-determined antigens, by the establishment of permanent immunity to re-infection, and by a harbouring of the virus which persists for the rest of the individual's life. This harbouring takes two forms: firstly, a small number of circulating B lymphocytes can be shown to carry the virus genome; and secondly, there is a productive infection somewhere in the oropharynx or its adnexa, with shedding of virus into the buccal fluid. It is, of course, this shedding of infectious virus which is responsible by horizontal transmission for all natural primary infections. In developed countries about 75–80% of the population ultimately become infected, whereas in developing countries 99.9% of children are already infected by about the age of 3. Each of these findings has been documented in full elsewhere (Epstein & Achong, 1979a; W. Henle & G. Henle, 1979; G. Henle & W. Henle, 1979).

If natural primary infection is delayed to adolescence or young adulthood there is a 50% chance that it will be accompanied by the clinical manifestations of infectious mononucleosis (IM). Such delayed infection is more frequent in the privileged classes of Western societies enjoying high standards of hygiene, than in the lower socio-economic groups, thus explaining the characteristic association of IM with affluence. The uniformly early age of EB virus infection in developing countries, which leaves no adolescents or young adults susceptible, accounts for the virtual absence there of IM. These epidemiological observations concerning EB virus and IM have recently been reviewed in detail (Epstein & Achong, 1977; G. Henle & W. Henle, 1979).

Why natural primary infection should usually remain clinically silent in childhood yet carry a 50% risk of IM amongst adolescents and young adults, almost certainly relates to the size of viral dose, mode of infection, and host physiological and immunological response in the two age groups. Young children appear to be readily infected by casual contamination with a small dose of virus shed into the environment whereas, in contrast, adolescents and young adults can only be infected with considerable difficulty. IM has long been recognized for its association with kissing among young people (Hoagland, 1955) and it would appear that the disease is only induced where a healthy seropositive individual shedding virus passes large quantities of the agent in buccal fluid to a seronegative (and therefore susceptible) partner during direct osculatory contact.

Indeed, there is very good evidence that infection does not occur where susceptible young adults merely share a room and washing facilities with those in the acute stage of IM (Hallee *et al.*, 1974) when the maximum shedding of virus occurs (Niederman *et al.*, 1976). However, although almost all IM is seen in those aged 15–25, it is worth noting that primary infection in very young children and in quite old adults can very occasionally be accompanied by the fully developed disease (Ginsberg, Henle, Henle & Horwitz, 1977; Horwitz *et al.*, 1976). In any event, the ultimate consequences of IM are exactly the same as with a silent primary infection – seroconversion, solid immunity to subsequent re-infection with EB virus, and life-long harbouring of the virus.

Immunological control of infection

During primary infection accompanied by IM, the spread of infectious virus within the patient becomes restricted as soon as virus-neutralizing antibodies develop. These first appear with the onset of symptoms and reach a peak by the end of the acute phase or at the start of convalescence, and it is interesting that thereafter they persist for life at the same level (Epstein & Achong, 1977). With silent natural primary infection, it is obviously not possible to monitor the development of neutralizing antibodies, but serial tests for them in seropositive individuals after primary infection show that here too, they remain constant (W. Henle & G. Henle, 1979).

It has been known for some time that IM gives rise to a remarkable T cell response (Sheldon, Papamichael, Hemsted & Holborow, 1973; Virolainen, Andersson, Lalla & von Essen, 1973; Yata, Desgranges, de-Thé & Tachibana, 1973) and that functionally, several different types of T cell activity are involved during the acute phase of the disease (Svedmyr & Jondal, 1975; Royston, Sullivan, Periman & Perlin, 1975; Hutt, Huang, Dascomb & Pagano, 1975; Rickinson, Crawford & Epstein, 1977; Tursz *et al.*, 1977; Bakacs, Svedmyr & Klein, 1978; Lipinski *et al.*, 1979; Svedmyr, Klein & Wiland, 1979; Tosato *et al.*, 1979). More recently, longitudinal studies have shown that specific memory T cells are constantly present in the peripheral circulation of all seropositive individuals and that with each given individual, the 'strength' of this specific T cell surveillance remains stable over months and years (Rickinson *et al.*, 1981).

PERSISTENCE OF EB VIRUS

Productive infection and virus shedding

It has already been pointed out that one component of the life-long carrier state which always follows infection with EB virus consists of a rather low-grade production of the virus somewhere in the oropharynx or its adnexa. This productive infection manifests itself by the shedding of virus particles in the buccal fluid. The incidence of individuals shedding EB virus from the mouth varies in different populations and under different conditions. In developed countries about 20% of healthy seropositive individuals are shedders at any one time and the dependence of this on a delicate immunological balance is clearly indicated by the fact that the rate rises significantly to more than 50% of seropositives amongst those subjected to immunosuppressive therapy. A similarly raised incidence of shedders has been observed in the general population of developing countries and the explanation has been put forward that this may result from the immunosuppressive effects of malarial infection and infestation with higher parasites. During acute IM, shedding of EB virus into the oropharynx occurs in almost all patients. The various aspects of virus shedding have been discussed by Epstein & Achong (1979a) and G. Henle & W. Henle (1979).

In early studies it was usually assumed that the productive infection in the oropharynx occurred exclusively in B lymphocytes within Waldeyer's ring, since only B cells were known to have demonstrable receptors for EB virus. However, the finding of EB virus at the orifice of Stensen's duct (Niederman *et al.*, 1976) and the demonstration that cannulated duct fluid contained considerably higher levels of virus than the buccal fluid (Morgan *et al.*, 1979) has led to the suggestion that salivary gland epithelial cells may also be involved.

The concept of EB virus replication in epithelial cells has become more attractive in recent years following experiments in which virus production was demonstrated in the EB virus genome-containing squamous epithelial cells from nasopharyngeal carcinomas (NPC) both in nude mice and *in vitro* (Trumper, Epstein & Giovanella, 1976; Trumper, Epstein, Giovanella & Finerty, 1977; Crawford *et al.*, 1979a), and the possibility therefore remains that the source of virus shed into the buccal fluid of normal seropositive individuals may ultimately prove to be in some special type of oro/nasopharyngeal epithelial cell. In this connection it is of interest that a number of different types of cell, in addition to primate B lymphocytes, have recently proved capable of expressing the earlier antigens of the virus replication cycle when provided experimentally with an artificial means of entry for the virus (Graessmann, Wolf & Bornkamm, 1980; Volsky, Shapiro & Klein, 1980).

On the other hand, even if EB virus replication leading to shedding does take place in salivary glands, it need not necessarily involve the epithelial elements; for, it has recently been shown that although horizontal transmission of the lymphotropic Herpesvirus sylvilagus takes place in the natural host (cotton-tail rabbit) through infected buccal fluid, virus replication in all the tissues of the mouth, pharynx and adjacent structures is restricted to infiltrating lymphocytes (Hinze & Lee, 1980). But wherever the oro/nasopharyngeal site of EB virus replication may be, it is undoubtedly dependent on immunological constraints, since where immunodepression occurs reactivation or augmentation of virus production results (see Epstein & Achong, 1979a).

EB virus infection of circulating B lymphocytes

The presence of EB virus-infected lymphocytes in the peripheral cirulation was first suspected when it was shown that EB virus-carrying continuous lymphoblastoid cell lines could be established *in vitro* from blood (Gerber, Whang-Peng & Monroe, 1969; Chang, Hsieh & Blankenship, 1971) or lymph node biopsies (Nilsson, Klein, Henle & Henle, 1971) from all seropositive individuals. The form in which the virus persists in these genome-carrying B lymphocytes remains a matter of controversy. It has been suggested that during the acute phase of IM, and at a lower level after the disease and in all silently infected individuals, the virus is present in a small number of B lymphocytes which it has transformed (Klein, 1979a) in a manner analogous to the transformation seen when infectious virus is added to B cells *in vitro* (Pope, 1979). A corollary of this view postulates that these cells are prevented from continuous multiplication by an immunological surveillance mechanism which holds them in check (Klein, 1973) and that unlimited malignant cell division leading to the development of lymphoma only supervenes when an additional cytogenetic change, $t(8q- : 14q+)$, takes place (Klein, 1979a). The demonstration of EB virus nuclear antigen (EBNA)-positive cells in the peripheral circulation of acute IM patients (Klein, Svedmyr, Jondal & Persson, 1976), the finding that such cells can undergo mitosis (Robinson, Smith & Niederman, 1980b), and the seemingly lymphomatous polyclonal multiplication of EBNA-positive cells in certain cases of fatal IM (Robinson *et al.*, 1980a) have all been adduced as evidence supporting this view.

On the other hand, the masses of EBNA-positive cells which infiltrate many organs in fatal IM could equally well be B lymphocytes supporting uncontrolled virus replication (Ernberg, Masucci & Klein, 1976), rather than a malignant proliferation. Such virus replication has been demonstrated in fatal IM (Crawford *et al.*, 1979b) and this interpretation would certainly explain the polyclonality of the infiltrate (Crawford *et al.*, 1979b; Robinson *et al.*, 1980a). In the same way, the small numbers of EBNA-positive cells in the blood of some IM patients early in the disease could also be concerned with virus replication.

In either case, it is likely that the controversy over the meaning of EBNA-positive cells in the blood is only of significance in relation to the acute phase of IM and has little relevance to the main issue concerning the nature of the cell–virus relationship between EB virus and some circulating B cells of all infected individuals. If one considers the lymphotropic herpesviruses of animals, there is good evidence now that in several systems such viruses persist in lymphoid target cells in a state of latency – that is to say, the cells are not transformed, infectious virus cannot be extracted directly from them in any known way, and the presence of the viral genetic information can only be demonstrated by activating it to virus production in tissue culture; in some cases this can only be done after the application of additional stimuli. Examples are provided by herpesvirus saimiri in its natural squirrel monkey host (Falk, Wolfe & Deinhardt, 1972), herpesvirus sylvilagus in the cotton-tail rabbit (Wegner & Hinze, 1974), Marek's disease virus in both T and B cells of the chicken spleen (Calnek, Shek & Schat, 1981) and mouse cytomegalovirus (CMV) (Olding, Kingsbury & Oldstone, 1976; Oldstone *et al.*, 1976); with each of these agents non-productive infections are established in lymphocytes from which infectious virus can only be recovered if the undisrupted cells are co-cultivated with a susceptible monolayer, and in the case of mouse CMV, activation is enhanced if the co-cultivation involves the stimulus of an immunological barrier (Olding, Jensen & Oldstone, 1975).

The EB virus genome-containing B cells present in the blood of all seropositive individuals appear to behave in an analogous manner. Thus, infectious virus cannot be detected when the cells are freshly removed and disrupted, but only becomes manifest if intact cells are cultured (Rickinson, Epstein & Crawford, 1975) whereupon the latent infection is activated, liberating infectious virus into the medium where it infects and transforms co-resident uninfected B lymphocytes to give rise to continuous cell lines (Rickinson, Jarvis, Crawford & Epstein, 1974; Rickinson, Finery & Epstein 1977a, b), the so-called two-step mechanism.

To complicate this issue further, it has been reported that in IM EBNA-negative EB virus-containing cells are also present in the circulation (Crawford, Rickinson, Finerty & Epstein, 1978) and a proposal has recently been made that the expression of EBNA may relate exclusively to transformation, leaving a different, EBNA-negative pathway leading to the virus productive cycle (Volsky, Klein, Volsky & Shapiro, 1981). It is quite clear, however, that this difficult and important question of the interaction between EB virus and the seemingly latently infected target cell will not prove easy to unravel until it becomes possible selectively to collect EB virus genome-containing cells from the peripheral blood of normal seropositive individuals in sufficient quantity for proper investigation. It is also necessary to determine the whereabouts in the infected individual of the immunologically privileged site from which such cells are released into the circulation throughout life.

EB virus in malignant cells

EB virus is associated with two, and only two, human malignancies; namely endemic (sometimes termed African) Burkitt's lymphoma (BL) (Burkitt, 1963) and undifferentiated or poorly differentiated NPC (Shanmugaratnam, 1971). It is most important that the exact nature of these two cancers should be understood and defined, since the restriction of the association to these tumours alone is of itself highly significant. Apart from BL, other lymphomas are quite unrelated to the virus, and tumours of the post-nasal space other than undifferentiated NPC or anywhere in the head and neck outside that narrow topographical region, are likewise unrelated to the virus.

Both BL and NPC are monoclonal tumours, BL being of B lymphocyte origin and NPC, however undifferentiated, always showing fine-structural evidence for the presence of keratin in its cells. In addition, BL cells consistently carry a characteristic chromosomal abnormality, t(8q– : 14q+). These findings have been reviewed elsewhere (Epstein & Achong, 1979b; Klein, 1979b).

The association of EB virus with BL and NPC cells bears striking similarities to the relationships long recognized between known oncogenic DNA viruses of animals and the cells of the tumours they cause (Tooze, 1973). Thus, almost every authenticated case of endemic BL carries multiple copies of the EB virus genome in every cell, and there is evidence that at least one of these genome copies is linearly integrated into the host cell DNA. Although all tumour cells from every appropriately examined example of undifferentiated NPC likewise contain multiple EB virus DNA copies, the state of this viral DNA has not yet been determined. With both tumours, the viral genetic information is responsible for the expression of EBNA which can be demonstrated in the cells of freshly removed biopsies, together with EB virus-determined membrane antigen (MA) in the case of endemic BL cells; so far MA has not been found on the undifferentiated squamous epithelial cells of NPC. Documentation of these observations has been given by Epstein & Achong (1979b) and by Klein (1979b). Experimental activation of the viral genome in the cells of both endemic BL and undifferentiated NPC appears to follow a similar pattern. With some tumours of each kind, the virus productive cycle is activated in a proportion of cells merely by growing biopsy material in tissue culture or by implanting it in the nude mouse. With other tumours, virus production can only be induced by treatment with such activators as halogenated pyrimidines or certain phorbol esters; finally, there is a third group of each type of tumour where virus activation cannot be brought about at all (Trumper et al., 1976, 1977; Crawford et al., 1979a).

Other important points in the association of EB virus with BL and NPC clearly reflect special virus-cell interactions but do not indicate what these might be. Thus, with both tumours there is a characteristic pattern of serological responses which does not occur in other EB virus-related conditions. Patients with BL have unusually high levels of antibody to the virus capsid antigen (VCA) as well as uniquely high titres of antibody to early antigen (EA) of the restricted (R) type; with NPC there are also raised levels of antibody to VCA, but in this disease additional antibodies are directed against EA of the diffuse (D) type and there is a curious production of IgA to VCA. In both tumours changes in antibody levels follow closely on, or precede, clinical changes in the patient. Full details of these serological data have been presented elsewhere (Epstein & Achong, 1979b; Klein, 1979b). It is also clear that individuals destined to develop BL show a peculiar and unusual response to infection by the virus long before the onset of malignancy (de-Thé *et al.*, 1978), but again, the nature of the special virus–cell interactions underlying this have not been determined.

In this connection, it is of interest to note that where malignant reticuloproliferative disease has been induced experimentally in animals by the injection of EB virus (Shope, Dechairo & Miller, 1973; Epstein, Hunt & Rabin, 1973), the tumours are of B cell origin and the relationship of the virus to the tumour cells (Miller, 1979) parallels that seen in BL and NPC. There is also some evidence that the experimental tumours are of monoclonal origin (Rabin, Neubauer, Hopkins, & Levy, 1977).

DISCUSSION

Of the various types of infection which EB virus can bring about, productive infection presents the least difficulty since it can be studied *in vitro* in the small number of cells of 'producer' lines which express virus replicative functions (Hampar, 1979) and because it has close similarities to the 'lytic' infections of most ordinary herpesviruses. However, why only some cells in 'producer' lines make virus, why some cells in other lines are inducible, and why in yet other lines all cells remain stably transformed and totally uninducible, is not entirely understood. It seems likely that these differences are determined by interactions between virus genome expression and various blocks or bottle-necks in metabolic pathways which operate at several points in the virus replicative cycle (Epstein & Achong, 1979a) and which vary in different classes of B cells within the target population. There is certainly good evidence that different types of B cell are capable of exerting differing constraints on EB virus expression since B lymphocytes of foetal, adult seronegative, and marmoset origin transformed with the same EB virus isolate consistently give cell lines with different biological behaviours (Miller & Lipman, 1973; Crawford *et al.*, 1979a).

As for the persistence of productive infection throughout the life of some individuals as evidenced by virus shedding in the buccal fluid, here immunological constraints are clearly as important as constraints at the cellular level. The effect on shedding of immuno-suppressive therapy has already been mentioned and, in addition, recent studies have shown that cyclosporin A greatly depresses EB virus-specific T cell function both *in vitro* and *in vivo* (Crawford *et al.*, 1981; Bird, McLachlan & Britton, 1981).

It seems likely that cellular and immunological constraints are equally important in relation to latent infection and malignant transformation by EB virus. The crucial significance of hyperendemic malaria as a co-factor in the development of endemic BL has long been recognized (Burkitt, 1969) and has been explained in terms of the lymphostimulatory and the immunodepressive effects of the condition. The lymphostimulation could well provide unusual numbers of B lymphocytes in various stages of differentiation, with the possibility that some might be especially responsive to the transforming functions of the virus, while the immunodepression must obviously influence T cell control of transformed cells. It has sometimes been suggested that very early infection with EB virus, or some special temporal relationship between EB virus and malaria infection, play a part in determining the development of BL and each of these phenomena could also clearly mediate relaxation of cellular constraints or immunological control mechanisms. A full discussion of this topic has been given elsewhere (Epstein & Achong, 1979b).

The most difficult of the phenomena of persistence with EB virus is, of course, that of the virus genome-containing B cell in the peripheral circulation of the normal seropositive individual. By every kind of analogy with lymphotropic herpesviruses of animals, this should be a straight-forward state of latency, although to say that is merely to categorize rather than explain. The exact virus–cell relationships of latency are little understood in any animal virus system, but to distinguish it from some type of partial or controlled transformation is important both conceptually and biologically. However, as already pointed out, elucidation of the nature of this form of EB virus persistence is not likely to take place until it becomes possible to collect and study EB virus genome-containing cells from the circulation. But despite the many lacunae in our understanding of the life-long persistence of EB virus in the normal individual and of the exact relationship of the virus to the tumour cells of BL and NPC, sufficient information has been accumulated over the past few years for it to be evident that these phenomena are biologically important and in many ways unique. If progress in the next few years continues at the present pace, explanations are likely to be forthcoming sooner rather than later.

REFERENCES

BAKACS, T., SVEDMYR, E. & KLEIN, G. (1978). EBV-related cytotoxicity of Fc receptor negative T lymphocytes separated from the blood of infectious mononucleosis patients. *Cancer Letters*, **4**, 185–9.

BIRD, A. G., McLACHLAN, S. M. & BRITTON, S. (1981). Cyclosporin A promotes spontaneous outgrowth *in vitro* of Epstein-Barr virus-induced B-cell lines. *Nature*, **289**, 300–1.

BURKITT, D. (1963). A lymphoma syndrome in tropical Africa. In *International Review of Experimental Pathology*, ed. G. W. Richter & M. A. Epstein, vol. 2, pp. 67–138. Academic Press, New York & London.

BURKITT, D. P. (1969). Etiology of Burkitt's lymphoma – an alternative hypothesis to a vectored virus. *Journal of the National Cancer Institute*, **42**, 19–28.

CALNEK, B. W., SHEK, W. R. & SCHATT, K. A. (1981). Latent infections with Marek's disease virus and turkey herpesvirus. *Journal of the National Cancer Institute*, (in press).

CHANG, R. S., HSIEH, M. W. & BLANKENSHIP, W. (1971). Initiation and establishment of lymphoid cell lines from the blood of healthy persons. *Journal of the National Cancer Institute*, **47**, 469–77.

CRAWFORD, D. H., EPSTEIN, M. A., BORNKAMM, G. W., ACHONG, B. G., FINERTY, S. & THOMPSON, J. L. (1979a). Biological and biochemical observations on isolates of EB virus from the malignant epithelial cells of two nasopharyngeal carcinomas. *International Journal of Cancer*, **24**, 294–302.

CRAWFORD, D. H., EPSTEIN, M. A., ACHONG, B. G., FINERTY, S., NEWMAN, J., LIVERSEDGE, S., TEDDER, R. S. & STEWART, J. W. (1979b). Virological and immunological studies on a fatal case of infectious mononucleosis. *Journal of Infection*, **1**, 37–48.

CRAWFORD, D. H., RICKINSON, A. B., FINERTY, S. & EPSTEIN, M. A. (1978). Epstein-Barr (EB) virus genome-containing EB nuclear antigen-negative B-lymphocyte populations in blood in acute infectious mononucleosis. *Journal of General Virology*, **38**, 449–60.

CRAWFORD, D. H., SWENY, P., EDWARDS, J., JANOSSY, G. & HOFFBRAND, A. V. (1981). Long-term T-cell-mediated immunity to Epstein-Barr virus in renal allograft recipients receiving Cyclosporin A. *Lancet*, **i**, 10–13.

DE-THÉ, G., GESER, A., DAY, N. E., TUKEI, P. M., WILLIAMS, E. H., BERI, D. P. SMITH, P. G., DEAN, A. G., BORNKAMM, G. W., FEORINO, P. & HENLE, W. (1978). Epidemiological evidence for causal relationship between Epstein-Barr virus and Burkitt's lymphoma: results of the Ugandan prospective study. *Nature*, **274**, 756–61.

EPSTEIN, M. A. & ACHONG, B. G. (1977). Pathogenesis of infectious mononucleosis. *Lancet*, **ii**, 1270–3.

EPSTEIN, M. A. & ACHONG, B. G. (1979a). Introduction and general biology of the virus. In *The Epstein-Barr Virus*, ed. M. A. Epstein & B. G. Achong, pp. 1–22. Springer, Berlin, Heidelberg & New York.

EPSTEIN, M. A. & ACHONG, B. G. (1979b). The relationship of the virus to Burkitt's lymphoma. In *The Epstein-Barr Virus*, ed. M. A. Epstein & B. G. Achong, pp. 321–37. Springer, Berlin, Heidelberg & New York.

EPSTEIN, M. A., ACHONG, B. G. & BARR, Y. M. (1964). Virus particles in cultured lymphoblasts from Burkitt's lymphoma. *Lancet*, **i**, 702–3.

EPSTEIN, M. A., HUNT, R. D. & RABIN, H. (1973). Pilot experiments with EB virus in owl monkeys (*Aotus trivirgatus*). I. Reticuloproliferative disease in an inoculated animal. *International Journal of Cancer*, **12**, 309–18.

ERNBERG, I., MASUCCI, G. & KLEIN, G. (1976). Persistence of Epstein-Barr nuclear antigen (EBNA) in cells entering the EB viral cycle. *International Journal of Cancer*, **17**, 197–203.

FALK, L., WOLFE, L. G. & DEINHARDT, F. (1972). Isolation of *Herpesvirus saimiri* from blood of squirrel monkeys (*Saimiri sciureus*). *Journal of the National Cancer Institute*, **48**, 1499–1505.

GERBER, P., WHANG-PENG, J. & MONROE, J. H. (1969). Transformation and chromosome changes induced by Epstein-Barr virus in normal human leukocyte cultures. *Proceedings of the National Academy of Sciences, USA*, **63**, 740–7.

GINSBERG, C. M., HENLE, W., HENLE, G. & HORWITZ, C. A. (1977). Infectious mononucleosis in children. Evaluation of Epstein-Barr virus-specific serological data. *Journal of the American Medical Association*, **237**, 781–5.

GRAESSMANN, A., WOLF, H. & BORNKAMM, G. W. (1980). Expression of Epstein-Barr virus genes in different cell types after microinjection of viral DNA. *Proceedings of the National Academy of Sciences, USA*, **77**, 433–6.

HALLEE, T. J., EVANS, A. S., NIEDERMAN, J. C., BROOKS, C. M. & VOEGTLY, J. H. (1974). Infectious mononucleosis at the United States Military Academy. A prospective study of a single class over four years. *Yale Journal of Biological Medicine*, **47**, 182–95.

HAMPAR, B. (1979). Activation of the viral genome *in vitro*. In *The Epstein-Barr Virus*, ed. M. A. Epstein & B. G. Achong, pp. 283–95. Springer, Berlin, Heidelberg & New York.

HENLE, G. & HENLE, W. (1979). The virus as the etiologic agent of infectious mononucleosis. In *The Epstein-Barr Virus*, ed. M. A. Epstein & B. G. Achong, pp. 297–320. Springer, Berlin, Heidelberg & New York.

HENLE, W. & HENLE, G. (1979). Seroepidemiology of the virus. In *The Epstein-Barr Virus*, ed. M. A. Epstein & B. G. Achong, pp. 61–78. Springer, Berlin, Heidelberg & New York.

HINZE, H. C. & LEE, K. W. (1980). Mechanism of oropharyngeal excretion of the rabbit mononucleosis lymphoma agent, Herpesvirus sylvilagus. In *Viruses in Naturally Occurring Cancers*, Book A., ed. M. Essex, G. Todaro & H. zur Hausen, pp. 163–9. Cold Spring Harbor.

HOAGLAND, R. J. (1955). The transmission of infectious mononucleosis. *American Journal of Medical Science*, **229**, 262–72.

HORWITZ, C. A., HENLE, W., HENLE, G., SEGAL, M., ARNOLD, T., LEWIS, F. B. ZANICK, D. & WARD, P. C. J. (1976). Clinical and laboratory evaluation of elderly patients with heterophil-antibody positive infectious mononucleosis. *American Journal of Medicine*, **61**, 333–9.

HUTT, L. M., HUANG, Y. T., DASCOMB, H. E. & PAGANO, J. S. (1975). Enhanced destruction of lymphoid cell lines by peripheral blood leukocytes taken from patients with acute infectious mononucleosis. *Journal of Immunology*, **115**, 243–8.

JONDAL, M. & KLEIN, G. (1973). Surface markers on human B and T lymphocytes. II. Presence of Epstein-Barr virus receptors on B lymphocytes. *Journal of Experimental Medicine*, **138**, 1365–78.

KLEIN, G. (1973). Tumor immunology. *Transplantation Proceedings*, **5**, 31–41.

KLEIN, G. (1979a). Lymphoma development in mice and humans: diversity of initiation is followed by convergent cytogenetic evolution. *Proceedings of the National Academy of Sciences, USA*, **76**, 2442–6.

KLEIN, G. (1979b). The relationship of the virus to nasopharyngeal carcinoma. In *The Epstein-Barr Virus*, ed. M. A. Epstein & B. G. Achong, pp. 339–50. Springer, Berlin, Heidelberg & New York.

KLEIN, G., SVEDMYR, E., JONDAL, M. & PERSSON, P. O. (1976). EBV-determined nuclear antigen (EBNA) positive cells in the peripheral blood of infectious mononucleosis patients. *International Journal of Cancer*, **17**, 21–6.

LIPINSKI, M., FRIDMAN, W. H., TURSZ, T., VINCENT, C., PIOUS, D. & FELLOUS, M. (1979). Absence of allogeneic restriction in human T-cell-mediated cytotoxicity to Epstein-Barr virus-infected target cells. Demonstration of an HLA-linked control at the effector level. *Journal of Experimental Medicine*, **150**, 1310–22.

MILLER, G. (1979). Experimental carcinogenicity by the virus *in vivo*. In *The Epstein-Barr Virus*, ed. M. A. Epstein & B. G. Achong, pp. 351–72. Springer, Berlin, Heidelberg & New York.

MILLER, G. & LIPMAN, M. (1973). Release of infectious Epstein-Barr virus by transformed marmoset leukocytes. *Proceedings of the National Academy of Sciences, USA*, **70**, 190–4.

MORGAN, D. G., NIEDERMAN, J. C., MILLER, G., SMITH, H. W. & DOWALIBY, J. M. (1979). Site of Epstein-Barr replication in the oropharynx. *Lancet*, **ii**, 1154–7.

NIEDERMAN, J. C., MILLER, G., PEARSON, H. A., PAGANO, J. S. & DOWALIBY, J. M. (1976). Infectious mononucleosis: Epstein-Barr virus shedding in saliva and the oropharynx. *New England Journal of Medicine*, **294**, 1355–9.

NILSSON, K., KLEIN, G., HENLE, W. & HENLE, G. (1971). The establishment of lymphoblastoid lines from adult and fetal human lymphoid tissue and its dependence on EBV. *International Journal of Cancer*, **8**, 443–50.

OLDING, L. B., JENSEN, F. C. & OLDSTONE, M. B. A. (1975). Pathogenesis of cytomegalovirus infection. I. Activation of virus from bone marrow derived lymphocytes by *in vitro* allogencic reaction. *Journal of Experimental Medicine*, **141**, 561–72.

OLDING, L. B., KINGSBURY, D. T. & OLDSTONE, M. B. A. (1976). Pathogcnesis of cytomegalovirus infection. Distribution of viral products, immune complexes and autoimmunity during latent murine infection. *Journal of General Virology*, **33**, 267–80.

OLDSTONE, M. B. A., HASPEL, M. V., PELLEGRINO, M. A., KINGSBURY, D. T. & OLDING, L. (1976). Histocompatibility complex and virus infection latency and activation. *Transplantation Reviews*, **31**, 225–39.

PATTENGALE, P. K., SMITH, R. W. & GERBER, P. (1973). Selective transformation of B lymphocytes by EB virus. *Lancet*, **ii**, 93.

POPE, J. H. (1979). Transformation by the virus *in vitro*. In *The Epstein-Barr Virus*, ed. M. A. Epstein & B. G. Achong, pp. 205–23. Springer, Berlin, Heidelberg & New York.

RABIN, H., NEUBAUER, R. H., HOPKINS III, R. F. & LEVY, B. M. (1977). Characterization of lymphoid cell lines established from multiple Epstein-Barr virus (EBV-)induced lymphomas in a cotton-topped marmoset. *International Journal of Cancer*, **20**, 44–50.

RICKINSON, A. B., CRAWFORD, D. & EPSTEIN, M. A. (1977). Inhibition of the *in vitro* outgrowth of Epstein-Barr virus-transformed lymphocytes by thymus-dependent lymphocytes from infectious mononucleosis patients. *Clinical and Experimental Immunology*, **28**, 72–9.

RICKINSON, A. B., EPSTEIN, M. A. & CRAWFORD, D. H. (1975). Absence of infectious Epstein-Barr virus in blood in acute infectious mononucleosis. *Nature*, **258**, 236–8.

RICKINSON, A. B., FINERTY, S. & EPSTEIN, M. A. (1977a). Comparative studies on adult donor lymphocytes infected by EB virus *in vivo* or *in vitro*: Origin of transformed cells arising in co-cultures with foetal lymphocytes. *International Journal of Cancer*, **19**, 775–82.

RICKINSON, A. B., FINERTY, S. & EPSTEIN, M. A. (1977b). Mechanism of the establishment of Epstein-Barr virus genome-containing lymphoid cell lines from infectious mononucleosis patients: studies with phosphonoacetate. *International Journal of Cancer*, **20**, 861–8.

RICKINSON, A. B., JARVIS, J. E., CRAWFORD, D. H. & EPSTEIN, M. A. (1974). Observations on the type of infection by Epstein-Barr virus in peripheral lymphoid cells of patients with infectious mononucleosis. *International Journal of Cancer*, **14**, 704–15.

RICKINSON, A. B., MOSS, D. J., WALLACE, L. E., MISKO, I. S., EPSTEIN, M. A. & POPE, J. H. (1981). Long-term T cell-mediated immunity to Epstein-Barr virus. *Cancer Research*, (in press).

ROBINSON, J. E., BROWN, N., ANDIMAN, W., HALLIDAY, K., FRANCKE, U., ROBERT, M. F., ANDERSSON-ANVRET, M., HORSTMANN, D. & MILLER, G. (1980a). Diffuse polyclonal B-cell lymphoma during primary infection with Epstein-Barr virus. *New England Journal of Medicine*, **302**, 1293–7.

ROBINSON, J., SMITH, D. & NIEDERMAN, J. (1980b). Mitotic EBNA-positive lymphocytes in peripheral blood during infectious mononucleosis. *Nature*, **287**, 334–5.

ROYSTON, I., SULLIVAN, J. L., PERIMAN, P. O. & PERLIN, E. (1975). Cell-mediated immunity to Epstein-Barr virus-transformed lymphoblastoid cells in acute infectious mononucleosis. *New England Journal of Medicine*, **293**, 1159–63.

SHANMUGARATNAM, K. (1971). Studies on the etiology of nasopharyngeal carcinoma. In *International Review of Experimental Pathology*, ed. G. W. Richter & M. A. Epstein, vol. 10, pp. 361–413. Academic Press, New York & London.

SHELDON, P. J., PAPAMICHAIL, M., HEMSTED, E. H. & HOLBOROW, E. J. (1973). Thymic origin of atypical lymphoid cells in infectious mononucleosis. *Lancet*, **i**, 1153–5.

SHOPE, T., DECHAIRO, D. & MILLER, G. (1973). Malignant lymphoma in cottontop marmosets after inoculation with Epstein-Barr virus. *Proceedings of the National Academy of Sciences*, *USA*, **70**, 2487–91.

SVEDMYR, E. & JONDAL, M. (1975). Cytotoxic effector cells specific for B cell lines transformed by Epstein-Barr virus are present in patients with infectious mononucleosis. *Proceedings of the National Academy of Sciences*, *USA*, **72**, 1622–6.

SVEDMYR, E., KLEIN, G. & WEILAND, O. (1979). The EBV-carrying ß2M/HLA deficient Burkitt lymphoma line Daudi is sensitive to EBV-specific killer T cells of mononculeosis patients. *Cancer Letters*, **7**, 15–20.

TOOZE, J. (1973). *The Molecular Biology of Tumour Viruses*. Cold Spring Harbor Laboratory.

TOSATO, G., MAGRATH, I., KOSKI, I., DOOLEY, N. & BLAESE, M. (1979). Activation of suppressor T cells during Epstein-Barr virus-induced infectious mononucleosis. *New England Journal of Medicine*, **301**, 1133–7.

TRUMPER, P. A., EPSTEIN, M. A. & GIOVANELLA, B. C. (1976). Activation *in vitro* by BUdR of a productive EB virus infection in the epithelial cells of nasopharyngeal carcinoma. *International Journal of Cancer*, **17**, 578–87.

TRUMPER, P. A., EPSTEIN, M. A., GIOVANELLA, B. C. & FINERTY, S. (1977). Isolation of infectious EB virus from the epithelial tumour cells of nasopharyngeal carcinoma. *International Journal of Cancer*, **20**, 655–62.

TURSZ, T., FRIDMAN, W. H., SENIK, A., TSAPIS, A. & FELLOUS, M. (1977). Human virus-infected target cells lacking HLA antigens resist specific T-lymphocyte cytolysis. *Nature*, **269**, 806–8.

VIROLAINEN, M., ANDERSSON, L. C., LALLA, M. & VON ESSEN, R. (1973). T-lymphocyte proliferation in mononucleosis. *Clinical Immunology and Immunopathology*, **2**, 114–20.

VOLSKY, D. J. KLEIN, G., VOLSKY, B. & SHAPIRO, I. M. (1981). Production of infectious Epstein-Barr virus in mouse lymphocytes. *Nature*, **293**, 399–401.

VOLSKY, D. J., SHAPIRO, I. M. & KLEIN, G. (1980). Transfer of Epstein-Barr virus receptors to receptor-negative cells permits virus penetration and antigen expression. *Proceedings of the National Academy of Sciences, USA*, **77**, 5453–7.

WEGNER, D. L. & HINZE, H. C. (1974). Virus–host cell relationship of *Herpesvirus sylvilagus* with cotton tail rabbit leukocytes. *International Journal of Cancer*, **14**, 567–75.

YATA, J., DESGRANGES, C., DE-THÉ, G. & TACHIBANA, T. (1973). Lymphocytes in infectious mononucleosis: Properties of atypical cells and origin of the lymphoblastoid lines. *Biomedicine*, **19**, 479–83.

YEFENOF, E. & KLEIN, G. (1977). Membrane receptor stripping confirms the association between EBV receptors and complement receptors on the surface of human B lymphoma lines. *International Journal of Cancer*, **20**, 347–52.

YEFENOF, E., KLEIN, G., JONDAL, M. & OLDSTONE, M. B. A. (1976). Surface markers on human B and T lymphocytes. IX. Two-color immunofluorescence studies on the association between EBV receptors and complement receptors on the surface of lymphoid cell lines. *International Journal of Cancer*, **17**, 693–700.

VIRUS PERSISTENCE AND AVOIDANCE OF IMMUNE SURVEILLANCE: HOW MEASLES VIRUSES CAN BE INDUCED TO PERSIST IN CELLS, ESCAPE IMMUNE ASSAULT AND INJURE TISSUES

MICHAEL B. A. OLDSTONE AND ROBERT S. FUJINAMI

Department of Immunopathology, Research Institute of Scripps Clinic, Scripps Clinic and Research Foundation, La Jolla, California, USA

Contemporary virologic and biomedical research is much concerned with how, after acute infection, viruses sometimes persist in the infected host and evade immune surveillance, yet participate later in causing cellular injury. Virus persistence may be initiated and maintained by a variety of means. For convenience, these can be grouped into virus-induced and host-induced mechanisms, as presented in Table 1.

Several years ago while focusing our efforts on mapping viral and host gene products on the surfaces of infected cells (Joseph, Perrin & Oldstone, 1976; Holland *et al.*, 1976; Yefenof, Klein, Jondal & Oldstone, 1976; Welsh & Oldstone, 1977; Oldstone, Fujinami & Lampert, 1980), we were impressed by the decrease or absence of viral glycoprotein expression on cells' surfaces during persistent infection compared to the large amount expressed during acute infection. Further, when studying persistence of noncytopathic viruses in differentiated cells, we noted that the differentiated or luxury function of infected cells was dampened without severely or significantly altering the cells' vital functions. Persistent infection of neuroblastoma cells with lymphocytic choriomeningitis virus (LCMV) led to significant alterations in those cells' ability to make or degrade acetylcholine, a major neurotransmitter, but failed to alter the infected cells' growth rate, cloning efficiency, or total synthesis of RNA, DNA or protein (Oldstone, Holmstoen & Welsh, 1977). Recently, in a similar vein, we have found that hybridoma cells persistently infected with LCMV produce significantly less antibody than uninfected hybridomas cultured under similar conditions.

Table 1. *Escape from immune surveillance – establishment of virus persistence*

I. Virus property
  1. Generation of defective interfering virus
  2. Generation of temperature-sensitive mutants
  3. Generation of other mutants
  4. Generation of recombinant virus
  5. Integration
  6. Nonimmunogenic
  7. Infection and alteration of function of cells of immune system

II. Immune property
  1. Antibody-induced antigenic modulation
  2. Immune selection
  3. Blocking factors (antigen, immune complexes)
  4. Suppressor T cells
  5. Generation of immune interferon

III. Host cell property
  1. Generation of interferon
  2. Lack of enzyme(s) for complete viral replication
  3. Generation of mutant viruses
  4. Asymmetric budding (apical v. basilar)

LCMV persistence is probably initiated and maintained by the generation of mutant defective interfering virus (Lehmann-Grube, Slenczka & Tees, 1969; Welsh & Pfau, 1972; Welsh & Oldstone, 1977; reviewed Popescu, Schaefer & Lehmann-Grube, 1976; Welsh & Buchmeier, 1979; Buchmeier, Welsh, Dutko & Oldstone, 1980). Persistent LCMV infection *in vivo* occurs in virus excess. However, animals persistently infected with LCMV are fully competent immunologically at levels of both the B lymphocytes and T helper lymphocytes (Buchmeier & Oldstone, 1978) and make specific anti-LCMV immune responses.

In contrast to LCMV persistent infection, many other persistent viral infections, especially those in man, occur in antibody excess. One sees infections with measles virus, herpes virus, rubella virus, etc., continuing chronically despite vigorous immune responses by the host. For example, persistent measles virus infection of man can cause the neurologic disease subacute sclerosing panencephalitis (SSPE), in which cytotoxic antibodies and lymphocytes abound in the patient's circulation (Joseph, Cooper & Oldstone, 1975; Kreth, Kackell & ter Meulen, 1975; Perrin, Tishon & Oldstone, 1977), yet viruses persist in both neuronal and lymphoid cells (Horta-Barbosa *et al.*, 1971; reviewed in ter Meulen, Katz & Muller, 1972).

To account for the occurrence of virus persistence by two such differing pictures as in LCMV compared to measles virus infection,

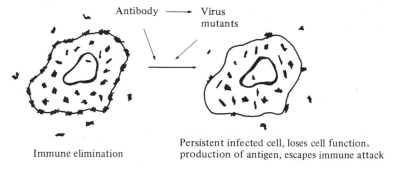

Fig. 1. Hypothesis presented to explain virus persistence and escape from immunologic surveillance. See text for discussion.

both of which bypass immune surveillance, we have proposed the hypothesis shown schematically in Fig. 1. Theoretically, as virus persistence begins, the number of viral antigens expressed on the surfaces of infected cells decreases so that insufficient material is present to cause lysis by cytotoxic immune reactants, thereby ensuring that cells remain alive and yet retain viral genome. Dampening of viral antigenic expression on the cell surface may be a function of the virus itself, enacted by deleting that part of its genome that encodes viral glycoproteins normally inserted and expressed on the cell surface. Alternatively, antiviral antibodies may bind to viral antigens on the cell surface and remove them from the cell's plasma membrane. Regardless of the mechanism employed, the end result is a cell relatively denuded of viral antigens. Further, these infected cells continue to replicate viral polypeptides, which can eventually disrupt the cell's biologic or physiologic function which we term loss of luxury functions. Implicit in this hypothesis are the notions that persistent virus infections occurring ordinarily in nature stem from noncytopathic rather than lytic viruses and that the related injury is caused, in part, by dysfunction of a differentiated cell. Hence, an attractive feature of this hypothesis is the prediction that disorders associated with hormonal aberrations may result from persistent infection of hormonal cells; persistent infection of immunocompetent lymphocytes may enhance (autoimmunity) or diminish (immunosuppression) their synthesis of immunoglobulins; persistent infection of oligodendrocytes may lead to demyelination; and involvement of neurons may lead to faulty synapses. This hypothesis is not meant to encompass all the variable mechanisms of virus persistence listed in Table 1.

Table 2. *Evidence for antibody-induced antigen modulation inducing virus persistence* in vitro *and* in vivo

Virus	Reference
In vitro studies	
Gross virus induced leukaemia virus	Aoki, T. & Johnson, P. A. (1972). Suppression of Gross leukemia cell-surface antigens: A kind of antigenic modulation. *Journal of the National Cancer Institute*, **49**, 183–92.
	Ioachim, H. L. & Sabbath, M. (1979). Redistribution and modulation of Gross murine leukemia virus antigens induced by specific antibodies. *Journal of the National Cancer Institute*, **62**, 169–80.
Friend leukaemia virus	Genovesi, E. V., Marx, P. A. & Wheelock, E. F. (1977). Antigenic modulation of Friend virus erythroleukemic cells *in vitro* by serum from mice with dormant erythroleukemia. *Journal of Experimental Medicine*, **146**, 520–34.
Mammary tumour virus	Calafat, J., Hilgers, J., von Blitterswijk, W. J., Verbeet, M. & Hageman, P. C. (1976). Antibody-induced modulation and shedding of mammary tumour virus antigens on the surfaces of GR ascites leukemia cells as compared with normal antigens. *Journal of the National Cancer Institute*, **56**, 1019–29.
	Yagi, M. J., Blair, P. B. & Lane, M.-A. (1978). Modulation of mouse mammary tumor virus production in the MJY-alpha cell line. *Journal of Virology*, **28**, 611–23.
Measles virus	Joseph, B. S. & Oldstone, M. B. A. (1975). Immunologic injury in measles virus infection. II. Suppression of immune injury through antigenic modulation. *Journal of Experimental Medicine*, **142**, 864–76.
	Fujinami, R. S. & Oldstone, M. B. A. (1979). Antiviral antibody reacting on the plasma membrane alters measles virus expression inside the cell. *Nature*, **279**, 529–30.
	Fujinami, R. S. & Oldstone, M. B. A. (1980). Alterations in expression of measles virus polypeptides by antibody: Molecular events in antibody-induced antigenic modulation. *Journal of Immunology*, **125**, 78–85.

In vivo studies

Friend leukaemia virus

Doig, D. & Chesebro, B. (1978). Antibody-induced loss of Friend virus leukemia cell surface antigens occur during progression of erythroleukemia in F_1 mice. *Journal of Experimental Medicine*, **148**, 1109–21.

Calafat, J., Hilgers, J., von Blitterswijk, W. J., Verbeet, M. & Hageman, P. C. (1976). Antibody-induced modulation and shedding of mammary tumour virus antigens on the surface of GR ascites leukemia cells as compared with normal antigens. *Journal of the National Cancer Institute*, **56**, 1019–29.

Measles virus

Rammohan, K. W., McFarland, H. E. & McFarlin, D. E. (1981). Induction of subacute murine measles encephalitis by monoclonal antibody to virus hemagglutinin. *Nature*, **290**, 588–9.

Herpes simplex virus

Stevens, J. G. & Cook, M. L. (1974). Maintenance of latent herpetic infection: An apparent role for antiviral IgG. *Journal of Immunology*, **113**, 1685–93.

Related studies

Parasites

Dwyer, D. M. (1976). Antibody-induced modulation of Lieshmania donovani surface membrane antigens. *Journal of Immunology*, **117**, 2081–91.

Hormone receptors

Heineman, S., Merlie, J. & Lindstrom, J. (1978). Modulation of acetylcholine receptor in rat diaphragm by anti-receptor sera. *Nature*, **274**, 55–8.

Tata, J. R. (1977). Modulation of hormone receptors. *Nature*, **269**, 757–8.

Immunologically mediated tissue injury, especially immune complex formation and its related diseases, frequently accompanies persistent infections by many viruses. In LCMV and measles virus persistent infection, viral glycoproteins react with their respective antibodies in the fluid phase or on the plasma membranes of infected cells to form complexes. Hence, viral proteins released from surfaces of infected cells have the opportunity to elicit immune responses and to form virus–antibody complexes. The deposition of these complexes may lead to tissue injury, especially causing glomerulonephritis, arteritis and choroiditis. In addition, persistently infected cells that fluctuate by abruptly expressing increased amounts of viral glycoproteins on their surfaces are subject to attack by the previously primed immune reactants. Our recent work has focused on, first, describing and, second, understanding the mechanism(s) whereby antibodies may induce virus persistence. The remainder of this chapter deals with work in this area.

In the course of infections by many types of viruses, antibodies to those viruses modulate or remove viral antigen from cell surfaces (Table 2). In its original sense, antibody-induced antigenic modulation referred to experiments with thymus leukaemia (TL) antigens expressed on normal thymus cells, as first reported by Boyse, Old and their associates (Boyse, Old & Luell, 1963; Boyse, Old & Stockert, 1965; Boyse, Stockert & Old, 1967; Old, Stockert, Boyse & Kim, 1968). These investigators noted that the phenotypic expression of TL was suppressed by TL antibodies both *in vivo* and *in vitro*. For this decreased expression of TL to occur, active metabolism of the infected cells was necessary. Subsequently, several virus systems with comparable events became apparent (Table 2). In general, with the exception of herpes simplex virus (HSV), the infecting agents used to study these events do not shut off host protein synthesis sufficiently to kill the infected cells. Furthermore, with the HSV model, an alternative role for antibody in manipulating the persistent state has recently been advanced by Notkins and his colleagues (Sekizawa, Openshaw, Wohlenberg & Notkins, 1980, 1981).

Experimentally, measles virus, *per se*, does not significantly halt host protein synthesis (Haspel, Pellergrino, Lampert & Oldstone, 1977); instead, cell death is associated with the function of a viral glycoprotein that allows cells to fuse, giant cells to form and syncytia to develop. Modulation, during which antibodies remove active measles virus fusion protein from cells (Fujinami & Oldstone, 1979;

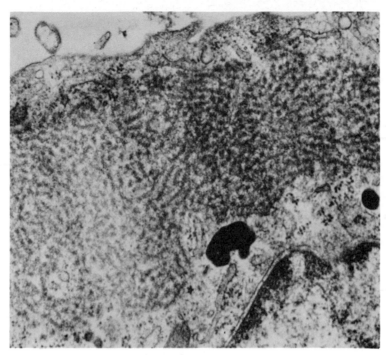

Fig. 2. Electron micrograph showing packed measles virus nucleocapsids arranged at random in the cytoplasm of a HeLa cell infected with measles virus. This cell was cultured in the presence of antibodies to measles virus. Note the absence of measles virus on the cell's surface. This picture, the result of antibody-induced modulation of measles virus antigens off the cell's surface *in vitro*, closely resembles the biopsies of patients with SSPE in which nucleocapsids are distributed and arranged similarly in the cytoplasm and no viral antigen appears on the surfaces of cells. This lack of viral antigen expression on the cell's surface allows such cells to escape ordinarily effective immune surveillance effector mechanisms (see Figs. 1 and 4).

Fujinami & Oldstone, 1980; Merz, Scheid & Choppin, 1980; Fujinami, Sissons & Oldstone, 1981), minimizes or prevents cell-to-cell fusion and, while allowing cell survival, also enables virus to persist. Modulation continues as long as sufficient antiviral antibody is in the culture medium to bind to viral glycoproteins expressed on the cell surface and strip these antigens. Once the initial phase of modulation has begun, infected cells can be cultured in serum containing specific antibodies and a functional complement source and/or cytotoxic lymphocytes without being lysed. Quantitative studies have shown that modulation diminishes the amount of viral antigens expressed on the cell surface to a point that binding of antibody and complement or immune lymphocytes is inadequate for lysis (Joseph & Oldstone, 1975; Oldstone & Tishon, 1978). During

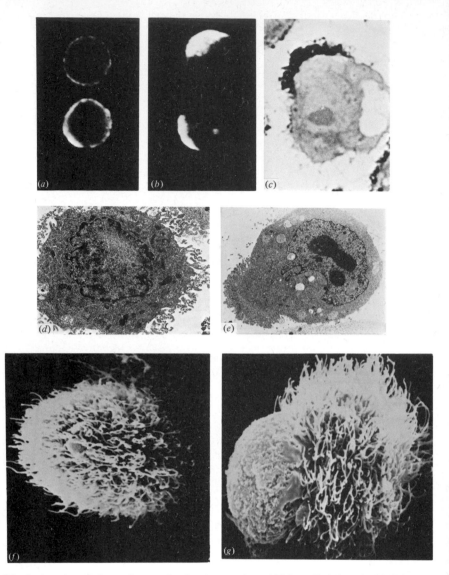

Fig. 3. A series of photomicrographs showing capping of HeLa cells infected with Edmond-ston strain of measles virus and exposed to anti-measles virus antibody. (*a*) and (*b*) utilize immunofluorescent microscopy; (*c*), immunoperoxidase technique with light microscopy; (*d*) and (*e*), transmission electron microscopy; and (*f*) and (*g*), scanning microscopy. In (*a*), infected cells are incubated at 4 °C with fluoresceinated human IgG antibody to measles virus. Note random circumferential distribution of viral antigens along the cell's surface. The same pattern is seen by transmission electron microscopy in (*d*) and by scanning electron microscopy in (*f*). In contrast, when living cells expressing measles virus antigens are incubated with anti-measles virus antibody at 37 °C, unipolar redistribution (capping) of measles virus antigens is noted by fluorescence microscopy (*b*), immunoperoxidase and light microscopy (*c*), transmission electron microscopy (*e*) and scanning electron microscopy (*g*). Photomicrographs published previously, in part, in *Journal of Immunology*, **113**, 1205, (1974); *Journal of Virology*, **15**, 1248, (1975); and *Progress in Medical Virology*, **26**, 45, (1980).

modulation, infected cells cultured with specifically reactive anti-viral antibodies grow at similar rates as uninfected cells (Joseph & Oldstone, 1975), but viral nucleocapsids dramatically increase in number inside infected cells and are positioned randomly, a picture closely resembling the distribution of nucleocapsids in cells obtained by biopsy from patients with SSPE (Fig. 2) (Iwasaki & Koprowski, 1974; Joseph & Oldstone, 1975).

Measles virus codes for six polypeptides, of which only the haemagglutinin (HA) and fusion (F) protein reach the surfaces of infected cells; both proteins are glycosylated. Cell surface HA is a dimer of 160 000 molecular weight, and under reducing conditions migrates electrophoretically as a monomer. The F protein is expressed on the cell surface as F0, denoting two disulphide-linked polypeptides, F1 (42 000 molecular weight) and F2 (24 000 molecular weight). The F2 fragment contains most of the carbohydrate in the intact F0 protein. The major glycoprotein on the surfaces of most human cells is the histocompatibility antigen complex (HLA). Mapping of viral glycoproteins and HLA antigens expressed on the surfaces of infected cells indicates that HLA antigens are both functionally and structurally distinct from the HA and F polypeptides (Haspel et al., 1977; Fujinami et al., 1981).

Initially, when antiviral antibodies are added to measles virus infected cells in culture, viral antigens expressed on the cells' surfaces become redistributed (capped) (reviewed in Oldstone et al., 1980, Fig. 3). Capping is well documented in vitro, and quite conceivably occurs in vivo, because the conditions needed to promote efficient capping are intrinsic: multivalent antibody, physiologic temperatures and living cells.

Polyvalent or monoclonal antibodies directed against measles virus antigens remove these antigens from the surfaces of infected cells during both acute and persistent infections. In both instances, the amount of viral antigens diminishes notably within 6 hours and greatly by 12 hours (Joseph & Oldstone, 1975; Oldstone & Tishon, 1978; Fujinami & Oldstone, 1979; Fujinami & Oldstone, 1980). This acute modulation of surface measles virus antigens is completely reversible by removing antibody from the culture medium (Fig. 4). However, as exposure to antiviral antibody lengthens before its removal, the rate at which measles virus antigens return to the cell's surface is extended. Thus, after one day's incubation with antiviral antibody, viral antigens reappear on 50% of the cells within 24 hours. After 5 days' incubation, 6 days pass before surface viral

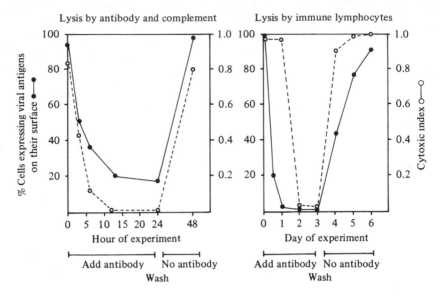

Fig. 4. Antibody-induced modulation of measles virus antigens off the surfaces of infected cells allows these infected cells to escape immune-mediated attack by either cytotoxic lymphocytes or cytotoxic antibodies and complement. Left panel: alteration of surface expression of measles virus antigens (●——●) and the ability of measles virus antibody and complement (○————○) to specifically lyse measles virus infected cells. Right panel: alteration of surface expression of measles virus antigens (●——●) and the ability of cytotoxic lymphocytes (○————○) to specifically lyse measles virus infected cells. Reprinted with permission from *Progress in Medical Virology*, **26**, 45–93, (1980). For experimental details see Joseph & Oldstone (1975) and Oldstone & Tishon (1978).

antigens emerge on 50% of the cells, and after 6 weeks of culture with antibody, it is difficult to detect viral antigens on the cells' surfaces after removing the antibody (Joseph & Oldstone, 1975).

Three major biological events occur as a result of antibody-induced modulation of measles virus antigens. First, as surface measles virus antigens are stripped from the surfaces of infected cells, neither antibody and complement nor cytotoxic lymphocytes can lyse the cells as they did when the antigens were present (Joseph & Oldstone, 1975; Oldstone & Tishon, 1978; Fig. 4). Such experiments confirm that the ability of immune reactants to kill measles virus infected cells directly parallels the quantity of viral antigens expressed on the cells' surfaces. Quantitative studies indicate that, whereas 1×10^5 antibody molecules are required to modulate viral antigens from the cell surface, at least five-fold greater antigen expression is needed for lysis by cytotoxic lymphocytes and ten-fold greater expression is needed for antibody and complement-associated immune lysis. The second biological event related to infected cells being antigenically modulated by anti-measles virus antibodies in culture is that the cells survive and grow as efficiently and at the same rate as uninfected cultured cells. Thus, removal of surface viral glycoproteins, specifically the F protein, prevents cell-to-cell fusion despite the continuing replication of measles virus polypeptides. The third biological consequence of interest is the distribution of viral polypeptides. Light and transmission electron microscopy indicate that little if any virus buds from the plasma membrane during antibody-induced modulation, yet there is an increased number of nucleocapsids in the cytoplasm (Fig. 2). Unlike their usual orderly pattern before modulation, these nucleocapsids are misaligned and disarrayed. It is useful to recall that the electron micrograph displayed in Fig. 2 was obtained by culturing measles virus infected cells with anti-measles virus antibody and is reminiscent of the disordered pattern of measles virus polypeptides observed in biopsied tissues taken directly from patients with SSPE.

So far, information regarding the molecular events associated with antibody-induced antigenic modulation of measles virus infected cells has been restricted to events occurring during the first 48 hours after infection. Polyvalent antibodies directed against HA and F protein or monoclonal antibodies against the HA of measles virus reproducibly alter the expression of these external viral polypeptides, as well as two internal polypeptides, P and M, within 6 hours after their addition to cultures of the infected cells (Fujinami & Oldstone, 1979; Fujinami & Oldstone, 1980). The decreases in numbers of F1, HA, P and M viral polypeptides are specific, because only antibodies directed against the measles virus antigens expressed on the cell surface cause this effect. Neither antibodies against nonviral surface antigens, such as HLA determinants, nor against viral antigens not expressed on the cell's surface, like measles virus nucleocapsid, are effective in this respect. Conceptually, the loss of F1 molecules early during antibody-induced antigenic modulation accounts for the lack of cell–cell fusion and giant cell formation that allows the survival of the infected cells. The associated loss of HA and F1 molecules enables virus-infected cells to resist immunologic attack and subsequent lysis by both humoral and cell-mediated immunologic reagents. Loss of P molecules may be important in regulating viral transcription, by analogy with other virus systems, and might be involved in the overproduction of nucleocapsid antigens. The alteration in M proteins probably relates

Table 3. *Anti-measles virus antibodies reacting with HA or F measles virus proteins expressed on the cell's surface alter the expression of measles virus P and M polypeptides found inside the cell*[a]

Measles virus polypeptide[b]	Label	HeLa cells acutely infected with measles virus and incubated with						
		Monoclonal no. 1 to HA[c]	Monoclonal no. 2 to HA[c]	Polyvalent (HA, F, L, NC, P, M)	Monoclonal to NC[c]	Polyvalent cell surface	Polyvalent HLA	FCS
HA	^{125}I	ND	ND	Slight ↓[d]	ND	ND	ND	No change
HA	^{35}S	No change	Slight ↑	Slight ↓	No change	No change	No change	No change
F	^{125}I	ND	ND	Marked ↓	ND	ND	ND	No change
F	^{35}S	Slight ↓	Slight ↓	Marked ↓	No change	No change	No change	No change
P	^{35}S	Marked ↓	Marked ↓	Marked ↓	No change	No change	No change	No change
P	^{32}P	ND	ND	Marked ↓	ND	No change	ND	No change
M	^{35}S	Slight ↓	Slight ↓	Slight ↓	No change	No change	No change	No change
M	^{32}P	ND	ND	Marked ↑	ND	No change	ND	No change

[a] HA and F are glycosylated polypeptides expressed on the plasma membrane. The L, NP, P and M viral polypeptides are not found on the plasma membrane but on the inner cytoplasmic side of the plasma membrane (M protein) or in the cytoplasm (L, NP and P). The L polypeptide was not studied. All the ^{35}S and ^{32}P data were corrected to NP expression to insure an internal control. Living cells were cultured with the various antibody sources, then labelled with appropriate radioisotopes. See Fujinami & Oldstone (1979) and Fujinami & Oldstone (1980) for experimental details.

[b] Iodine-125 incorporated into proteins expressed on the cell's surface using lactoperoxidase technique. NP, P and M, although not labelled by this technique, have tyrosine residues and can be labelled in situ with ^{125}I. HA and F polypeptides do not incorporate ^{32}P, while NP, P and M are phosphoproteins (Fujinami & Oldstone, 1980).

[c] Unpublished data, R. Fujinami, manuscript in preparation.

[d] Slight ↓, < 50% change decrease; moderate, > 50% but < 75%; marked, > 75% change. Comparison of cells modulated (antibody treated) to those non-modulated (foetal calf sera (FCS) incubated). ND, not done.

to the improper nucleocapsid alignment at the plasma membrane. The defect in P and M polypeptides observed under varying experimental conditions is shown in Table 3.

How well does the antibody-induced antigenic modulation, *in vitro* model, account for the events of naturally occurring persistent measles virus infection *in vivo*? Measles virus infection in man usually follows an acute, self-limiting course as the patient mounts an immune response that clears the virus from his tissues. During convalescence, low titres of antibody to measles virus and immune lymphocytes circulate and remain throughout life. By comparison, in SSPE, the chronic form of measles virus infection, patients have high titres of anti-measles virus antibodies and effective cytotoxic lymphocytes, yet virus persists. Although the virus can be isolated from cells of these patients' central nervous systems, antibody titres are 100- to 10 000-fold above normal in their sera and other body fluids, and cytotoxic lymphocytes in their peripheral blood are fully effective in killing measles virus infected targets *in vitro*. Sera from such patients are a source of fully functional and active complement. However, in normal persons and in patients with SSPE, the cerebral spinal fluid contains relatively little complement. Therefore, cerebral spinal fluid with these high titres of antibodies that bind to HA and F measles virus polypeptides, but having limited amounts of complement, can strip viral antigens from the surfaces of measles virus infected neurons, leaving these cells refractory to immune lysis. Such cells retain viral genetic information and a dramatically rising number of nucleocapsids scattered in a random arrangement. Cells of this type are in contrast to cells examined during acute virus infection with their rigid alignment of internal viral material. Recently, Rammohan *et al.* (Rammohan, McFarland & McFarlin, 1981) demonstrated that the inoculation of either polyvalent antibodies to measles virus or monoclonal antibody to measles virus HA caused chronic infection in mice, whereas inoculation of equivalent amounts of antibody to measles virus nucleocapsid did not modify the ordinarily acute infection. Other experiments also indicate that measles virus antibodies may initiate persistent infection *in vivo*. For example, Albrecht *et al.* (1977) showed that measles virus or SSPE virus inoculated into monkeys did not yield a persistent type infection unless the monkeys already had antibodies to measles virus. Similarly, Wear & Rapp (1971) found that persistent measles virus infection developed in newborn hamsters suckled by mothers containing antibodies to measles virus. Hence, a substantial body of *in vitro* and *in vivo* evidence suggests that initiation of measles virus persistence may be associated with measles virus antibodies, presumably by virtue of their modulation of antigens.

None of the studies mentioned here have probed the mechanisms associated with maintaining measles virus persistence. Choppin and his colleagues (reviewed Hall & Choppin, 1981) recently presented evidence of a specific defect in M protein in SSPE. Hence, it will be of interest to determine the effect of chronic antibody-induced measles virus antigenic modulation on the M protein. This is especially so as results of our experiments described above in which measles antibody-induced modulation of virus during acute infection suggested that M protein became defective.

The alterations we observed in P and M protein were noted in repeated experiments. Because we took care to avoid proteolytic enzyme activity in our cytosol preparations and used hybridoma antibodies to HA to show marked and significant alterations of P and M internal viral polypeptides, we view these findings as nonartifactual and likely to provide insight into regulation and maintenance of measles virus persistence. Current experiments are focusing on the role of P protein and other host phosphoproteins during the initial stage of persistent infection by measles viruses.

Studies of antibodies directed to TL differentiation antigens, followed by numerous experimental models of virus infection, have been used to establish the role of antibody in altering the surfaces of virus infected cells, leading to virus persistence both *in vitro* and *in vivo*. With this background, a flurry of recent interest indicates that antibodies may also play a significant role in receptor diseases like some forms of diabetes, myasthenia gravis and thyroid disorders (Table 2) as antibodies to insulin, acetylcholine and thyroid hormone receptors, respectively, have been blamed. Theoretically, these antibodies bind to their unique antigens on the cell surface and through clustering, aggregation or shedding alter important plasma membrane–cytoplasm signalling mechanisms. Hence, experimental results with measles virus antigens, infected cells and antibodies to measles virus may ultimately shed light on several poorly understood but potentially important observations concerning unrelated human diseases of both viral and nonviral origins.

SUMMARY

It is of interest to recall that many persistent infections *in vivo* occur despite a vigorous host immune response. For example, in humans persistent infections with measles virus, herpes viruses, rubella virus, etc., endure despite vigorous and specific antiviral immune responses by the host. In fact, these persistent infections most frequently occur in antiviral antibody excess. Other persistent infections, i.e. lymphocytic choriomeningitis virus and retrovirus infections in mice, occur in antigen (virus) excess although again the host makes antiviral immune responses. To explain these events we have hypothesized that during virus persistence the molecules of viral antigens expressed on the surfaces of infected cells decrease so that insufficient material is present for lysis by cytotoxic immune constituents. By this means, viral genetic information persists within the cell, yet such infected cells escape immunological surveillance. The decrease of viral antigens expressed on the cell surface may be a function of a virus itself, perhaps by deleting that part of its genome that codes for the viral glycoprotein(s) normally inserted and expressed on the cell's surface. Alternatively, the decrease in viral antigens expressed on the cell's surface may be a function of antibody. In this instance specific antiviral antibody binds to viral antigens on the cell's surface and strips such antigens off the plasma membrane. The additional point in this hypothesis is that an infected cell lacking viral antigens on its surface and escaping immunologic assault continues to carry and express viral information that may lead to an alteration in the differentiated cell's function. This may be expressed as an aberrant synapse by a neuron, failure of myelination by an oligodendrocyte, failure to make insulin by a beta islet cell, or the turn off or enhancement of production of immunoglobulin by B lymphocytes.

Our hypothesis allows several predictions to be made and experimentally tested. First, inherent is the notion that persistent virus infections occurring ordinarily in nature stem from noncytopathic rather than lytic viruses. Noncytopathic viruses can cause lifelong, relatively nonlytic states in which infected cells show normal or virus altered morphology. Second, the related cellular injury, in part, may be caused by a dysfunction or loss of a luxury function of a differentiated cell. Third, a common finding in persistence regardless of the initiating virus is a decrease or lack of viral antigen expression on the surfaces of infected cells. Fourth, there are several mechanisms either virus or host induced that cause a decreased expression of viral antigens on the cell's surface. For example, one of these host mechanisms is antibody-induced antigenic modulation or stripping. Fifth, antiviral host immune responses frequently occur in persistent infections. Their characteristics are common to immune responses occurring during acute infection. Lastly, cell and tissue injury associated with persistent infection may also be due to the host's antiviral immune response with the most frequent manifestation being the formation and deposition of virus–antiviral antibody immune complexes.

This is Publication No. 2497 from the Department of Immunopathology, Scripps Clinic and Research Foundation, La Jolla, California 92037.

This research was supported by USPHS grants AI-09484, AI-07007 and NS-12428.

REFERENCES

ALBRECHT, P., BURNSTEIN, T., KLUTCH, M. J., HICKS, J. J. & ENNIS, F. A. (1977). Subacute sclerosing panencephalitis: Experimental infection in primates. *Science*, **195**, 64–6.

BOYSE, E. A., OLD, L. J. & LUELL, S. (1963). Antigenic properties of experimental leukemias. II. Immunological studies *in vivo* with C57BL/6 radiation-induced leukemias. *Journal of the National Cancer Institute*, **31**, 987–95.

BOYSE, E. A., OLD, L. J. & STOCKERT, E. (1965). The TL (thymus leukemia) antigen: a review. In *Immunopathology*, ed. P. Grabar & P. Miescher, pp. 23–40. Basel: Schwabe & Co.

BOYSE, E. A., STOCKERT, E. & OLD, L. J. (1967). Modification of the antigenic structure of the cell membrane by thymus leukemia (TL) antibody. *Proceedings of the National Academy of Sciences, USA*, **58**, 954–7.

BUCHMEIER, M. J. & OLDSTONE, M. B. A. (1978). Virus-induced immune complex disease: Identification of specific viral antigens and antibodies deposited in complexes during chronic lymphocytic choriomeningitis virus infection. *Journal of Immunology*, **120**, 1297–304.

BUCHMEIER, M. J., WELSH, R. M., DUTKO, F. J. & OLDSTONE, M. B. A. (1980). The virology and immunobiology of lymphocytic choriomeningitis virus infection. *Advances in Immunology*, **30**, 275–331.

FUJINAMI, R. S. & OLDSTONE, M. B. A. (1979). Antiviral antibody reacting on the plasma membrane alters measles virus expression inside the cell. *Nature*, **279**, 529–30.

FUJINAMI, R. S. & OLDSTONE, M. B. A. (1980). Alterations in expression of measles virus polypeptides by antibody: Molecular events in antibody-induced antigenic modulation. *Journal of Immunology*, **125**, 78–85.

FUJINAMI, R. S., SISSONS, J. G. P. & OLDSTONE, M. B. A. (1981). Immune reactive measles virus polypeptides on the cell's surface: Turnover and relationship of the glycoprotein to each other and to HLA determinants. *Journal of Immunology*, (in press).

HALL, W. W. & CHOPPIN, P. W. (1981). Measles-virus proteins in the brain tissue of patients with subacute sclerosing panencephalitis. *New England Journal of Medicine*, **304**, 1152–5.

HASPEL, M. V., PELLEGRINO, M. A., LAMPERT, P. W. & OLDSTONE, M. B. A. (1977). Human histocompatibility determinants and virus antigens: Effect of measles virus infection on HLA expression. *Journal of Experimental Medicine*, **146**, 146–56.

HOLLAND, J. J., VILLARREAL, L. P., WELSH, R. M., OLDSTONE, M. B. A., KOHNE, D., LAZZARINI, R. & SCOLNICK, E. (1976). Long term persistent vesicular stomatitis virus and rabies virus infection of cells *in vitro*. *Journal of General Virology*, **33**, 193–211.

HORTA-BARBOSA, L., HAMILTON, R., WITTEG, B., FUCCILLO, D. A. & SEVER, J. L. (1971). Subacute sclerosing panencephalitis: Isolation of suppressed measles virus from lymph node biopsies. *Science*, **173**, 840–1.

IWASAKI, Y. & KOPROWSKI, H. (1974). Cell to cell transmission of virus in the central nervous system. I. Subacute sclerosing panencephalitis. *Laboratory Investigation*, **31**, 187–96.

JOSEPH, B. S., COOPER, N. R. & OLDSTONE, M. B. A. (1975). Immunologic injury of cultured cells infected with measles virus. I. Role of IgG antibody and the alternative complement pathway. *Journal of Experimental Medicine*, **141**, 761–74.

JOSEPH, B. S. & OLDSTONE, M. B. A. (1975). Immunologic injury in measles virus infection. II. Suppression of immune injury through antigenic modulation. *Journal of Experimental Medicine*, **142**, 864–76.

JOSEPH, B. S., PERRIN, L. H. & OLDSTONE, M. B. A. (1976). Measurement of virus antigens on the surfaces of HeLa cells persistently infected with wild type and vaccine strains of measles virus by radioimmune assay. *Journal of General Virology*, **30**, 329–37.

KRETH, W. H., KACKELL, M. Y. & TER MEULEN, V. (1975). Demonstration of *in vitro* lymphocyte-mediated cytotoxicity against measles virus in SSPE. *Journal of Immunology*, **114**, 1042–6.

LEHMANN-GRUBE, F., SLENCZKA, W. & TEES, R. (1969). A persistent and inapparent infection of L cells with the virus of lymphocytic choriomeningitis. *Journal of General Virology*, **5**, 63–81.

MERZ, D. C., SCHEID, A. & CHOPPIN, P. W. (1980). Importance of antibodies to the fusion glycoprotein of paramyxoviruses in the prevention of spread of infection. *Journal of Experimental Medicine*, **151**, 275–88.

OLD, L. J., STOCKERT, E., BOYSE, E. A. & KIM, J. H. (1968). Antigenic modulation. Loss of TL antigen from cells exposed to TL antibody. Study of the phenomenon *in vitro*. *Journal of Experimental Medicine*, **127**, 523–39.

OLDSTONE, M. B. A., FUJINAMI, R. S. & LAMPERT, P. W. (1980). Membrane and cytoplasmic changes in virus infected cells induced by interactions of antiviral antibody with surface viral antigen. In *Progress in Medical Virology*, vol. 26, ed. J. L. Melnick, pp. 45–93. Basel: S. Karger.

OLDSTONE, M. B. A., HOLMSTOEN, J. & WELSH, R. M., JR (1977). Alterations of acetylcholine enzymes in neuroblastoma cells persistently infected with lymphocytic choriomeningitis virus. *Journal of Cellular Physiology*, **91**, 459–72.

OLDSTONE, M. B. A. & TISHON, A. (1978). Immunologic injury in measles virus infection. IV. Antigenic modulation and abrogation of lymphocyte lysis of virus-infected cells. *Clinical Immunology and Immunopathology*, **9**, 55–62.

202   MICHAEL B. A. OLDSTONE AND ROBERT S. FUJINAMI

PERRIN, L. H., TISHON, A. & OLDSTONE, M. B. A. (1977). Immunologic injury of measles virus infection. III. Presence and characterization of human cytotoxic lymphocytes. *Journal of Immunology*, **118**, 282–90.

POPESCU, M., SCHAEFER, H. & LEHMANN-GRUBE, F. (1976). Homologous interference of lymphocytic choriomeningitis virus. Detection and measurement of interference focus-forming units. *Journal of Virology*, **20**, 1–8.

RAMMOHAN, K. W., McFARLAND, H. F. & McFARLIN, D. E. (1981). Induction of subacute murine measles encephalitis by monoclonal antibody to virus haemagglutinin. *Nature*, **290**, 588–9.

SEKIZAWA, T., OPENSHAW, H., WOHLENBERG, C. & NOTKINS, A. L. (1980). Latency of herpes simplex virus in absence of neutralizing antibody: Model for reactivation. *Science*, **210**, 1026–8.

SEKIZAWA, T., OPENSHAW, H., WOHLENBERG, C. & NOTKINS, A. L. (1981). Reactivation of latent herpes simplex virus. *Immunologic Abstracts*, (in press).

TER MEULEN, V., KATZ, M. & MULLER, D. (1972). Subacute sclerosing panencephalitis: A review. *Current Topics in Microbiology and Immunology*, **57**, 1–38.

WEAR, D. J. & RAPP, F. (1971). Latent measles virus infection of the hamster central nervous system. *Journal of Immunology*, **107**, 1593–8.

WELSH, R. M., JR & BUCHMEIER, M. J. (1979). Protein analysis of defective interfering lymphocytic choriomeningitis virus and persistently infected cells. *Virology*, **96**, 503–15.

WELSH, R. M. & OLDSTONE, M. B. A. (1977). Inhibition of immunologic injury of cultured cells infected with lymphocytic chorimeningitis virus: Role of defective interfering virus in regulating viral antigenic expression. *Journal of Experimental Medicine*, **145**, 1449–68.

WELSH, R. M. & PFAU, C. J. (1972). Determinants of lymphocytic choriomeningitis interference. *Journal of General Virology*, **14**, 177–87.

YEFENOF, E., KLEIN, G., JONDAL, M. & OLDSTONE, M. B. A. (1976). Surface markers on human B- and T-lymphocytes. IX. Two-color immunofluorescence studies on the association between EBV receptors and complement receptors on the surface of lymphoid cell lines. *International Journal of Cancer*, **17**, 693–700.

A MODEL FOR THE CONTROL OF NON-SEGMENTED NEGATIVE STRAND VIRUS GENOME REPLICATION

DANIEL KOLAKOFSKY AND
BENJAMIN M. BLUMBERG

Department of Microbiology, Medical School, University of Geneva, 64 Avenue de la Roseraie, 1205 Geneva, Switzerland

When single-stranded (ss) RNA viruses infect their host cells, the infecting viral genome must serve at least two separate and sequential functions. The first function is to provide viral mRNAs for viral protein synthesis, and the second function is to provide a template for viral genome replication via a full-length complementary strand. These two functions of the infecting viral genome are mutually exclusive in both plus-strand and minus-strand RNA virus infections. The switching of the viral genome between these two functions, a central control feature in these viral replication schemes, is mediated through the synthesis of viral proteins.

In the case of plus-strand RNA viruses, such as polio virus or the bacteriophage Qß, the infecting virus genome is itself the template mRNA for protein synthesis. The viral genome is therefore first associated with ribosomes in a polysome which synthesizes the viral proteins. However, since ribosomes translate the viral genome by moving in the 5' to 3' direction, whereas the viral replicase must synthesize a complementary minus-strand by reading the genome template in the opposite direction, a means must be found to clear the infecting genome template of ribosomes to allow the viral replicase to pass. In the case of the bacteriophage Qß, this mechanism is effected through the synthesis of the viral replicase itself, which has the added function of acting as a repressor of protein synthesis (Kolakofsky & Weissman, 1971a, b). The replicase acts as a repressor by specifically attaching to the ribosome initiation site of the coat protein gene, thereby preventing additional ribosomes from entering the polysome. The genome template is thus cleared of ribosomes by their detachment after the completion of the proteins being synthesized, i.e. run-off. It is worth noting that the virus-coded replicase polypeptide in the Qß system does not act as a polymerase or a ribosome repressor by itself, but carries out its function in concert with three additional host cell proteins (Kamen, 1970; Kondo, Gallerani & Weissmann, 1970).

In the case of negative strand viruses such as Sendai virus or vesicular stomatitis virus (VSV), the infecting viral genome is itself the complement of the viral mRNAs. These viral genomes are, however, never present as free RNA, but are found only in the form of viral nucleocapsids. The viral nucleocapsid is a linear structure of helical symmetry, morphologically analogous to tobacco mosaic virus (TMV), but less rigid and less compact. It is composed, like TMV, of mostly a single protein, the viral N protein, whose attachment to the viral RNA is exceedingly stable. However, unlike TMV (a plus-strand virus), these negative strand virus nucleocapsids also contain two additional proteins (NS and L for VSV, and P and L for Sendai virus), which can be separated from the core nucleocapsid (RNA and N protein) by high salt treatment, and which function with the core nucleocapsid as the viral polymerase (Emerson & Wagner, 1972; Choppin & Compans, 1975; Wagner, 1975). Like naked plus-strand virus RNAs, the viral minus-strand nucleocapsid is infectious (Szilagyi & Uryvyev, 1973) as it contains within itself all the enzymatic activities required for the production of active monocistronic mRNA, i.e. capping, methylation, and polyadenylation, as well as RNA polymerase activity (Banerjee, Abraham & Colonno, 1977). Purified virions in the presence of non-ionic detergents are also found to contain these same enzymatic activities.

RNA synthesis in animal cells infected with negative strand viruses such as VSV can be followed most simply by the addition of ^{3}H-uridine in the presence of the drug actinomycin D, which prevents DNA-dependent host cell RNA synthesis, but does not affect viral RNA-dependent RNA synthesis. By use of this procedure, VSV infected cells have been found to synthesize virus-specific RNA in a sequential manner. Shortly after infection, capped and polyadenylated RNAs complementary to the viral genome (the viral mRNAs), which sediment as two main groups at 12–17 S and 30 S on SDS-sucrose gradients, are synthesized (primary transcription), and associate with ribosomes to synthesize viral proteins (early protein synthesis). Primary transcription, like the *in vitro* virion polymerase activity, takes place equally well in the presence of inhibitors of protein synthesis such as cycloheximide, suggesting that no newly synthesized proteins are required in this process. However, in the absence of protein synthesis, only primary transcripts are made (Huang & Manders, 1972; Perrault & Holland, 1972). When protein synthesis is allowed to proceed, the infecting viral genome can now be switched to its second function, that of serving as a template for genome RNA replication, and about 2 h after infection newly synthesized viral genome RNAs are observed, which sediment at 42 S in SDS-sucrose gradients (Soria, Little & Huang, 1974; Simonsen, Batt-Humphries & Summers, 1979).

The genome RNAs differ from the viral mRNAs in several important ways. The genome RNAs are full-length copies of the viral genome and exist as both plus-strands (antigenomes) and minus-strands (genomes) which are neither capped nor polyadenylated, whereas the mRNAs are subgenomic segments of the antigenome and exist only as capped and polyadenylated plus-strands. In addition, the viral genome and antigenome RNAs are found only in nucleocapsids, whereas the viral mRNAs, although associated with other proteins, as mRNPs, are not found in nucleocapsids. A simple way of demonstrating this difference is to centrifuge cytoplasmic extracts of infected cells on CsCl density gradients. Viral nucleocapsids are sufficiently stable in the high salt conditions of the gradient to band at a buoyant density indicative of their protein and RNA composition (1.34 gm/ml), whereas the viral mRNPs, along with other host cell RNA–protein complexes, are disaggregated during centrifugation and their RNAs sediment into the gradient pellet as free RNA (Leppert et al., 1979).

A further important difference between mRNA synthesis (transcription) and genome synthesis (replication) in VSV infected cells is that on-going protein synthesis is required not only for switching the infecting viral genome from primary transcription to genome replication, but is equally required for genome replication late in infection, after amplification of the infecting viral genome (Perlman & Huang, 1973; Wertz & Levine, 1973). Although these experiments, which are based on cycloheximide inhibition of protein synthesis, do not allow us to distinguish whether the newly synthesized proteins required for genome replication are viral or host, other experiments do offer suggestive evidence in this regard. Kang & Allen (1978) have reported experiments in which BHK cells were pretreated with actinomycin D for 24 h before infection, and upon VSV infection still produced normal amounts of virus progeny. These experiments would seem to exclude host cell proteins whose synthesis is continuously required for genome replication. However, it should be noted that if the mRNAs for these putative host proteins were unusually stable (so that even 24 h after new host mRNA synthesis was inhibited enough of this mRNA remained active), then some of the proteins, whose on-going synthesis are required for genome replication, could in fact be of host origin. With this possibility in mind, we shall assume that those proteins whose continuous synthesis is required for genome replication are viral in origin. This, of course, does not exclude the participation of stable host cell proteins in genome replication.

Although minus-strand RNA viruses, in which mRNA and genome RNA are on complementary strands, do not face the same problem of plus-strand RNA viruses in which translating ribosomes and the viral replicase must traverse the same RNA strand in opposite directions, minus-strand RNA viruses face a somewhat analogous problem. In minus-strand RNA virus infections, the viral polymerase must copy the same genome template in two different ways, which are also mutually exclusive. During mRNA transcription, the polymerase must synthesize subgenomic segments of plus-strand which are not assembled into nucleocapsids, whereas during genome replication, the viral polymerase must synthesize an uninterrupted copy of the genome template, which is assembled into nucleocapsids. It should be noted that the precise mechanism whereby viral mRNAs are synthesized, i.e. whether the mono-cistronic mRNAs are the result of the polymerase terminating and reinitiating RNA synthesis along the genome template, or whether they result from endonucleolytic processing of a polycistronic precursor, has yet to be clarified despite an enormous effort in this direction. In the model outlined below we have arbitrarily chosen the stop-start mechanism of mRNA synthesis for purposes of discussion. However, this choice is not of great consequence since the model can easily be adapted to the alternative processing mechanism of mRNA synthesis.

A possible mechanism for how this switch from genome transcription to replication takes place through the synthesis of the viral proteins was suggested by the discovery of the viral leader RNAs. Leader RNAs were first noted by Reichmann et al. (1974) as the products of the in vitro VSV defective-interfering (DI) particle polymerase reaction. At that time, the significance of this RNA product (minus-strand leader RNA) was unclear, since its origin on the DI-genome template had not been determined. Plus-strand leader RNAs were then noted by Banerjee and co-workers (Colonno & Banerjee, 1976, 1977, 1978) in the non-defective (ND) VSV virion polymerase reaction. These workers demonstrated that this 48-nucleotide-long product was the first RNA synthesized in the *in vitro* reaction, and more importantly, that it was the exact complement of the 3' end of the minus-strand genome template. Thus, contrary to previous expectation (Abraham & Banerjee, 1976; Ball & White, 1976), the N protein mRNA was not the first RNA synthesized by the viral polymerase *in vitro*, but its synthesis was dependent on the prior synthesis of the leader RNA. Subsequent sequence determination of the 46-nucleotide-long minus-strand leader RNA synthesized *in vitro* demonstrated that this RNA was the exact complement of the 3' end of the DI-genome and the ND-antigenome (Schubert, Keene, Lazzarini & Emerson, 1978; Semler, Perrault, Abelson & Holland, 1978).

The model for the control of VSV genome transcription and replication based on leader RNAs is somewhat analogous to how the attenuation signal in the tryptophan operon in prokaryotes controls the relative amount of transcription from this operon (Yanofsky, 1981). The bacterial RNA polymerase has a single entry site into this operon, at the promoter, and must transcribe the template for a certain distance (the leader RNA region) before it arrives at the first structural gene. Whether the polymerase terminates RNA synthesis at the end of the leader RNA sequence, at the attenuation signal, or whether it reads through this termination signal to produce the polycistronic mRNA for the structural genes, is controlled through the intracellular level of trp-tRNA. When the intracellular level of trp-tRNA is low, due to an insufficiency of tryptophan, the attenuation signal is suppressed and the polymerase reads through into the structural genes providing the mRNA for these proteins. When the intercellular level of trp-tRNA is adequate, the attentuation signal is observed, and RNA synthesis terminates to produce only leader RNA.

The analogies to transcription and replication of the minus-strand VSV genome are as follows. The VSV minus-strand genome ('the operon') also has a single entry site for the viral polymerase, at the exact 3' end of the template (Colonno & Banerjee, 1977), and the polymerase must transcribe the template for a certain distance, the leader region, before it arrives at the first structural gene. At this junction of the leader region and the first structural gene lies the leader RNA termination site, the attenuator, which is thought to control the choice between transcription and replication. Whether the viral polymerase observes the termination signal to produce leader RNA, or reads through this signal, is controlled by the intracellular level of the viral N protein. N protein is thought to control the attenuation signal because of its ability to bind to the leader RNA sequences, which in this model contains the site for the initiation of nucleocapsid assembly. Transcription and replication are started by the same viral polymerase at the same promoter on the genome template (the exact 3' end), but the fate of the nascent strand depends on the intracellular level of N protein. In the absence of adequate amounts of the viral N protein, e.g. during primary transcription, the attenuation signal is observed and the nascent RNA chain terminates as leader RNA. During genome replication, in the presence of adequate cytoplasmic levels of the viral N protein, nucleocapsid assembly is initiated on the leader RNA sequences of the nascent genome chain, the leader termination signal is suppressed, and RNA synthesis and nucleocapsid assembly proceed uninterruptedly in the 5' to 3' direction along the nascent genome, preventing the synthesis of mRNAs and leading to the synthesis of a full-length antigenome nucleocapsid.

A basic difference in the models for the control of VSV genome replication and that for tryptophan operon transcription is inherent in the different ways that eukaryotic and prokaryotic ribosomes translate mRNAs; eukaryotic ribosomes can only initiate protein synthesis at the 5' ends of mRNAs, whereas prokaryotic ribosomes can also initiate at internal sites (Kozak, 1979). In the model for VSV replication, the minus-strand genome is thought to contain individual promoters for each virus structural gene, although access to these promoters is gained only through the single-entry promoter at the 3' end. During genome transcription, the termination of the leader RNA chain allows the viral polymerase to initiate mRNA synthesis on the succeeding structural genes. The consequent separation of the leader RNA sequence from the viral mRNAs prevents the viral N protein from assembling these mRNAs into nucleocapsid structures, since the mRNAs themselves lack the initiation site for nucleocapsid assembly. Thus, in contrast to the tryptophan attenuator, which simply controls read-through of the polymerase into the structural gene region of its operon, the VSV attenuator causes its operon to synthesize one of two alternate RNA products, both of which derive from the structural gene region. Expression of the VSV attenuation signal leads to the synthesis of monocistronic mRNAs for the succeeding structural genes, whereas suppression of the attenuation signal also allows transcription of the structural gene sequences, but these sequences are transcribed as one continuous chain which is simultaneously assembled into an antigenome nucleocapsid, i.e. genome replication.

In spite of the differences just noted, tryptophan mRNA synthesis and VSV genome replication are both controlled by the same basic mechanism, that of feedback regulation. In the absence of adequate intracellular levels of N protein, VSV genomes are locked into transcription to provide templates for the synthesis of additional viral proteins. When adequate amounts of N protein become available, VSV genomes switch to genome replication, which utilizes large amounts of N protein during nucleocapsid assembly (there are approximately 2500 N protein molecules per genome nucleocapsid (Wagner, 1975)). In this way, N protein controls both its own synthesis and utilization.

Although we have used the prokaryotic tryptophan operon as an analogy because of the presence of a suppressible termination signal, other analogies of self-regulatory proteins exist which are just as striking. For example, the coordinate expression of prokaryotic ribosomal protein genes also takes place through a similar feedback mechanism, controlled in this case by the availability of unassembled rRNA (Yates & Nomura, 1981). The primary binding or core ribosomal proteins can bind both to rRNA, and to ribosome initiation sites on the ribosomal protein mRNAs, the binding to rRNA being preferred. In the presence of unassembled rRNA, these ribosomal proteins bind to the rRNA, leaving the ribosomal protein mRNAs open to translation. However, in the absence of unassembled rRNA, these proteins bind to their own mRNAs and repress their own synthesis. Another analogous example is found in the control of bacteriophage T4 gene 32 protein synthesis (Lemaire, Gold & Yarus, 1978). This protein binds preferentially to ss-DNA, but it can also bind to its own mRNA and inhibit its translation. In the presence of adequate amounts of DNA replication and therefore ssDNA, gene 32 protein binds to ssDNA, leaving its own synthesis unimpaired. In the absence of ssDNA, gene 32 protein similarly shuts down its own synthesis. In both of the above examples, a protein simultaneously controls its own synthesis and utilization by feedback regulation.

If one of the functions of the VSV plus-strand leader RNA is to contain the site for the initiation of nucleocapsid assembly and thereby to control transcription and replication of the minus-strand genome template, the question remains as to why a similar leader

RNA (minus-strand leader RNA) is transcribed from the 3' end of the ND-antigenome. Antigenome nucleocapsids are templates only for genome synthesis, no mRNAs are transcribed from this strand. A clue to a possible explanation of the minus-strand leader RNA comes from the way RNA synthesis on the minus-strand genome template is controlled. One of the features of the model outlined above is that no full-length antigenome strands can be synthesized until the intracellular concentration of N protein reaches a critical level. If the concentration of N protein required for the initiation of nucleocapsid assembly were higher than that required for elongation, this would ensure that nucleocapsid assembly would take place only under conditions where sufficient N protein is available to completely assemble the antigenome chain. Since this would similarly hold for genome synthesis from antigenome templates, the attenuation signal near the 3' end of the antigenome template could prevent genome RNA from being synthesized under conditions where it could not be assembled into nucleocapsids. Unassembled genome RNA could theoretically anneal to the viral mRNAs and not only impair the translation of these mRNAs, but increase the cytoplasmic concentration of dsRNA, thereby increasing the antiviral state induced by interferon.

The model for the control of VSV genome replication was formulated on the discovery of plus- and minus-strand leader RNAs in standard VSV infected cells (Leppert et al., 1979). However, several additional experiments have now provided supporting evidence.

(1) Like ND-antigenomes, DI-genomes, which are derived from the 5' end of the minus-strand genome by a copy-back mechanism (Leppert, Kort & Kolakofsky, 1977), are templates only for replication; no mRNAs are transcribed from these DI-genomes. The question can therefore be asked whether minus-strand leader RNA synthesis from these templates, like DI-genome replication, also requires continuous protein synthesis. Treatment of mixed virus infected cells with cycloheximide has clearly shown that intracellular minus-strand leader RNA synthesis from DI-genomes is not restricted under these conditions (Blumberg, Leppert & Kolakofsky, 1981). The initiation of these DI-genome chains, therefore, does not require on-going protein synthesis. The requirement for on-going protein synthesis for DI-genome replication must

therefore act elsewhere, presumably at the level of read-through of the attenuation signal.

(2) Since the model predicts that the site for the initiation of nucleocapsid assembly is contained within the leader RNA sequence, it is reasonable to inquire whether the leader RNA might not itself be encapsidated into a nucleocapsid structure. Recent evidence demonstrates that this is indeed the case. Intracellular plus- and minus-strand leader RNAs sediment as an 18S complex in non-deproteinizing sucrose gradients (Blumberg *et al.*, 1981; Blumberg & Kolakofsky, 1981). Free leader RNA sediments at 4 S. In the case of plus-strand leader RNA, this complex was shown to contain the viral N protein, and to be sensitive to SDS and proteinase, but not to ribonuclease (Blumberg *et al.*, 1981). Thus the leader RNA in this complex, like genome RNA in genome nucleocapsids, is protected against ribonuclease digestion. In addition, the leader RNA-N protein complex bands in CsCl density gradients at a similar buoyant density (1.33 gm/ml) as genome nucleocapsids.

(3) In the electron microscope, genome nucleocapsids often appear to be composed of a series of discs, analogous to TMV, but less rigid and less compact. When examined in the electron microscope, the 18 S leader complex appears as a single disc of similar morphology (Blumberg & Kolakofsky, 1981).

(4) N protein, which was extracted from genome nucleocapsids, can assemble leader RNA *in vitro* into structures which band in CsCl density gradients at the buoyant density of genome nucleocapsids, and are also resistant to ribonuclease digestion (B. M. Blumberg & D. Kolakofsky, unpublished data).

The above experiments demonstrate that intracellular leader RNAs are found in nucleocapsid structures. Leader RNAs must therefore contain the site for the initiation of nucleocapsid assembly, a crucial prediction of the model. What remains to be done is to examine whether, *in vitro*, addition of sufficient amounts of the viral N protein alone can convert transcribing nucleocapsid structures into replicative intermediates, or whether additional viral and/or host proteins are required.

REFERENCES

ABRAHAM, G. & BANERJEE, A. K. (1976). Sequential transcription of the genes of vesicular stomatitis virus. *Proceedings of the National Academy of Sciences, USA*, **73**, 1504–8.

BALL, L. A. & WHITE, C. N. (1976). Order of transcription of genes of vesicular stomatitis virus. *Proceedings of the National Academy of Sciences, USA*, **73**, 442–6.

BANERJEE, A. K., ABRAHAM, G. & COLONNO, R. J. (1977). Vesicular stomatitis virus; mode of transcription (a review). *Journal of General Virology*, **34**, 1–8.

BLUMBERG, B. M. & KOLAKOFSKY, D. (1981). Intracellular VSV leader RNAs are found in nucleocapsid structures. *Journal of Virology*, (submitted).

BLUMBERG, B. M., LEPPERT, M. & KOLAKOFSKY, D. (1981). Interaction of VSV leader RNA and nucleocapsid protein may control VSV genome replication. *Cell*, **23**, 837–45.

CHOPPIN, P. W. & COMPANS, R. W. (1975). Reproduction of paramyxoviruses. In *Comprehensive Virology*, vol. 4, ed. H. Fraenkel Conrat and R. Wagner, pp. 95–178. New York: Plenum Press.

COLONNO, R. J. & BANERJEE, A. K. (1976). A unique RNA species involved in initiation of vesicular stomatitis virus RNA transcription *in vitro*. *Cell*, **8**, 197–204.

COLONNO, R. J. & BANERJEE, A. K. (1977). Mapping and initiation studies on the leader RNA of vesicular stomatitis virus. *Virology*, **77**, 260–8.

COLONNO, R. J. & BANERJEE, A. K. (1978). Complete nucleotide sequence of the leader RNA synthesized *in vitro* by vesicular stomatitis virus. *Cell*, **15**, 93–101.

EMERSON, S. U. & WAGNER, R. R. (1972). Dissociation and reconstitution of the transcriptase and template activities of vesicular stomatitis B and T virions. *Journal of Virology*, **10**, 297–309.

HUANG, A. S. & MANDERS, E. (1972). Ribonucleic acid synthesis of vesicular stomatitis virus. IV. Transcription by standard virus in the presence of defective interfering particles. *Journal of Virology*, **9**, 909–16.

KAMEN, R. (1970). Subunit composition of Qß replicase. *Nature*, **228**, 527.

KANG, C. Y. & ALLEN, R. (1978). Host function-dependent induction of defective interfering particles of vesicular stomatitis virus. *Journal of Virology*, **25**, 240–9.

KOLAKOFSKY, D. & WEISSMANN, C. (1971a). Possible mechanism for transition of viral RNA from polysome to replication complex. *Nature New Biology*, **231**, 42–6.

KOLAKOFSKY, D. & WEISSMANN, C. (1971b). Qß replicase as repressor of Qß RNA-directed protein synthesis. *Biochimica Biophysica Acta*, **246**, 596–9.

KONDO, M., GALLERANI, R. & WEISSMANN, C. (1970). Subunit structure of Qß replicase. *Nature*, **228**, 525–7.

KOZAK, M. (1979). Inability of circular messenger RNAs to attach to eukaryotic ribosomes. *Nature*, **280**, 82–5.

LEMAIRE, G., GOLD, L. & YARUS, M. (1978). Autogenous translational repression of bacteriophage T4 gene 32 expression *in vitro*. *Journal of Molecular Biology*, **126**, 73–90.

LEPPERT, M., KORT, L. & KOLAKOFSKY, D. (1977). Further characterization of Sendai virus DI-RNAs: a model for their generation. *Cell*, **12**, 539–52.

LEPPERT, M., RITTENHOUSE, L., PERRAULT, J., SUMMERS, D. F. & KOLAKOFSKY, D. (1979). Plus and minus-strand leader RNAs in negative strand virus infected cells. *Cell*, **18**, 735–47.

PERLMAN, S. M. & HUANG, A. S. (1973). RNA synthesis of vesicular stomatitis virus. V. Interaction between transcription and replication. *Journal of Virology*, **12**, 1395–1400.

PERRAULT, J. & HOLLAND, J. J. (1972). Absence of transcriptase activity or transcription-inhibiting ability in defective interfering particles of vesicular stomatitis virus. *Virology*, **50**, 159–170.

REICHMANN, M. E., VILLARREAL, L. P., KOHNE, D., LESNAW, J. & HOLLAND, J. J. (1974). RNA polymerase activity and poly(A) synthesizing activity in defective T particles of vesicular stomatitis virus. *Virology*, **58**, 240–9.

SCHUBERT, M., KEENE, J. D., LAZZARINI, R. A. & EMERSON, S. U. (1978). The complete sequence of a unique RNA species synthesized by a DI particle of VSV. *Cell*, **15**, 103–12.

SEMLER, B. L., PERRAULT, J., ABELSON, J. & HOLLAND, J. J. (1978). Sequence of a RNA templated by 3'-OH RNA terminus of defective interfering particles of vesicular stomatitis virus. *Proceedings of the National Academy of Sciences, USA*, **75**, 4704–8.

SIMONSEN, C. C., BATT-HUMPHRIES, S. & SUMMERS, D. F. (1979). RNA synthesis of vesicular stomatitis virus-infected cells: *in vivo* regulation of replication. *Journal of Virology*, **31**, 124–32.

SORIA, M., LITTLE, S. P. & HUANG, A. S. (1974). Characterization of vesicular stomatitis nucleocapsids. I. Complementary 40 S RNA molecules in nucleocapsids. *Virology*, **61**, 270–80.

SZILAGYI, J. F. & URYVYEV, L. (1973). Isolation of an infectious ribonucleoprotein from vesicular stomatitis virus containing an active RNA transcriptase. *Journal of Virology*, **11**, 279–86.

WAGNER, R. R. (1975). Reproduction of rhabdoviruses. In *Comprehensive Virology*, vol. 4, H. Fraenkel Conrat and R. Wagner, pp. 1–94. New York: Plenum Press.

WERTZ, G. W. & LEVINE, M. (1973). RNA synthesis by vesicular stomatitis virus and a small plaque mutant: effects of cycloheximide. *Journal of Virology*, **25**, 202–6.

YANOFSKY, C. (1981). Attenuation in the control of expression of bacterial operons. *Nature*, **289**, 751–8.

YATES, J. L. & NOMURA, N. (1981). Feedback regulation of ribosomal protein synthesis in *E. coli*: localization of the mRNA target sites for repressor action of ribosomal protein L1. *Cell*, **24**, 243–9.

ANTIGENIC VARIATION IN THE INFLUENZA A (HONG KONG) VIRUSES

J. J. SKEHEL*, A. R. DOUGLAS*, I. A. WILSON†
AND D. C. WILEY†

*Division of Virology, National Institute for Medical Research, Mill Hill, London NW7, UK
†Gibbs Laboratory, Harvard University, Cambridge, Mass., USA

Of the biological properties of influenza viruses which influence the frequent recurrence of influenza, antigenic variation is the best defined genetically and has received the most attention in research. The antigens of importance are the virus membrane glycoproteins, the haemagglutinin and the neuraminidase, and differences in the antigenic properties of these components of different viruses form the basis of the division of influenza A viruses into subtypes (WHO memorandum, 1980). Although the antigenicities of both haemagglutinin and neuraminidase are known to change independently it is generally considered that variation in the former is more important since antibodies against the haemagglutinin neutralize virus infectivity (Webster & Laver, 1975). As a consequence in the last few years a considerable amount of information has been obtained on the structure of the haemagglutinins of a number of viruses in attempts to understand the molecular basis of the antigenic differences between viruses of different subtypes and of the variation which occurs within subtypes often referred to as antigenic drift. In the latter case particular attention has been focused on the haemagglutinins of viruses of the H3N2, 'Hong Kong', subtype and this is a review of information accumulated on antigenic drift in these viruses which were isolated between 1968 and 1980. It contains summaries of the antigenic and primary structure differences between the haemagglutinins and considers the consequences of these variations with reference to the three-dimensional structure of a 1968 haemagglutinin and in terms of the survival of the Hong Kong viruses during this period in the human population.

ANTIGENIC VARIATION

Analyses of antigenic differences in haemagglutinins are routinely made in haemagglutination-inhibition tests using post-infection fer-

Table 1. *Antigenic analysis by haemagglutination-inhibition tests of representative Hong Kong viruses isolated between 1968 and 1981*

	A/HK/1/68	A/Eng/878/69	A/HK/107/71	A/Eng/42/72	A/PC/1/73	A/Scot/840/74	A/Vic/3/75	A/Tok/1/75	A/Eng/864/75	A/Vic/112/76	A/Tex/1/77	A/BK/1/79	A/BK/2/79	A/Belg/2/81
A/Hong Kong/1/68	2560	640	640	160	320	160	40	80	160	40	40	<40	<40	<40
A/England/878/69	1280	1280	640	160	320	80	80	80	80	80	<40	<40	<40	<40
A/Hong Kong/107/71	640	640	2560	320	320	80	160	160	80	160	40	40	<40	<40
A/England/42/72	640	320	160	2560	640	640	160	160	160	80	80	160	80	80
A/Pt. Chalmers/1/73	320	160	320	640	1280	640	160	160	80	80	160	160	160	80
A/Scotland/840/74	160	160	160	640	320	2560	320	320	160	160	160	160	80	80
A/Victoria/3/75	80	80	160	320	320	640	1280	320	160	640	160	40	40	80
A/Tokyo/1/75	40	80	<40	160	320	320	320	5120	320	320	320	160	160	160
A/England/864/75	<40	<40	40	320	320	160	160	640	2560	320	1280	1280	320	320
A/Victoria/112/76	40	<40	40	80	80	160	80	160	320	1280	320	320	160	160
A/Texas/1/77	<40	<40	<40	320	160	160	80	640	1280	320	2560	1280	640	640
A/Bangkok/1/79	<40	<40	<40	80	160	80	160	640	640	480	320	2560	640	480
A/Bangkok/2/79	<40	<40	<40	<40	80	40	80	320	320	160	320	640	5120	320
A/Belgium/2/81	<40	<40	<40	40	40	40	80	320	640	320	1280	1280	320	2560

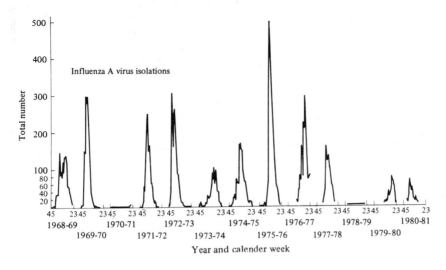

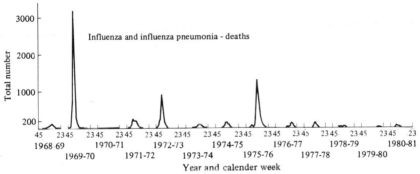

Figs. 1 & 2. The results of influenza surveillance in the United Kingdom from 1968–81. The graphs were drawn from information provided by the Communicable Disease Surveillance Centre of the Public Health Laboratory Service, Colindale, London.

ret antisera which are highly strain specific. The results in Table 1 are presented as examples of the type of information available for the Hong Kong viruses and give an indication of the degree of difference between selected viruses observed in influenza surveillance studies. The viruses in the Table were in the main the most frequently isolated in the particular year although for example in 1974 viruses like A/Port Chalmers/1/73 and not A/Scotland/864/74 were more common and in 1976 viruses like A/Victoria/3/75 were more frequently isolated than A/Victoria/112/76. In addition it may be noted that evidence of the co-circulation of different viruses was frequently obtained and this was observed most strikingly in 1975 with the viruses A/England/864/75, A/Victoria/3/75 and A/Tokyo/ 1/75.

The numbers of Hong Kong viruses isolated each year in the United Kingdom since 1968 are shown in Fig. 1 to indicate variations in the level of influenza activity from year to year and as an index of the impact of influenza in different years figures for mortality due to influenza in this period are shown in Fig. 2. Clearly the numbers of isolates may vary considerably from year to year even though viruses of apparently significant antigenic difference are detected and this may be a reflection of differences in the viruses unrelated to their antigenicity. Moreover in certain years the number of viruses isolated is disproportionate to the amount of disease indicated by the mortality figures, e.g. in 1976–77. However, although since 1975 there appears to have been little influenza in the United Kingdom, and this coincides with experience in the majority of countries, it contrasts with the situation in North America where in the USA 1980–81 was the most serious year for influenza since 1968. The reason for this difference from country to country is not known but since the viruses isolated in all countries were antigenically similar these observations would seem to indicate differences in immunity in the respective populations.

Nevertheless with these reservations it can be concluded that Hong Kong influenza viruses have circulated nearly every year since 1968 and that internationally since the initial epidemics of 1968 and 1969 there have been at least two others, in 1972 and 1975, and that these were caused by viruses (A/England/42/72 and A/Victoria/3/75) which were antigenically distinct from those previously identified.

HAEMAGGLUTININ STRUCTURE

The haemagglutinin of the 1968 Hong Kong virus A/Aichi/2/68 (X–31, Kilbourne, 1969) has a molecular weight of about 225 000 and is a trimer of identical subunits each consisting of two polypeptide chains, HA_1 and HA_2 (Wiley, Skehel & Waterfield, 1977). HA_1 contains 328 amino acids and HA_2 221 and the two polypeptides are linked in each subunit by a single disulphide bond between residues 14 of HA_1 and 137 of HA_2. The amino acid sequences of both chains are known (Verhoeyen et al., 1980; Ward & Dopheide, 1980) and are summarized in Fig 3. Both polypeptides are glycosylated: at six sites in HA_1, at asparagine residues 8, 22, 38,

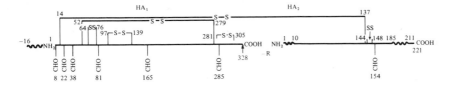

Fig. 3. A summary of the primary structure of the haemagglutinin subunit.

81, 165 and 285; and at one site in HA_2, asparagine 154. In virus particles the molecule is associated with the lipid membrane by a hydrophobic region near the carboxyl terminus of HA_2 between residues 185 and 210; the projecting portion of the molecule consisting of the HA_1 polypeptides and the HA_2 (residues 1–175) polypeptides, of molecular weight about 210 000 is released as a soluble glycoprotein, BHA, by digesting viruses with bromelain. X-ray crystallographic analyses of BHA crystals indicate that the amino terminus of HA_1 is also near the lipid membrane of the virus. The HA_1 chain extends from the base of the molecule through a fibrous stem into a peripheral β-structure-rich region, and then returns to the fibrous region and terminates about 30 Å from the virus membrane. The most prominent features of the part of the subunit composed of HA_2 residues are two antiparallel α-helices, one 29 Å long which proceeds distally from the membrane end of the molecule to connect through an extended chain with the other helix which stretches 76 Å back towards the membrane. A stereo drawing of a tracing of the α-carbon atoms of a subunit is shown in Fig. 4 together with a schematic diagram. Details of the structure have been published (Wilson, Skehel & Wiley, 1981).

VARIATIONS IN STRUCTURE

Extensive immunochemical information has not been reported for the haemagglutinin; that which is available indicates that strain-specific antibody binding sites are located on the HA_1 polypeptide chain. Of the isolated polypeptides, only HA_1 was observed to react in immunodiffusion tests with antibodies against the complete molecule (Brand & Skehel, 1972); following cyanogen bromide cleavage, antibody binding activity was found to be restricted to the 168 residue amino terminal fragment of HA_1 (Jackson et al., 1979);

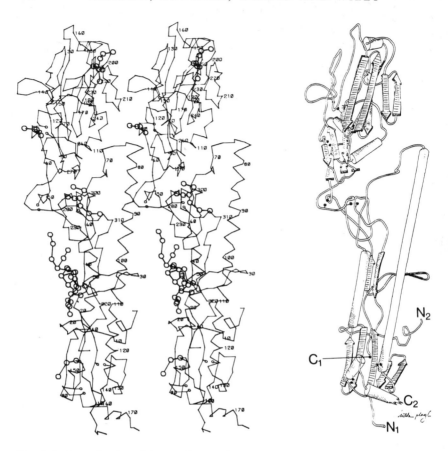

Fig. 4. A stereo drawing of the α-carbon tracing of a haemagglutinin subunit and on the right a schematic diagram indicating the terminal amino acids. ○——○ indicates carbohydrate side chains.

and haemagglutination-inhibition activity by a variety of monoclonal antibodies was found to be blocked by a large fragment of HA_1, residues 28–328, obtained by tryptic digestion of haemagglutinin after incubation at pH 5.0 (unpublished). These findings are consistent with the observations of limited variation in the amino acid sequences of the HA_2 components since only four amino acid differences at residues 2, 132, 150 and 212 were detected between the haemagglutinins of 1968 and 1975 viruses (Min Jou *et al.*, 1980; Verhoeyen *et al.*, 1980). In contrast, the data in Table 2 indicate that in the same haemagglutinins there were 21 amino acid substitutions in the HA_1 components. This Table also contains the amino acid sequence of the HA_1 polypeptides of eight Hong Kong viruses

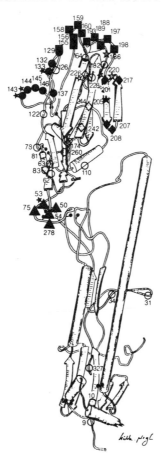

Fig. 5. Schematic diagram of a subunit of the 1968 haemagglutinin showing the positions of amino acid substitutions observed in the haemagglutinin sequence of A/Bangkok/1/79. Stars indicate the positions of amino acid substitutions observed in the haemagglutinins of antigenic variants selected using monoclonal antibodies. Amino acid sequence data communicated by G. W. Both and M. J. Sleigh and from Webster & Laver (1980) and Laver et al. (1979).

considered in Table 1 and indicates that in comparison with the 1968 haemagglutinin the number of amino acid substitutions has increased from 12 in 1972 to 21 in 1975 and 33 in 1979 (Sleigh, Both, Underwood & Bender, 1981). The locations of these substitutions are not randomly distributed throughout the polypeptide; there appear to be three main regions of variation between residues 50 and 63, 122 and 160, and 182 and 201, and two long sequences of conserved amino acids between residues 84 and 121, and 279 and 326. The consequences of the changes in amino acid side chain size, charge, or both size and charge for the structure of the molecule

Table 2. *Amino acid sequences of HA₁ polypeptides from respresentati*

```
                      10        20        30        40        50
X-31               QDLPGNDNSTATLCLGHHAVPNGTLVKTITDDQIEVTNATELVQSSSTGKICNNPH
A/England/878/69                                 N
A/Hong Kong/107/71          K                    N             T
A/England/42/72                                  N
A/Memphis/102/72      F                          N
A/Pt. Chalmers/1/73                              N
A/Victoria/3/75                                  N
A/Texas/1/77        N                            N                      R   DS
A/Bangkok/1/79      N                            N                      R   DS


                     130       140       150       160       170
X-31               ITEGFTWTGVTQNGGSNACKRGPGSGFFSRLNWLTKSGSTYPVLNVTMPNNDNFD
A/England/878/69
A/Hong Kong/107/71          E   E             D                   R
A/England/42/72    N                          D             Y
A/Memphis/102/72   N                          D             Y
A/Pt. Chalmers/1/73 N                         D             Y    A
A/Victoria/3/75    N    N              S       D             Y      Q            S
A/Texas/1/77       N    N              Y      DNS            Y  E              G
A/Bangkok/1/79     N    N          S   Y      DNS            YE E  K           G


                     250       260       270       280       290
X-31               DVLVINSNGNLIAPRGYFKMRTGKSSIMRSDAPIDTCISECITPNGSIPNDKPFQNVN
A/England/878/69
A/Hong Kong/107/71
A/England/42/72    I                                          G
A/memphis/102/72   I                                          G
A/Pt. Chalmers/1/73 I                                         G
A/Victoria/3/75    I                                          G   S
A/Texas/1/77       I  L                         I            G   S
A/Bangkok/1/79     I  L                         I            G   S
```

Data from Sleigh, Both, Underwood & Bender (1981), and communicated by M. J. Sleigh and G. W. Both.

have been considered before (Wiley, Wilson & Skehel, 1981). The substitution of threonine 83 in A/Victoria/3/75 haemagglutinin removes the carbohydrate attachment site at asparagine 81 and the aspartic to asparagine change at residue 63 creates another site for glycosylation.

Estimates of the significance of these changes in amino acid sequence for antigenic variation are presently the main approach to defining the antigenically important regions of the molecule. They have been considerably assisted by the results of experiments involving analyses of the haemagglutinins of antigenic mutants selected by growth of viruses in the presence of monoclonal antibodies (Laver *et al.*, 1979; Webster & Laver, 1980; Moss, Underwood, Bender & Whittaker, 1980). These results indicate

Hong Kong viruses

```
  60        70        80        90       100       110       120
:ILDGIDCTLIDALLGDPHCDVFQNETWDLFVERSKAFSNCYPYDVPDYASLRSLVASSGTLEF
   N                 D
                 G
                 G
   N             G
   N             G      K
   KN            G      K
   KN            G      K


 180       190       200       210       220       230       240
.LYIWGIHHPSTNQEQTSLYVQASGRVTVSTRRSQQTIIPNIGSRPWVRGLSSRISIYWTIVKPG
      V      D        T
      V                        KG
      V      D                 K
      V      D    N    T       K              Q
      V      DK   N        K   K       V
      V      DK   N            K
      V      DK   N    R       K


 00       310       320
ITYGACPKYVKQNTLKLATGMRNVPEKQT

                       R
```

that amino acid substitutions which affect the binding of different groups of antibodies are located at amino acid residue 54, between residues 133 and 144 and between residues 201 and 226. At least amino acid substitutions in these regions of the haemagglutinins of the viruses shown in Table 2 may, therefore, be implicated in antigenic variation.

Further assessment of the involvement of different regions of the haemagglutinin in antibody binding comes from consideration of the positions of amino acid substitutions in the three-dimensional structure of the molecule. This was discussed recently for Hong Kong viruses up to A/Victoria/3/75 (Wiley *et al.*, 1981) and in Fig. 5 this sort of analysis is extended to the A/Bangkok/1/79 virus by illustrating the sites of amino acid substitution in this haemagglutinin in comparison with the X-31 1968 haemagglutinin for which the structure is known. In agreement with observations made with the earlier isolates of the subtype, the amino acid substitutions are clustered in four main regions: from amino acids 122 to 146 including the protruding loop from residues 139 to 146; at the top of the molecule, residues 155–160 of a loop structure, and in the region containing α-helix 188–193 between residues 188 and 198; in the vicinity of the disulphide bond between residues 52 and 277 and, contrasting with these three areas which are exposed on the haemagglutinin surface, in an interface region between subunits which includes residues 201–220. The suggested importance of these regions in antigenic variation is again consistent with the locations of the amino acid substitutions, mentioned above, which were observed in the haemagglutinins of antigenic variants selected using monoclonal and avid antibodies. In these cases haemagglutination-inhibition and neutralization of virus by the antibodies were prevented by amino acid substitutions at, for example, residues 143, 186, 54 and 205.

The relationship between amino acid substitutions in these regions, antibody binding and the neutralization of virus infectivity, has not been determined. The proximity of the first two areas described to the proposed sialic acid binding pocket (Wilson *et al.*, 1981) suggests that antibody binding in these areas would directly prevent interaction of the haemagglutinin with cellular receptors by steric hindrance. Since the monoclonal antibodies used in the selection of all the antigenic variants considered inhibit haemagglutination, it is also possible that binding of antibody at the sites influenced by amino acid substitution in the last two areas described also effects virus neutralization by preventing interaction with the sialic acid-containing receptors. Alternatively such antibodies may primarily influence the participation of the haemagglutinin in the membrane fusion activity of the virus; such alternative mechanisms of neutralization, however, remain to be established.

What is the contribution of these changes in haemagglutinin structure to the survival of Hong Kong influenza viruses in the human population since their introduction in 1968? On the basis that survival necessitates infection of non-immune individuals it is achieved by viruses infecting the newly born exclusively or by viruses with haemagglutinins sufficiently novel antigenically to allow re-infections as well as primary infections. Obviously the Hong Kong influenza viruses followed the latter course, as did influenza viruses of the H_1 and H_2 subtypes before, between 1918 and 1957 and 1957 and 1968 respectively. Assessment of the extent of the variation in haemagglutinin structure required to allow re-infection can be made by examining the haemagglutinins of the viruses which were able to cause the second and third world-wide epidemics of this period in 1972 and 1975. As discussed in detail previously (Wiley *et al.*, 1981) amino acid substitutions in all four of the regions of the haemagglutinin molecule described above occur in the haemagglutinins of both A/England/42/72 and A/Victoria/3/75 viruses. Whether all of these changes are required for a virus to achieve the potential to re-infect or whether changes in structure at certain antigenic sites are more important than at others is not known. It is possible that analyses of the human immune response to influenza infections will resolve these questions but at present it can only be concluded that since binding of antibodies at sites influenced by amino acid substitutions in any one of the four regions prevents infection, modification in all four regions will be required to avoid neutralization and ensure virus survival.

REFERENCES

BRAND, C. M. & SKEHEL, J. J. (1972). Crystalline antigen from the influenza virus envelope. *Nature, New Biology*, **238**, 145–7.

JACKSON, D. C., DOPHEIDE, T. A., RUSSELL, R. J., WHITE, D. O. & WARD, C. W. (1979). Antigenic determinants of influenza virus haemagglutinin. II. Antigenic reactivity of the isolated N-terminal cyanogen bromide peptide of A/Memphis/72 haemagglutinin heavy chain. *Virology*, **93**, 458–65.

KILBOURNE, E. D. (1969). Future influenza vaccines and the use of genetic recombinants. *Bulletin of the World Health Organization*, **41**, 643–5.

LAVER, W. G., AIR, G. M., WEBSTER, R. G., GERHARD, W., WARD, C. W. & DOPHEIDE, T. A. A. (1979). Antigenic drift in type A influenza virus. Sequence differences in the haemagglutinin of Hong Kong (H3N2) variants selected with monoclonal hybridoma antibodies. *Virology*, **98**, 226–37.

MIN JOU, W., VERHOEYEN, M., DEVOS, R., SAMAN, E., FANG, R., HUYLEBROECK, D. & FIERS, W. (1980). Complete structure of the haemagglutinin gene from the human influenza A/Victoria/3/75 (H3N2) strain as determined from cloned DNA. *Cell*, **19**, 683–96.

MOSS, B. A., UNDERWOOD, P. A., BENDER, V. J. & WHITTAKER, R. G. (1980). Antigenic drift in the haemagglutinin from various strains of influenza virus A/Hong Kong/68 (H3N2). In *Structure and Variation in Influenza Virus*, ed. W. G. Laver & G. M. Air, pp. 329–38. New York: Elsevier.

SLEIGH, M. J., BOTH, G. W., UNDERWOOD, P. A. & BENDER, V. J. (1981). Antigenic drift in the haemagglutinin of the Hong Kong influenza subtype: Correlation of amino acid changes with alterations in viral antigenicity. *Journal of Virology*, **37**, 845–53.

VERHOEYEN, M., FANG, R., MIN JOU, W., DEVOS, R., HUYLEBROECK, D., SAMAN, E. & FIERS, W. (1980). Antigenic drift between the haemagglutinin of the Hong Kong influenza strains A/Aichi/2/68 and A/Victoria/3/75. *Nature*, **286**, 771–6.

WARD, C. W. & DOPHEIDE, T. A. A. (1980). Completion of the amino acid sequence of a Hong Kong influenza haemagglutinin heavy chain: sequence of cyanogen bromide fragment CNI. *Virology*, **103**, 37–53.

WEBSTER, R. G. & LAVER, W. G. (1975). Antigenic variation of influenza viruses. In *Influenza Viruses and Influenza*, ed. E. D. Kilbourne, pp. 270–314. New York: Academic Press.

WEBSTER, R. G. & LAVER, W. G. (1980). Determination of the number of nonoverlapping antigenic areas on Hong Kong (H3N2) influenza virus haemagglutinin with monoclonal antibodies and the selection of variants with potential epidemiological significance. *Virology*, **104**, 139–48.

WHO MEMORANDUM (1980). A revision of the system of nomenclature for influenza viruses: a WHO memorandum. *Bulletin of the World Health Organization*, **58**.

WILEY, D. C., SKEHEL, J. J. & WATERFIELD, M. (1977). Evidence from studies with a cross-linking reagent that the haemagglutinin of influenza virus is a trimer. *Virology*, **79**, 446–8.

WILEY, D. C., WILSON, I. A. & SKEHEL, J. J. (1981). Structural identification of the antibody-binding sites of Hong Kong influenza haemagglutinin and their involvement in antigenic variation. *Nature*, **289**, 373–8.

WILSON, I. A., SKEHEL, J. J. & WILEY, D. C. (1981). Structure of the haemagglutinin membrane glycoprotein of influenza virus at 3 Å resolution. *Nature*, **289**, 366–73.

HOW DO PAPOVA VIRUSES PERSIST IN THEIR HOSTS?

BEVERLY E. GRIFFIN

Imperial Cancer Research Fund, Lincoln's Inn Fields, London WC2A 3PX, UK

INTRODUCTION

The papova viruses are small DNA viruses which act as causative agents for tumour formation in a variety of species. The classification 'papova' is derived from a combination of the names denoting *pap*illoma, *pol*yoma and *va*culating viruses. The papillomas are wart viruses, the best studied among them being the bovine viruses. Since they cannot be propagated in tissue culture, although of undoubted importance, papillomas remain the least well defined of the papova viruses. Not only are they larger than the other members of this class of viruses, but it is not clear whether there is any relation between the papillomas and the other viruses classified under the papova heading. For these reasons, this chapter will concentrate on the latter viruses (the 'povas'). Of these, polyoma virus has the mouse as natural host and the vacuolating virus, simian virus 40 (SV40) has monkey as its natural host. In the past decade, human variants of SV40 have been discovered, the best characterized of which is the BK virus, first isolated from an immunosuppressed person. In their natural hosts, among healthy specimens, none of these viruses is apparently tumorigenic. In tissue culture systems, on the other hand, they probably all are, as will be discussed in more detail later. Data on species specificities of the three papova viruses, polyoma, SV40 and BKV are summarized in Table 1.

PERSISTENCE

While our knowledge of the molecular biology of papova viruses surpasses that of any other group of animal DNA viruses, our understanding of infection *in vivo* is very limited. It is clear that the majority of individuals become seroconverted early in life and that primary infection is not associated with major illness. The high frequency with which papova viruses can be isolated from the urine

Table 1. *Species specificities of the three papova viruses*

Virus	Natural host	Species range for transformation *in vitro*
Polyoma virus	Mouse	Mouse, rat, hamster, calf, (man?)[a]
SV40	Monkey	Mouse, rat, hamster, guinea pig, rabbit, pig, calf, horse, monkey, man[a]
BK virus	Man	Mouse, hamster, man, rat, rabbit, monkey[b]

[a] Data adapted from Pontén (1971, p. 44). [b] Data taken from Tooze (1980, p. 358).

of immunosuppressed individuals is best explained by the 'reactivation' of pre-existing virus, but the nature of this viral persistence is entirely unknown. Infection *in vitro* of cells of the natural host results in virus growth and cell death or in cell transformation (see Table 1). It seems unlikely that the virus persists *in vivo* through continuous cycles of lytic infection of host cells, and no neoplastic disease has yet been associated with papova viruses in their natural hosts. Whatever the mechanism of persistence it seems that the maintenance of the virus in a quiescent state is achieved, at least in part, by immune mechanisms, since interference with these mechanisms commonly results in virus excretion and, much more rarely, results in clinical disease. There is evidence implicating one human papova virus, JC virus, as the cause of a rare demyelinating disease, progressive multifocal leucoencephalopathy (PML) in patients with immune deficiencies due either to disease or to immunosuppressive therapy (Gardner, 1977). This disease has also been seen in monkeys, in association with SV40 (Holmberg *et al.*, 1977), and in recent studies, we have found that nude (*nu/nu*) mice challenged with polyoma virus produce symptoms characteristic of PML, that is, brain and spinal cord demyelination, as well as posterior paralysis (McCance, Sebesteny, Griffin, Balkwill & Tilly, unpublished results). Recently, D. J. McCance (personal communication) has extended these studies to outbred Swiss mice. His results show that a small population of the mice (<1%) exhibit symptoms corresponding to those found in the nude mice. However, he has not been able to make direct association between the disease and polyoma virus. Nonetheless, these studies suggest that persistence and expression of polyoma virus *in vivo* perhaps warrants further investigation.

What happens to cells infected with papova viruses *in vitro* is much better defined. In general, mouse cells infected with polyoma virus, monkey cells with SV40, and human cells with BKV respond by producing more virus, with the ultimate death of the cell. Under special conditions, viral infection of these same cells, and a variety of others (see Table 1), results in cellular transformation. Cells so-transformed generally produce tumours in the relevant whole animals. Detailed investigations, as subsequently discussed, show that in some cases all, and in others a part, of the viral genome persists in the transformed cells although only a portion of the genome is expressed. This expression appears to be necessary for maintaining the transformed state (or those phenotypic responses associated with it).

MOLECULAR ORGANIZATION OF POLYOMA VIRUS, SV40 AND BKV

Before proceeding into a detailed discussion of transformed cells and viral sequences that persist within them, it seems relevant to outline what is known about the molecular organization of the three papova viruses being considered and the proteins expressed by them. Polyoma virus, SV40 and BKV are all of similar size, having the capacity to code within their DNAs for about 200 000 daltons of protein. This size in itself, particularly in regard to polyoma virus, was previously somewhat baffling since proteins corresponding to about 280 000 daltons had been identified as being coded, at least in part, by this virus. A solution to this dilemma could be suggested when the complete DNA sequence of the polyoma viral genome became available (Deininger *et al.*, 1980; Soeda *et al.*, 1980a). Similarly, a somewhat less complicated organization of the SV40 and BKV genomes was suggested by their DNA sequences (Fiers *et al.*, 1978; Reddy *et al.*, 1978; Yang & Wu, 1979; Sief, Khoury & Dhar, 1979). The arrangements of viral gene products in relation to the genomes of polyoma virus and SV40 are shown in Fig. 1. (BKV has an organization virtually indistinguishable from SV40). Polyoma virus is known to code for six proteins, three of them (the small, middle and large T-antigens) have been identified by their antigenic properties and are referred to as 'early proteins', since they do not depend for their expression upon DNA replication and appear in the cell before the onset of replication. The other three are the viral

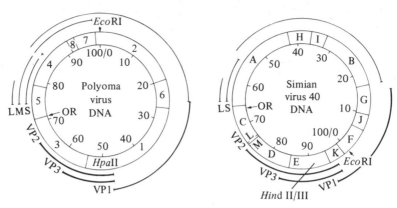

Fig. 1. Predictions allowed by the complete DNA sequence of polyoma virus and SV40. Left, a physical map of polyoma virus DNA based on the eight fragments produced by cleavage of the DNA with the restriction enzyme *Hpa*II. The map is divided into 100 units and on it are indicated the single *Eco*RI restriction site, the viral origin of replication (OR) and the location of the areas coding for the six known polyoma viral proteins, the large (L), middle (M) and small (S) T-antigens and the capsid proteins VP1, VP2 and VP3. The N-termini of the three T-antigens (located near the origin) appear to be identical; elsewhere on the genome, a single region codes for both middle and large T-antigens making use of two different reading frames. It can also be seen that the C-termini of VP2 and VP3, known to share common sequences, overlap the N-terminus of VP1. The latter is encoded within a different reading frame. Right, a physical map of SV40 DNA based on the thirteen fragments produced by cleavage of the DNA with restriction enzymes from *Haemophilus influenzae* (*Hind*II and III). The map is also divided into 100 units and is oriented so that homologies with polyoma virus are readily evident. SV40 apparently codes for no protein which corresponds to the middle T-antigen of polyoma virus. (BK virus shares a similar organization with SV40). As with polyoma virus, the N-termini of large and small T-antigens appear to be identical, but their C-termini are different. The organization of the late proteins is very similar to that seen with polyoma virus.

capsid proteins, VP1, VP2 and VP3, the 'late proteins', which are only expressed subsequent to DNA replication. SV40 and BKV have no viral equivalents of the polyoma virus middle T-antigen, and as will be discussed later, this is of primary importance in attempting to understand cellular transformation by the relevant viruses. The DNA sequence showed how the 'extra' protein could be accommodated by the polyoma virus DNA, without a concomitant expansion in size, by the use of overlapping sequences within the genome for encoding more than one protein, as shown in Fig. 1.

While the DNA sequence could suggest a solution for the coding dilemma set by polyoma virus, sequences, in themselves, were not proof. Although the ultimate proof can only come from a comparison of protein sequences (not yet available) with DNA sequence, evidence in support of the postulated organization comes in part from a consideration of protein data and in part from studies on viral mutants. Smart & Ito (1978) showed that all three of the

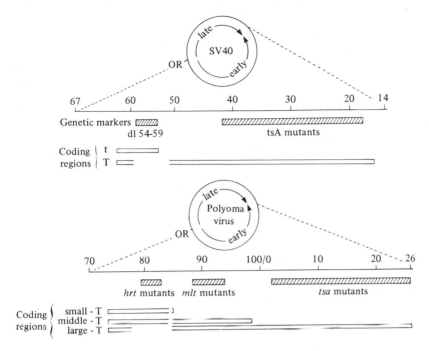

Fig. 2. A linearization of the 'early' regions of polyoma virus and SV40 (see Fig. 1), showing the respective viral origins of replication (OR), and the position of the viral early proteins, the large (T) and small (t) T-antigens of SV40 and the large, middle and small T-antigens of polyoma virus, relative to the positions where viral early mutants for both species have been mapped. The similarity in location of the 54–59 mutants of SV40 and the *hrt* mutants of polyoma virus should be noted; in both cases they lie within regions that are removed from the mRNAs of the large T-antigens by splicing. For phenotypes of these mutants, see the text. In the case of polyoma virus, it can be concluded that a functional large T-antigen *per se* is not sufficient to initiate transformation; in the case of SV40, the mutant data have been interpreted to suggest that small T-antigen serves some function in transformation, since its alteration affects the phenotype.

polyoma virus T-antigens had peptides in common which appeared to originate from the N-termini of the proteins. In addition, the viral large and middle T-antigens contained some unique peptides. The study of mutants is in general accord with these data. For polyoma virus, there are three known classes of viral early mutants, the *tsa*, *hrt* and *mlt* mutants. The lesion responsible for the temperature sensitivity in *tsa* mutants lies within the viral large T-antigen and by marker rescue experiments has been localized in the C-terminal portion of this protein (Miller & Fried, 1976). The *hrt* and *mlt* mutants are double mutants. The former have lesions in middle and small T-antigens and can be complemented by *tsa* mutants (Eckhart, 1977; Fluck, Staneloni & Benjamin, 1977). The latter have lesions in middle and large T-antigens (Ito, Spurr & Griffin, 1980). The location of these mutants is shown in Fig. 2, and is in agreement with coding assignments allowed by the DNA sequence (Fig. 1). Similarly, for SV40, the large and small T-antigens have been found to share common peptides (Paucha *et al.*, 1978; Simmons & Martin, 1978); *tsa* mutants have lesions in large T-antigen which map within the C-terminal portion of the protein (Rundell *et al.*, 1977) and the 54–59 mutants have lesions only in the small T-antigen (Shenk, Carbon & Berg, 1976; Sleigh, Topp, Hanich & Sambrook, 1978), as shown in Fig. 2. These data also agree with coding assignments allowed by the DNA sequence. For a more detailed discussion of genomic organization and additional references, see Tooze (1980).

TRANSFORMATION BY POLYOMA VIRUS AND SV40

Polyoma virus was first isolated by Gross (1953), and SV40 by Sweet & Hilleman (1960). The two viruses are organizationally similar in many respects: they replicate mainly by a bidirectional mode, starting from a single region (see Fig. 1); their capsid proteins are the same in number and similar in sizes and composition; their large and small T-antigens are similar in size and composition; the biological role of their large T-antigens, in initiating new rounds of cellular and viral synthesis, regulating expression of the other viral proteins, promoting viral integration into host cells, and possibly initiating integration into host chromosomes, is similar (Tooze, 1980; Griffin, Soeda, Barrell & Staden, 1980). These similarities, (which are possibly real with regard to lytic infection by the virus) have been a hindrance to studies on transformation, since notions on homologous behaviour derived from lytic studies have without doubt influenced studies on transformation. That studies on BKV have not been similarly hampered is probably due to its later discovery (Gardner, Field, Coleman & Hulme, 1971). Although a common mode for cellular transformation was earlier sought, it is now becoming clear that SV40 and polyoma virus probably transform cells by different pathways.

While the evidence for this notion has been around for a long time, it has not been given due consideration. For example, Pontén (1971) emphasized the fact that the earliest growth control exerted by SV40 resulted in *unrestrained* cellular growth, whereas polyoma virus seemed to induce *irregular* growth in cells. Pontén interpreted this to mean that SV40 might be altering the control of cell division, whereas the key alteration in the case of polyoma virus implicated the cell membrane. To him, the differences in cellular changes by the two viruses suggested different modes of action. To turn to yet another aspect of transformation, all attempts to find viable non-transforming mutants of SV40, comparable to the *hrt* mutants of polyoma virus isolated by Benjamin (1970), have met with no success. Although data like these pointed to differences, it seems fair to say that the critical data which finally resulted in a reassessment of the ideas on transformation by these two viruses were: (i) the discovery of the polyoma virus middle T-antigen (Ito, Brocklehurst & Dulbecco, 1977), and the subsequent failure in attempts to find a comparable virus-coded protein for SV40; and (ii) experiments which showed that whereas the entire early region of SV40 is required for the initiation and maintenance of cellular transformation (Graham *et al.*, 1974) and any interruption of this region of the DNA produced fragments that were non-transforming, only a portion of the early region of polyoma virus is required for transformation (Israel *et al.*, 1979; Hassel, Topp, Rifkin & Moreau, 1980; Novak, Dilworth & Griffin, 1980). In studies aimed at defining the minimum sequence required for the initiation and maintenance of transformation by polyoma virus, this region has been shown to be the part of the genome that encodes the viral middle T-antigen (Novak *et al.*, 1980; Novak & Griffin, 1981a).

There is now a wealth of evidence to support the concept that the middle T-antigen, a membrane-associated protein (Ito *et al.*, 1977), plays a pivotal role in transformation by polyoma virus and within this protein, two, and possibly three, specific regions are at present known to be functionally involved (for elaboration, see Griffin *et al.*, 1979). One lies at the C-terminus of the protein, which contains a long stretch of hydrophobic sequences, flanked by polar amino acids. Interference with the C-terminus results in a marked decrease in transformation (Novak & Griffin, 1981a, b). The other is the part of the protein which lies within the region where the *mlt* mutants map (see Fig. 2). In most cases, mutations within this region have been shown to have a detectable, and often marked, effect upon transformation. In one case (with mutant *dl* 8), cells express the transformed phenotype to a greater extent than seen with wild-type virus whereas with other mutants, notably mutant *dl* 23 and mutant 1015, the transformed phenotype is poorly expressed, if at all

(Griffin & Maddock, 1979; Griffin *et al.*, 1979; Magnusson & Berg, 1979). This region has been implicated in a protein kinase activity that appears to be associated with the middle T-antigen (Smith, Smith, Griffin & Fried, 1979). The third possible region involved is that found in the lesion of one of the *hrt* mutants, *NG* 59 (see Fig. 2). Although all the other host-range mutants have been found to be frame-shift mutants which ultimately lead to a premature termination of middle T-antigen, NG-59 is apparently an in-phase mutant which appears to result in only a minor modification of the protein (Benjamin, Carmichael & Schaffhausen, 1979). Nevertheless, this mutant is non-transforming. All these data point to changes in cellular transformation when the polyoma virus middle T-antigen is perturbed, and reemphasize its importance in transformation.

The DNA and predicted protein sequences of polyoma virus, SV40 and BKV suggest that they might have evolved from a common ancestor (Soeda, Maruyama, Arrand & Griffin, 1980b). Nonetheless, the transforming sequences within polyoma virus apparently have no counterpart in the other two viruses. How then do the other two viruses transform? And from where did the 'extra' sequences in polyoma virus originate? It is easier at the moment to speculate on the latter than the former question. The nearest analogy for transformation by polyoma virus appears to lie, not with its fellow papova viruses, but with the RNA retroviruses, because the retrovirus gene product pp60 *src* appears to be a protein essential to transformation which is also membrane associated, contains hydrophobic regions and has an associated protein kinase activity. Since the *src* gene appears to have been acquired by the virus from its host cell (Levinson, Courtneidge & Bishop, 1981), it seems reasonable to suggest that polyoma virus may also have acquired its transforming sequences from its host cell. There are no data that confirm or refute this. To the former question about transformation induced by SV40 or BKV, studies to date suggest that both the viral small and large T-antigens play some role in this process. Thus, the SV40 54–59 mutants, which affect only the small T-antigen, reduce, but do not abolish, transformation (Sleigh *et al.*, 1978; Bouck *et al.*, 1978) and *tsA* mutants which affect only the large T-antigen do not transform at the non-permissive temperature (Rundell *et al.*, 1977). Map positions of these mutants are shown in Fig. 2. These data, however, do not distinguish between a direct and an auxiliary role for SV40 in transformation. It is now well documented that a protein present in normal cells (53–54K) is present in much higher quantities in SV40 and other transformed cells, and complexes with the SV40 large T-antigen (Linzer & Levine, 1979; Melero, Stitt, Mangel & Carroll, 1979; Linzer, Maltzman & Levine, 1979a, b; Crawford et al., 1979). It is an open question whether SV40 (and presumably BKV) transform cells by virtue of a transforming viral gene product, by integrating into host cells in such a way that the production of a host protein (an 'oncogene' product) is modified and/or stimulated or by a combination of these or other mechanisms. In this connection, it is noteworthy that whatever the mode, polyoma virus produces a higher frequency of transformation than is generally observed with SV40.

PERSISTENCE OF VIRAL SEQUENCES WITHIN TRANSFORMED CELLS

It has been known for some time that polyoma virus and SV40 sequences persist within transformed cells. In the case of polyoma virus, only early sequences appeared to be expressed in transformed cells and the presence of the entire genome was in doubt, since numerous attempts to rescue infectious virus from transformed cells always met with failure. Nonetheless, a part, at least, of the viral genome appeared to persist indefinitely and to be transmitted to daughter cells. These sequences were thought to be the cause of the altered cellular state. The case for persistence of the SV40 genome was stronger, since infectious virus could be rescued from transformed cells, indicative of the presence of the entire genome. The early studies on viral persistence are reviewed by Pontén (1971, pp. 72–75, and 87–91).

The advent of one particular technique made possible more detailed studies of persisting viral sequences. This technique, commonly referred to as 'Southern blotting' (Southern, 1975), couples cleavage of total cellular DNA with restriction enzymes, separation of resulting fragments, transfer to nitrocellulose filters and analysis of viral sequences by means of hybridization with high specific activity radioactively labelled viral DNA probes, as outlined in Figs. 3 and 4. This technique is described (Fig. 3) in detail since it is the basis for most current analyses of viral sequences persisting within transformed cells.

The earliest studies aimed at defining the state of viral sequences were carried out on rat or mouse cells transformed with SV40. They

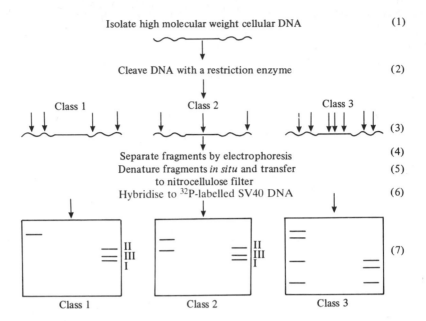

Fig. 3. Scheme for determining the arrangement of viral sequences in the DNA of transformed cells. (1 to 3) DNA from transformed cell lines is isolated and cleaved with three different kinds of restriction enzymes: those that do not cleave the viral DNA (Class 1), those which cleave viral DNA only once (Class 2), and those which cleave it more than once (Class 3). (Viral DNA is indicated by a straight line and host cell DNA by wavy lines.) (4) The DNA fragments are fractionated according to size by electrophoresis on an agarose gel. (Large fragments move slowly in the electrical field and smaller fragments move more rapidly.) (5) The fragments are denatured *in situ* by treatment with alkali and then transferred to a nitrocellulose filter, and (6) hybridized to 'probe' viral DNA ^{32}P-labelled to high specific activity. In this technique the labelled viral DNA is present in vast excess and provides the driving force for the reaction. Hybridization allows homologous DNA to be detected. Thus, probe ^{32}P-labelled SV40 DNA can be used to locate SV40 DNA sequences within cellular DNA, etc. (7) The distribution of viral sequences among the different sized fragments of transformed cellular DNA is determined by autoradiography. Fragments that contain as little as 10^{-13}g. of viral DNA can be detected using a high specific activity probe. For an indication of the size of the fragments obtained from cellular DNA, a control experiment shown on the right of each block in (7) is carried out using only viral DNA, supercoiled (I), circular (II) or full-length linear (III) for Class 1 and 2 experiments, or cleaved with an appropriate restriction enzyme for the Class 3 experiment.

With enzymes that do not cleave viral DNA (Class 1), all fragments of DNA detected must come from the cleavage of the host cell DNA, and the number of radioactive bands from any one cell line provides an estimate of the number of separate sites at which the virus is integrated.

Enzymes that cleave viral DNA once (Class 2) can be used to detect the presence in transformed cells of partially duplicated or tandem copies. These enzymes should cleave within repititious DNA to generate a whole copy of viral DNA that, if present, should migrate together with linear DNA. In the example shown, there are no tandem copies.

Enzymes that cleave viral DNA at more than one site (Class 3) can be used to catalogue viral DNA sequences within the transformed cell by comparing fragment sizes with those obtained by cleaving the viral DNA alone. (Adapted from Botchan *et al.*, 1976.)

suggested: (i) that in any particular cloned transformed cell line, the location of viral sequences within the cellular DNA remained stable; (ii) there could be multiple copies of intact or partial viral genomes which were integrated at different regions within the host DNA; (iii) the integration sites between viral and host DNAs mapped at different positions on the viral DNA (Botchan, Topp & Sambrook, 1976; Ketner & Kelly, 1976). Similar findings were subsequently reported for integrated SV40 sequences in a human transformed cell line (Hwang & Kucherlapati, 1980). Taken together, it would appear that at a gross level there is no apparent site specificity for the integration of SV40 into either rodent or human cells. More recent studies have aimed at defining the details within any one or all of the three points mentioned above. Finer analyses of SV40 viral sequences integrated into host chromosomes have served only to emphasize, however, the apparent randomness of the integration process. Rigby, Chia, Clayton & Lovett (1980), studying mouse transformed cells, found that in five out of six lines, although the viral sequences were tandemly duplicated, there was no apparent specificity for integration. In one line, the viral origin of replication was repeated three times, although there was only one complete copy of the early region, which led to the suggestion that a trimer of viral DNA was initially integrated, followed by subsequent deletion events. Other studies, in which fragments of SV40 DNA integrated into rat cells were cloned in *E. coli* and then sequenced, revealed convolutions of viral DNA sequences that could only be explained by assuming that they represented end-products of a number of recombination events (Sambrook *et al.*, 1979, 1980). Cell–viral junctions showed no sequence homology among different lines. More extensive studies of the same nature, using rat cells transformed by polyoma virus, have led to similar findings (E. Ruley, personal communication). To try to fathom this apparent disorder within transformed cell lines, Hiscott, Murphy & Defendi (1981) have recently carried out some interesting experiments that use newly transformed clones of mouse cell origin. Their studies showed post-integration rearrangements of the viral genome which they explain as resulting from labile interactions between the SV40 DNA and its host genome. What seems clear from their experiments is that integrated viral DNA is unstable and becomes altered during propagation of cells. Many of the previous results arise from studies on established cell lines passaged many times in culture and probably are a reflection of this instability. They, therefore, add

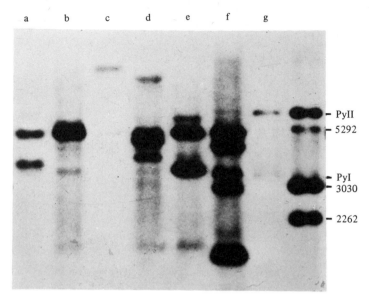

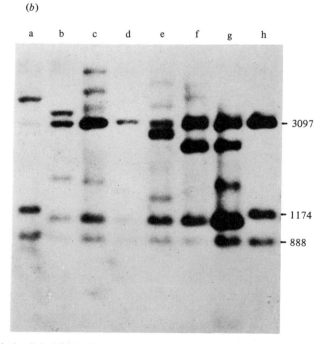

Fig. 4. Analysis of viral DNA from rat cells transformed by wild-type (A2 strain) and mutant (*dl* 8) polyoma viruses. Total cellular DNA was cleaved with restriction endonucleases and analysed by the general procedure of Southern (1975) (see Fig. 3).
(*a*) DNA cleaved with *Eco*RI (see Fig. 1) Lanes (a, b, c), DNA from cell lines established

little to our understanding of the initial event(s) that led to the transformed state of the cell. No-one, to my knowledge, has carried out studies to see if viral sequences in long-established cell lines continue to change, or whether a stable integrated state is ultimately reached.

From similar types of studies, the randomness of the integration event in polyoma virus transformed cells is evident (Birg, Dulbecco, Fried & Kamen, 1979; Lania *et al.*, 1979). For example, in some recent studies on rat cells transformed by the polyoma virus *mlt* mutant, *dl* 8 (see Fig. 2), a number of cell lines established from colonies grown in soft agar were examined. Two distinct morphological colony types were observed following cellular transformation with this mutant. Cell lines derived from either type produced tumours in syngeneic rats at a rate about twice as fast as that observed with animals injected with cells transformed by wild-type virus (Griffin *et al.*, 1979). In an attempt to correlate morphology with viral DNA sequences, we investigated total cellular DNA by the Southern blotting/hybridization method discussed previously (Fig. 3). Some of the data on cellular DNA cleaved with the restriction endonucleases *Eco*RI (Fig. 3, class 2), or with *Pvu*II (Fig. 3, class 3) are shown in Fig. 4. Our conclusions regarding the integration of polyoma virus DNA were roughly similar to those reached from studies on SV40 transformed cells. That is, many but not all cells contained free copies of the viral genome and most cells contained tandemly integrated copies. One cell line, *dl* 8/L2, however, fulfilled neither of these condition. No specific sites of integration were seen, and no patterns observed which could explain morphological differences. The general conclusions from our experiments and those of others with polyoma virus are that transformed cells may or may not contain free copies of the viral genome, they may contain single or multiple copies of integrated DNA, and they may contain either less than one whole genome or tandemly

Fig. 4 (*cont.*)
from large colonies produced by infection with the mutant *dl* 8 (lane c is *dl* 8/L2), lanes (d, e, f) DNA from cell lines established from small colonies, lane g, marker polyoma virus circular DNA (PyII) and supercoiled DNA (PyI), lane h, PyII, PyI and, for size markers, polyoma virus full-length linear DNA produced by cleavage with *Eco*RI (5292) and linear DNAs produced by cleavage with *Hind*III (3030 and 2262).

(*b*) DNA cleaved with the restriction endonuclease *Pvu*II (four sites in polyoma virus DNA). Lane, a, DNA from a wild-type virus transformed cell line, lanes (b, c, d) DNA from cell lines established from large colonies produced by infection with *dl* 8 (lane d is *dl* 8/L2), lanes (e, f, g) DNA from cell lines established from small colonies, and lane h, marker DNA showing the sizes of the three largest fragments produced by digestion of polyoma virus DNA with *Pvu*II.

duplicated copies. Data taken from studies of Lania *et al*. (1979), including the protein analysis of some transformed rat lines, illustrate the heterogeneity of the state of viral sequences within the host chromosome.

Since the coding information in the DNAs of these small oncogenic viruses is limited (see Fig. 1), it appears likely that integration into cellular chromosomes is mediated by cellular recombination enzymes. Extensive investigations such as those cited above all suggest that these recombination events occur at random sites in the host and viral sequences. Were these all the data, the question of integration could be tabled until such time as recombination becomes better understood. However, a recent paper by Mougneau, Birg, Rassoulzadegan & Cuzin (1980) keeps the issue open. They studied rat cells that had, in independent experiments, been transformed with SV40 under carefully controlled conditions, and then passaged only a few times in culture. Their data, the analysis of which has been limited to date to the Southern blotting technique, point to some integration specificity. These findings therefore differ significantly from those of others. In the first place, Mougneau *et al*. failed to detect any free (non-integrated) viral DNA. In the second, a number of cell lines produced viral fragments which were indistinguishable among the lines and virus–cell junction fragments which were also indistinguishable. An attractive interpretation of all the data is that there may be some specificity in the initial integration events which is 'lost' during subsequent passaging of the transformed cells. For example, due to instabilities such as those reported by Hiscott *et al*. (1981), sequences may become scrambled in such a way that they lead to the apparent randomness observed in all other studies. It should be noted, however, that the aforementioned studies by Mougneau *et al*. have not yet been confirmed by DNA sequencing and until this is done, no firm conclusions can be reached on this controversial issue.

CONCLUSIONS AND SPECULATIONS

When considering the viral sequences retained within cells transformed by polyoma virus and SV40 (and probably BKV and the other human variants), one main difference and several similarities emerge. The difference is that for the maintenance of the transformed state, the entire 'early' region of SV40 appears to be required whereas with polyoma virus this is definitely not the case. Numerous attempts to transform cells with less than the whole early region of SV40 on the one hand, or to find transformed cell lines which contain only portions of the early region on the other, have met with little success. This is consistent with studies on *tsA* mutants of SV40 that show a need for a functional large T-antigen probably to initiate and certainly to maintain cellular transformation. With polyoma virus, the minimum sequence required appears to be only that needed to express the middle T-antigen, or roughly one-half of the early region. In addition to studies on the DNA and RNA within transformed cells, which are in agreement with the above statement, Ito & Spurr (1979) have examined the proteins in a wide variety of cells transformed by polyoma virus. Without exception, all lines expressed the viral middle (and small) T-antigen. The presence of large T-antigen was highly variable. Some lines had full-sized large T-antigens, some had truncated versions of the protein and some contained no detectable amounts of any protein related to large T-antigen. With both SV40 and polyoma virus, the sequences that persist appear to reflect the need to express a particular viral protein.

Similarities that emerge are: (i) those that reflect the presence of free viral genomes within transformed cells, (ii) tandem arrays of viral sequences, and (iii) a general lack of specificity in the integration sites between viral and host cell DNAs. Although the human papova viruses are more difficult to study, preliminary data on BKV lead to similar conclusions (Chenciner *et al.*, 1980; Ter Schegget, Voves, Van Strein & Van der Noordaa, 1980). With regard to point (i), the persistence of extrachromosomal viral DNA is not a requirement for transformation and some experiments appear to militate against its presence. Studies by Zouzias, Prasad & Basilico (1977) showed that within a population of cells, not only did some of them contain no free genome copies, but further, that spontaneous induction of viral DNA replication occurred only in a minority of the population, in so-called 'virus factories'. Studies by Basilico *et al.* (1979) and Lania, Hayday & Fried (1981), among others, showed that free viral DNA disappeared from some lines upon passaging of the cells either in culture or *in vivo*. (ii) The presence of tandem head-to-tail copies of viral sequences, in the case of SV40 at least, may in the absence of specific integration, ensure the expression of the viral early functions. In the case of polyoma virus, the need for tandem integration is less clear and,

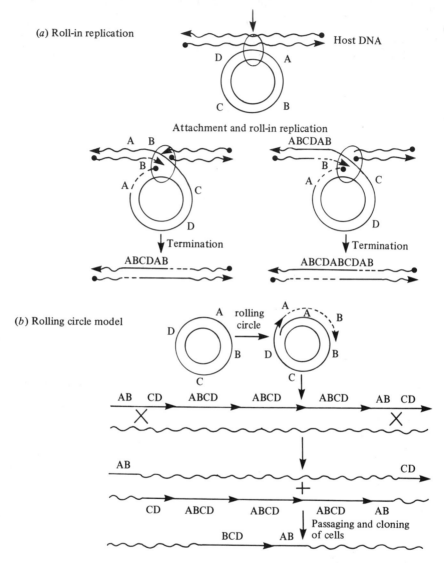

Fig. 5. Possible mechanisms by which 'head-to-tail' tandem copies of viral DNA sequences become integrated into host chromosomes. In both models, wavy lines represent host DNA sequences, and letters above non-wavy lines represent viral sequences. The *roll-in replication* model is based on one suggested for the phage mu by Harshey & Bukhari (1981). In this model, following attachment of the viral DNA to host DNA, one strand of the virus becomes integrated, and viral replication occurs as a post-integration event. Depending upon the nature of the termination, sequences corresponding all the way from less than a whole copy of viral DNA to multiple tandem copies could be integrated into the host chromosome. In a *rolling circle model*, replication precedes integration. The recombination event between viral and host DNAs would then determine the nature of the viral sequences found in the host chromosome. In both models, rearranged sequences could occur by intramolecular events that occur during the passaging and cloning of cells.

indeed, not all cell lines examined have tandem copies. The origin of tandem integration has been the subject of debate. It has generally been suggested that although polyoma virus and SV40 replicate mainly by a bidirectional mode originating from the single origin on their genomes (see Fig. 1), a minor rolling circle mode of replication may exist which could be responsible for the tandem copies of the genomes found within host cells (Bjursell, 1978). An alternative mode of replication that could also produce tandem sequences is the roll-in replication model proposed for the phage mu (Harshey & Bukhari, 1981). Either of the models (see Fig. 5) could lead to tandem arrays of sequences and provide integrated forms of DNA that could recombine to give apparent 'scrambled' sequences. In the latter connection, tandem copies, in general, have been found to be unstable (Basilico et al., 1979). To point (iii), if expression of a viral function(s) is important in cellular transformation, there is no prerequisite for specific integration events. Integration in such a way as to increase the efficiency of expression would appear to have obvious advantages, although this has not been shown to be so. In the case of SV40, it could be argued that as long as the large T-antigen could be expressed, and presumably the expression of at least one of the viral capsid proteins suppressed, no other requirement need be met. Although the same may hold true for polyoma virus, for it, the statement is probably only partly sufficient. Most cells infected with this virus are semi-permissive for its expression (see Table 1). In order for a full and stable expression of transformation in vitro, it may be necessary to suppress the machinery that leads to a lytic response. All studies to date have implicated the viral large T-antigen in the initiation of viral DNA synthesis. Therefore, for polyoma virus, the main constraint imposed on integration may be that it occurs in such a way that a functional large T-antigen is not expressed. Cells which do not fulfil this requirement may either die or be unstable with respect to the integrated viral DNA. An interesting, but unanswered question for polyoma virus, is what is the maximum amount of viral information that can be retained concomitant with the stable appearance of cellular transformation? For SV40, the host cell itself appears to have a controlling influence on expression.

REFERENCES

BASILICO, C., ZOUZIAS, D., DELLA-VALLE, G., GATTONI, S., COLANTUONI, V., FENTON, R. & BAILEY, L. (1979). Integration and excision of polyomavirus genomes. *Cold Spring Harbor Symposium in Quantitative Biology*, **44**, 611–20.

BENJAMIN, T. L. (1970). Host range mutants of polyoma virus. *Proceedings of the National Academy of Sciences, USA*, **67**, 394–9.

BENJAMIN, T. L., CARMICHAEL, G. G. & SCHAFFHAUSEN, B. S. (1979). The hr-t gene of polyoma virus. *Cold Spring Harbor Symposium in Quantitative Biology*, **44**, 263–70.

BIRG, F., DULBECCO, R., FRIED, M. & KAMEN, R. (1979). State and organization of polyoma virus DNA sequences in transformed rat cell lines. *Journal of Virology*, **29**, 633–48.

BJURSELL, G. (1978). Effects of 2′deoxy-2′-azidocytidine on polyoma virus DNA replication: evidence for rolling circle-type mechanism. *Journal of Virology*, **26**, 136–42.

BOTCHAN, M., TOPP, W. & SAMBROOK, J. (1976). The arrangement of Simian Virus 40 sequences in the DNA of transformed cells. *Cell*, **9**, 269–87.

BOUCK, N., BEALES, N., SHENK, T., BERG, P. & DIMAYORCA, S. (1978). New region of the Simian Virus 40 genome required for efficient viral transformation. *Proceedings of the National Academy of Sciences, USA*, **75**, 2473–7.

CHENCINER, N., MENEGUZZI, G., CORALLINI, A., GROSSI, M. P., GRASSI, M. P., BARBANTI-BRODANO, G. & MILANESI, G. (1980). Integrated and free viral DNA in hamster tumours induced by BK virus. *Proceedings of the National Academy of Sciences, USA*, **77**, 975–9.

CRAWFORD, L. V., LANE, D. P., DENHARDT, D. T., HARLOW, E. E., NICKLIN, P. M., OSBORN, K. & PIM, D. C. (1979). Characterization of the complex between SV40 large T antigen and the 53K host protein in transformed mouse cells. *Cold Spring Harbor Symposium in Quantitative Biology*, **44**, 179–87.

DEININGER, P. L., ESTY, A., LAPORTE, P., HSU, H. & FRIEDMANN, T. (1980). Nucleotide sequence and restriction enzymes sites of the polyoma genome. *Nucleic Acids Research*, **8**, 855–60.

ECKHART, W. (1977). Complementation between temperature sensitive (ts) and host range non-transforming (hr-t) mutants of polyoma virus. *Virology*, **77**, 589–97.

FIERS, W., CONTRERAS, R., HAEGEMAN, G., ROGIERS, R., VAN DE VOORDE, A., VAN HEUVERSWYN, H., VAN HERREWEGHE, J., VOLCKAERT, G. & YSEBAERT, M. (1978). Complete nucleotide sequence of SV40 DNA. *Nature*, **273**, 113–20.

FLUCK, M. M., STANELONI, R. J. & BENJAMIN, T. L. (1977). Hr-t and ts-a: two early gene functions of polyoma virus. *Virology*, **77**, 610–24.

GARDNER, S. D. (1977). The new human papova viruses; their nature and significance. In *Recent Advances in Clinical Virology*, vol. 1, ed. A. P. Waterson, pp. 93–115. Churchill Livingstone, Edinburgh.

GARDNER, S. D., FIELD, A. M., COLEMAN, D. V. & HULME, B. (1971). New human papova virus (BK) isolated from urine after renal transplantation. *Lancet*, **i**, 1253–7.

GRAHAM, F. L., ABRAHAMS, P. J., MULDER, C., HEIJNEKER, H. L., WARNAAR, S. O., DE VRIES, F. A. J., FIERS, W. & VAN DER EB, A. J. (1974). Studies on *in vitro* transformation by DNA and DNA fragments of human adenoviruses and Simian virus 40. *Cold Spring Harbor Symposium in Quantitative Biology*, **39**, 637–50.

GRIFFIN, B. E., ITO, Y., NOVAK, U., SPURR, N., DILWORTH, S. M., SMOLAR, N., POLLACK, R., SMITH, K. & RIFKIN, D. B. (1979). Early mutants of polyoma virus (dl8 and dl23) with altered transforming properties: is polyoma virus middle T antigen a transforming gene product? *Cold Spring Harbor Symposium in Quantitative Biology*, **44**, 271–83.

GRIFFIN, B. E. & MADDOCK, C. (1979). New classes of viable deletion mutants in the early region of polyoma virus. *Journal of Virology*, **31**, 645–56.

GRIFFIN, B. E., SOEDA, E., BARRELL, B. G. & STADEN, R. (1980). Sequence and analysis of polyoma virus DNA. In *Molecular Biology of Tumour Viruses*, vol. 2, 2nd edn, ed. J. Tooze, pp. 831–96. Cold Spring Harbor Laboratories.

GROSS, L. (1953). A filterable agent recovered from AK leukaemic extracts causing salivary gland carcinomas in C3H mice. *Proceedings of the Society for Experimental Biology and Medicine (New York)*, **83**, 414–21.

HARSHEY, R. M. & BUKHARI, A. I. (1981). A mechanism of DNA transposition. *Proceedings of the National Academy of Sciences, USA*, **78**, 1090–4.

HASSEL, J. A., TOPP, W. C., RIFKIN, D. B. & MOREAU, P. E. (1980). Transformation of rat embryo fibroblasts by cloned polyoma virus DNA fragments containing only part of the early region. *Proceedings of the National Academy of Sciences, USA*, **77**, 3978–82.

HISCOTT, J. B., MURPHY, D. & DEFENDI, V. (1981). Instability of integrated viral DNA in mouse cells transformed by Simian Virus 40. *Proceedings of the National Academy of Sciences, USA*, **78**, 1736–40.

HOLMBERG, C. A., GRIBBLE, D. H., TAKEMOTO, K. K., HOWLEY, P. M., ESPAMA, C. & OSBORN, B. I. (1977). Isolation of Simian Virus 40 from Rhesus monkeys (*Macaca mulatta*) with spontaneous progressive multifocal leukoencephalopathy. *Journal of Infectious Diseases*, **136**, 593–6.

HWANG, S.-P. & KUCHERLAPATI, R. (1980). Localization and organization of integrated Simian Virus 40 sequences in a human cell line. *Virology*, **105**, 196–204.

ISRAEL, M. A., SIMMONDS, D. T., HOURIHAN, S. L., ROWE, W. P. & MARTIN, M. A. (1979). Interrupting the early region of polyoma virus DNA enhances tumourigenicity. *Proceedings of the National Academy of Sciences, USA*, **76**, 3713–16.

ITO, Y., BROCKLEHURST, J. R. & DULBECCO, R. (1977). Virus specific proteins in the plasma membrane of cells lytically infected or transformed by polyoma virus. *Proceedings of the National Academy of Sciences, USA*, **74**, 4666–70.

ITO, Y. & SPURR, N. (1979). Polyoma virus T antigens expressed in transformed cells: significance of middle T antigen in transformation. *Cold Spring Harbor Symposium in Quantitative Biology*, **44**, 149–57.

ITO, Y., SPURR, N. & GRIFFIN, B. E. (1980). Middle T antigen as the primary inducer of full expression of the phenotype of transformation by polyoma virus. *Journal of Virology*, **35**, 219–32.

KETNER, G. & KELLY, T. J. (1976). Integrated Simian Virus 40 sequences in transformed cell DNA: analysis using restriction endonucleases. *Proceedings of the National Academy of Sciences, USA*, **73**, 1102–6.

LANIA, L., HAYDAY, A., BJURSELL, G., GANDINI-ATTARDI, D. & FRIED, M. (1979). Organisation and expression of integrated polyoma virus DNA sequences in transformed rodent cells. *Cold Spring Harbor Symposium in Quantitative Biology*, **44**, 597–603.

LANIA, L., HAYDAY, A. & FRIED, M. (1981). The loss of functional large T antigen and free viral genomes from Py *in vitro*-transformed cells after passage *in vivo* as tumour cells. *Journal of Virology*, **39**, 422–31.

LEVINSON, A. D., COURTNEIDGE, S. A. & BISHOP, J. M. (1981) Structural and functional domains of the Rous Sarcoma Virus transforming protein (pp60[src]). *Proceedings of the National Academy of Sciences, USA*, **78**, 1624–8.

LINZER, D. I. H. & LEVINE, A. J. (1979). Characterization of a 54K dalton cellular SV40 tumour antigen present in SV40-transformed cells and uninfected embryonal carcinoma cells. *Cell*, **17**, 43–52.

LINZER, D. I. H., MALTZMAN, W. & LEVINE, A. J. (1979a). The SV40 A gene product is required for the production of a 54 000 M.W. cellular tumour antigen. *Virology*, **98**, 308–18.

LINZER, D. I. H., MALTZMAN, W. & LEVINE, A. J. (1979b). Characterization of a murine cellular SV40 T antigen in SV40-transformed cells and uninfected embryonal carcina cells. *Cold Spring Harbor Symposium in Quantitative Biology*, **44**, 215–24.

MAGNUSSON, G. & BERG, P. (1979). Construction and analysis of viable deletion mutants of polyoma virus. *Journal of Virology*, **32**, 523–9.

MELERO, J. A., STITT, D. T., MANGEL, W. F. & CARROLL, R. B. (1979). Identification of new polypeptide species (48–55K) immunoprecipitable by antiserum to purified large T-antigen and present in SV40 infected and transformed cells. *Virology*, **93**, 466–80.

MILLER, L. K. & FRIED, M. (1976). Construction of the genetic map of the polyoma genome. *Journal of Virology*, **18**, 824–32.

MOUGNEAU, E., BIRG, F., RASSOULZADEGAN, M. & CUZIN, F. (1980). Integration sites and sequence arrangement of SV40 DNA in a homogenous series of transformed rat fibroblast lines. *Cell*, **22**, 917–27.

NOVAK, U., DILWORTH, S. M. & GRIFFIN, B. E. (1980). Coding capacity of a 35% fragment of the polyoma virus genome is sufficient to initiate and maintain cellular transformation. *Proceedings of the National Academy of Sciences, USA*, **77**, 3278–82.

NOVAK, U. & GRIFFIN, B. E. (1981a). Requirement for the C-terminal region of middle T antigen in cellular transformation by polyoma virus. *Nucleic Acids Research*, **9**, 2055–73.

NOVAK, U. & GRIFFIN, B. E. (1981b). Cellular transformation by polyoma virus. In *International Cell Biology 1980–1981*, ed. H. G. Schweiger, pp. 448–56. Springer-Verlag, Berlin.

PAUCHA, E., MELLOR, A., HARVEY, R., SMITH, A. E., HEWICK, R. M. & WATERFIELD, M. D. (1978). Large and small tumour antigens from Simian Virus 40 have identical amino termini mapping at 0.65 map units. *Proceedings of the National Academy of Sciences, USA*, **75**, 2165–9.

PONTÉN, J. (1971). Spontaneous and virus induced transformation in cell culture. *Virology Monographs*, **8.**

REDDY, V. B., THIMMAPAYA, B., DHAR, R., SUBRAMANIAN, K. N., ZAIN, B. S., PAN, J., CELMA, M. L., GHOSH, P. K. & WEISSMAN, S. M. (1978). The genome of Simian Virus 40. *Science*, **200**, 494–502.

RIGBY, P. W. J., CHIA, W., CLAYTON, C. E. & LOVETT, M. (1980). The structure and expression of the integrated viral DNA in mouse cells transformed by Simian Virus 40. *Proceedings of the Royal Society of London*, **B210**, 437–50.

RUNDELL, K., COLLINS, J. K., TEGTMEYER, P., OZER, H. L., LAI, C. & NATHANS, D. (1977). Identification of Simian Virus 40 protein A. *Journal of Virology*, **21**, 636–46.

SAMBROOK, J., BOTCHAN, M., HU, S. L., MITCHISON, T. & STRINGER, J. (1980). Integration of viral DNA sequences in cells transformed by adenovirus 2 or SV40. *Proceedings of the Royal Society of London*, **B210**, 423–35.

SAMBROOK, J., GREENE, R., STRINGER, J., MITCHISON, T., HU, S.-L. & BOTCHAN, M. (1979). Analysis of the sites of integration of viral DNA sequences in rat cells transformed by adenovirus 2 or SV40. *Cold Spring Harbor Symposium in Quantitative Biology*, **44**, 569–84.

SHENK, T. E., CARBON, J. & BERG, P. (1976). Construction and analysis of viable deletion mutants of Simian Virus 40. *Journal of Virology*, **18**, 664–71.

SIEF, I., KHOURY, G. & DHAR, R. (1979). The genome of human papova virus BKV. *Cell*, **18**, 963–77.

SIMMONS, D. T. & MARTIN, M. A. (1978). Common methionine-tryptic peptides near the amino-terminal end of primate papova virus tumour antigens. *Proceedings of the National Academy of Sciences, USA*, **75**, 1131–5.

SLEIGH, M. J., TOPP, W. C., HANICH, R. & SAMBROOK, J. (1978). Mutants of SV40 with an altered small t protein are reduced in their ability to transform cells. *Cell*, **14**, 79–88.

SMART, J. E. & ITO, Y. (1978). Three species of polyoma virus tumour antigens share common peptides probably near the amino termini of the proteins. *Cell*, **15**, 1427–37.

SMITH, A. E., SMITH, R., GRIFFIN, B. E. & FRIED, M. (1979). Protein kinase activity associated with polyoma virus middle T antigen in vitro. *Cell*, **18**, 915–24.

SOEDA, E., ARRAND, J. R., SMOLAR, N., WALSH, J. E. & GRIFFIN, B. E. (1980a). Coding potential and regulatory signals of the polyoma virus genome. *Nature*, **283**, 445–53.

SOEDA, E., MARUYAMA, T., ARRAND, J. R. & GRIFFIN, B. E. (1980b). Host-dependent evolution of three papova viruses. *Nature*, **285**, 165–7.

SOUTHERN, E. M. (1975). Detection of specific sequences among DNA fragments separated by gel electrophoresis. *Journal of Molecular Biology*, **98**, 503–17.

SWEET, B. H. & HILLEMAN, M. R. (1960). The vacuolating virus SV40. *Proceedings of the Society for Experimental Biology and Medicine*, **105**, 420–7.

TER SCHEGGET, J., VOVES, J., VAN STREIN, A. & VAN DER NOORDAA, J. (1980). Free viral DNA in BK virus-induced hamster tumour cells. *Journal of Virology*, **35**, 331–9.

TOOZE, J. (1980). *The Molecular Biology of Tumour Viruses*, 2nd edn, vol. 2, *DNA Tumour Viruses*. Cold Spring Harbor Laboratories.

YANG, C. A. & WU, R. (1979). BK virus DNA: complete nucleotide sequence of a human tumour virus. *Science*, **206**, 456–62.

ZOUZIAS, D., PRASAD, L. & BASILICO, C. (1977). State of the viral DNA in rat cells transformed by polyoma virus. *Journal of Virology*, **24**, 142–50.

ADENO-ASSOCIATED VIRUS LATENT INFECTION

K. I. BERNS, A. K.-M. CHEUNG*, J. M. OSTROVE† AND M. LEWIS

Department of Immunology and Medical Microbiology, University of Florida College of Medicine, Gainesville, Florida 32610, USA

The adeno-associated viruses (AAV) are defective parvoviruses which require co-infection with a helper virus for a productive infection (Atchison, Casto & Hammon, 1965; Melnick, Mayor, Smith & Rapp, 1965; Hoggan, Blacklow & Rowe, 1966). The natural helper is adenovirus but several types of herpesviruses, including herpes simplex (HSV), varicella zoster, cytomegalovirus, and Epstein-Barr virus (Atchison, 1970; Blacklow, Hoggan & McClanahan, 1970; Blacklow, Dolin & Hoggan, 1971), have also been found to function as at least partial helpers in cell culture. Although different serotypes of AAV have been isolated from a wide range of host species including man, monkeys, cows, dogs and birds (Mayor, Jamison, Jordan & Melnick, 1965; Hoggan *et al.*, 1966; Dutta & Pomroy, 1967; Domoto & Yanagawa, 1969; Luchsinger *et al.*, 1970), it appears that a given AAV serotype can productively infect cells of any species as long as there is concomitant infection by a helper virus able to replicate in the target cells (Casto, Atchison & Hammon, 1967). In the absence of co-infection by a helper the AAV virion can adsorb to the cell and penetrate to the nucleus where the DNA is uncoated but there is no virus-specific DNA, RNA, or protein synthesis detectable by traditional methods (Rose & Koczot, 1972). In a productive co-infection with adenovirus the first AAV-specific macromolecule whose synthesis can be detected is DNA. This occurs approximately 7 h post-infection (Rose & Koczot, 1972) or shortly after the onset of adenovirus DNA synthesis. Thus, it would appear that AAV codes for no early functions and is totally dependent on helper virus and host cell functions for DNA replication. However, more recent data suggest

Current addresses:
* Marjorie B. Kovler Viral Oncology Laboratories, University of Chicago, Chicago, Illinois 60637.
† Department of Microbiology, Johns Hopkins University School of Medicine, Baltimore, Maryland 21205.

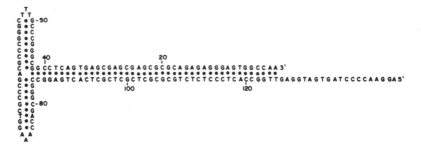

Fig. 1. Nucleotide sequence of the inverted terminal repetition in AAV 2 DNA shown in the folded configuration that permits maximum self base pairing.

that an AAV gene product may also be required to maintain and amplify DNA synthesis (Meyers & Carter, 1981) and that expression of this gene product is also dependent on the presence of helper virus. There have been several recent reviews of AAV biology and replication (Berns, 1974; Rose, 1974; Berns & Hauswirth, 1979).

Although the defective parvoviruses (AAV) are not readily distinguishable from the autonomous parvoviruses at the level of the intact virion (icosahedral, 18–24 nm diameter) or the number (3) and approximate sizes of the coat proteins, there are significant differences at the level of the genome. Both types have particles which contain linear single-stranded DNA with a molecular weight of 1.5–1.8×10^6 daltons (Crawford, Follett, Burdon, & McGeoch, 1969; Gerry, Kelly & Berns, 1973). Whereas the autonomous parvoviruses encapsidate only the plus strand, in the case of AAV strands of both polarities are encapsidated with equal frequency in different particles (Crawford *et al.*, 1969; Mayor, Torikai, Melnick, & Mandel, 1969; Rose, Berns, Hoggan & Koczot, 1969; Berns & Rose, 1970; Berns & Adler, 1972). The AAV genome has been the subject of detailed structural analysis. The pertinent finding with regard to latent infection is the arrangement of the nucleotide sequences at the ends of the DNA. AAV 2 DNA contains an inverted terminal repetition that is 145 nucleotides long (Lusby, Fife & Berns, 1980) (Fig. 1). The first 125 nucleotides of the terminal repetition form an overall palindromic or self-complementary sequence which can fold back on itself to form a base-paired hairpin structure. The structure is T or Y-shaped rather than a simple hairpin because the sequence involved is more complex than a simple palindrome. The first 41 nucleotides are perfectly complementary to nucleotides 85–125, in addition to which there are two independent palindromes of equal length between nucleotides 42–84. It is known that DNA replication begins nearly opposite the 3′ end of the parental strand (Hauswirth & Berns, 1977) and it is believed that the hairpin structure formed by folding over the first 125 nucleotides at the 3′ end of the parental strand serves as the primer for the initiation of DNA synthesis. The structure of the termini of AAV DNA also appears to be significant in the mechanism underlying the process of latent infection (Cheung, Hoggan, Hauswirth & Berns, 1980).

As a defective virus requiring a helper for productive infection, AAV is faced with an inherent problem *in vivo*: how to persist until the advent of the helper. This is apparently accomplished by establishing a latent infection. In the absence of helper co-infection, the AAV virion can penetrate to the nucleus of the cell where the DNA is uncoated and forms a stable relationship with the host genome. The viral genome can then be rescued from the latent state at some future time by co-infection of the cell with a competent helper virus. AAV latent infection was discovered by Hoggan, Thomas, Thomas & Johnson (1972) as a consequence of the programme in the United States to screen primary cell cultures to be used in vaccine production for the presence of endogenous viruses. None of the cell lots of human or monkey origin tested yielded infectious AAV or serological evidence for the presence of AAV proteins. However, when the primary cultures were challenged by infection with an appropriate adenovirus 1–2% of the human embryonic kidney cell lots and up to 20% of the lots of African green monkey kidney cells produced infectious AAV. Thus, AAV latent or cryptic infection appeared to be a rather common event in nature.

In order to be able to study in detail the mechanisms involved in the latent infection it was necessary to establish a model system in cell culture. This proved relatively easy to do. Infection of cells of human origin (Detroit 6, HeLa, or Hep-2) with a human serotype of AAV at a high multiplicity in the absence of helper virus allowed the ready detection of infectious AAV upon addition of adenovirus, even after many passages, and the subsequent isolation of stable clones of latently infected cells. After the initial infection, maintenance of the latent infection was not sensitive to AAV-neutralizing antibody and up to 30% of cloned cell cultures would yield AAV upon challenge by adenovirus (Hoggan *et al.*, 1972). Therefore, the concept that a sort of carrier culture had been established in which only a few cells were infected at any one time or that there was a constant yield of infectious virus at a low rate seemed unlikely. In the absence of helper challenge it was impossible to detect either infectious virus or viral antigen. Additionally, no evidence could be found for latent infection by either adenovirus or any of the herpesviruses (Berns, Pinkerton, Thomas & Hoggan, 1975).

Although it was relatively easy to establish such a latent infection in cell culture the efficiency with which this occurred was not possible to quantitate precisely. At a multiplicity of infection (m.o.i.) of 250 $TCID_{50}$ (tissue culture infectious doses/cell), the initial mass culture remained positive for AAV upon challenge with adenovirus for more than 100 passages. At passage 39 approximately 30% of the cells cloned were positive for a latent AAV infection. But the initial m.o.i. meant that the cells had each originally been exposed to at least several thousand particles, so that the efficiency of the system is not at all comparable, for example, to that seen with a temperate bacteriophage. Reduction of the m.o.i. by a factor of 100 still led to a detectable latent infection in cell culture, but AAV could not be detected by adenovirus challenge after 34 passages.

Detroit 6 cells which had been infected with AAV at a m.o.i. of 250 $TCID_{50}$ as described above were cloned after passage 39 and tested for AAV (Berns et al., 1975). One of the positive clones was recloned and two of the subclones utilized for further study to see if the presence of AAV DNA could be detected and the number of copies of the viral genome per cell determined. To do this the kinetics of reassociation in solution of a highly labelled viral DNA probe were determined in the manner first described by Gelb, Kohne & Martin (1971). This approach is based upon the fact that DNA renaturation is bimolecular and thus concentration dependent. The presence of even one or a few copies of the viral genome per cell will accelerate the rate of reannealing of a very small amount of highly radioactively labelled purified viral DNA added to a large amount of cellular DNA purified from latently infected cells. The increase in the rate of reannealing of the viral DNA used as a probe is a reflection of the number of copies of the viral genome in the cell. By means of this approach it was determined that there were 3–5 copies of the AAV genome per diploid amount of cell DNA in the two subclones.

Handa, Shiroki & Shimojo (1977) presented evidence that AAV DNA in latently infected HeLa cells was integrated into cellular

DNA. To test whether viral DNA was covalently associated with cellular DNA they made use of the 'network technique'. Total DNA was isolated from latently infected cells, denatured, and then allowed to reassociate for a very brief time. Under the conditions used, only highly repetitive sequences in cellular DNA can reanneal. Because these sequences are present in many thousands of copies throughout the genome, a large open network is formed which contains almost all of the cellular DNA. In high salt this network can then be pelleted by centrifugation at low speed. If the several copies of viral DNA present in latently infected cells are covalently asociated with the cellular DNA they will also pellet. On the other hand, if they exist in the cells as a plasmid they should remain in the supernatant fluid. In their experiments Handa *et al.* (1977) found that 80% of the cellular DNA pelleted and along with it 80% of the viral DNA sequences. Thus, at least 80% and perhaps all of the AAV sequences were covalently associated with cellular DNA, i.e. integration had occurred.

To characterize structural details of the latent AAV genome, Cheung *et al.* (1980) studied a subclone of latently infected Detroit 6 cells described above using the blotting technique of Southern (1975). Total DNA from latently infected cells was digested with various bacterial restriction endonucleases, the digestion products separated by electrophoresis in agarose gels which were then blotted to nitrocellulose paper and the positions of AAV-specific sequences detected by hybridization with highly radioactively labelled AAV virion DNA. Both an early passage of the clone (passage 9–10) and a late passage (118) were studied. When the cellular DNA either was not digested or was digested with a restriction enzyme which cleaved cellular sequences but had no recognition sequence within AAV DNA, none of the AAV sequences in the DNA from early passage cells migrated as free AAV DNA. In uncleaved DNA the AAV sequences all migrated as high molecular weight material. The apparent molecular weight of the AAV sequences was reduced when the cellular DNA was cleaved with enzymes which did not cut AAV DNA (*Hpa*I, *Pvu*I and *Bgl*II). Because only one band of AAV DNA could be detected after cleavage, it was concluded that the AAV DNA in early passage cells was covalently associated with cellular DNA and that the AAV DNA was located at only a very few, possibly only one, sites in the cellular DNA. In support of the notion that all of the AAV DNA was associated with cellular DNA the latter was precipitated using the technique of Hirt (1967). All of the AAV DNA sequences were found in the pellet, none was detectable in the supernatant fluid. When DNA from late passage cells was examined in the same manner, AAV sequences were found in the same high molecular weight species, seen in early passage DNA and were associated with cellular DNA. However, there was an additional band which represented a small fraction of the total AAV sequences seen in the blots and which migrated at the position of free AAV virion DNA. Further characterization with several restriction enzymes that cut within the AAV genome of the free AAV DNA seen in late passage cells demonstrated that, at this level of resolution, the free DNA was indistinguishable from virion DNA, both in size and in the overall orientation of the AAV DNA sequences. Thus, this free DNA apparently represented AAV sequences that had originally been integrated into cellular DNA (because there was no free AAV DNA in early passage cells) and had then 'popped out' in some manner with continued passage of the latently infected cells. Whether the free copies were the result of excision by recombination or of selective replication by copying of an integrated sequence is unknown. Further, it was not ascertained whether the AAV DNA present as a free copy in the latently infected cells could replicate or whether the low level of free DNA detected in the late passage cells represented a low level of 'popping out' with continued passage, followed by subsequent dilution of the free copies with cell division. Because of the defective nature of the AAV genome with regard to replication, the latter possibility would seem more likely.

After the demonstration that AAV DNA was covalently attached to cellular DNA in the latent state the next question was whether the linkage involved a specific site on the viral genome or was random as has been observed for the papovaviruses (Botchan, Topp & Sambrook, 1976; Ketner & Kelly, 1976). To determine this the DNA from early passage cells was cleaved with restriction enzymes which cut at three sites within AAV DNA (*Hinc*II and *Pst*I). Three types of fragments were observed. Both internal fragments were produced by cleavage with each enzyme. Some of the integrated sequences had lost one internal *Hinc*II site, but the fused fragment was of the expected size. No terminal fragments corresponding in size to those from virion DNA were observed. Instead all terminal sequences were found in two classes of larger fragments. The first class contained sequences from both termini and were apparently fusion fragments resulting from a head-to-tail tandem repeat of the

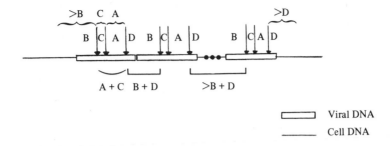

Fig. 2. Model for the integration of AAV DNA in Detroit 6 cells. Some of the copies of the genome appear to be tandemly repeated with no intervening sequences. Others have intervening sequences of unknown origin. The arrows represent *Hind*II cleavage sites along the AAV genome. B + D and >B + D represent fragments containing both B and D sequences. A + C represents a fragment equal in size to the sum of the A and C fragments and containing both A and C sequences.

viral genome. In some cases the fusion fragments were almost exactly the size expected from joining of the two terminal fragments and in other cases the fusion fragments were larger than this so that sequences in addition to those in the two normal terminal fragments must have been present. Whether this additional DNA represented viral or interposed cellular sequences is not known. The second class of terminal fragments contained sequences from only one end of the genome or the other in a species larger than a normal terminal fragment. It was concluded that these species represented fragments containing joints between cellular and viral DNA. Sequences from both the right and left ends of the genome were found in separate species of this type. For both internal and terminal fragments hybridization probes containing the specific sequences involved were used. Because internal sequences were never detected as being joined to nonviral sequences, but terminal sequences were, and because of fusion of terminal fragments from both ends of the genome, it was concluded that multiple copies of the AAV genome were integrated into the cellular DNA as a head-to-tail tandem repeat(s) and that joining between viral and cellular sequences specifically involved the termini of the viral DNA (Fig. 2). It is important to note that although the evidence strongly supported the notion that AAV DNA is integrated into cellular DNA in latently infected cells, the data did not allow the distinction to be made between incorporation into chromosomal DNA or into a large plasmid composed of both cellular and viral DNAs.

In all of the experiments cited above identical results were obtained for integrated AAV sequences in cells from both the early and late passages of the clone investigated. Because the data derived from this clone were consistent with the integration of the several copies of the AAV genome at a single site in the cellular genome, it was necessary to repeat most of these experiments using a second independently derived clone (M. Lewis, M. D. Hoggan & K. I. Berns, unpublished observations). Again evidence was obtained for integration of several copies of the viral DNA as a head-to-tail tandem repeat and again the linkage of viral to cellular DNA involved the termini of the viral DNA. However, the species containing viral sequences from one end of the DNA, presumably joined to cellular sequences, differed in size from those observed with the initial clone investigated. Therefore, at this level of discrimination the sites of integration apparently differed from clone to clone. Whether integration sites have common sequences over a more limited region will hopefully be determined as a result of ongoing experiments in which the joint fragments are being cloned in *E. coli*.

Cleavage of DNA with several restriction enzymes that cleave within the AAV genome (*Hind*III, *Eco*RI, *Bam*HI, *Sal*I, *Hinc*II and *Pst*I) gave identical results for the integrated AAV DNA from both early and late passage cells of the initial clone investigated. However, quite different results were obtained when DNAs from the early and late passages were digested with *Sma*I. *Sma*I only cleaves AAV DNA near the middle of the terminal palindromic sequences. In this case two of the four bands containing AAV sequences observed in digests of early passage DNA had significantly reduced electrophoretic mobilities when late passage DNA was digested with *Sma*I. Apparently two *Sma*I sites present in early passage DNA were lost upon prolonged passage. This suggested a dynamic state for the terminal AAV sequences and suggests continued recombination at these sites during prolonged passage. Such activity would support the hypothesis that integration specifically involves the ends of the viral genome because of the special structure of the terminal sequences. The data also are in accord with the notion that the intact free viral DNA observed in late passage cells might have been a consequence of this type of recombination.

The steps in the pathway leading to a latent infection are unknown. One obvious question is whether the single-stranded AAV genome is converted to a duplex structure prior to integration. This seems to be a likely step although the possibility of recombination between the single-stranded form of the viral genome and cellular DNA followed by gap repair synthesis cannot be ruled out *a priori*. Two possible routes exist for conversion of the virion DNA to a double-stranded structure prior to integration. Because strands of both polarities are encapsidated into AAV virions, it is conceivable that after uncoating in the nucleus the strands could anneal to form duplex DNA. The high m.o.i. required to establish latent infection in a significant fraction of the cells in a culture would be consistent with this notion. The other possibility is that the viral genome, which is self-priming in *in vitro* assays by virtue of its 3' terminal hairpin structure (W. W. Hauswirth & K. I. Berns, unpublished data), could undergo one round of synthesis to produce a complementary strand even in the absence of helper virus co-infection. This limited extent of DNA synthesis would not be detected by traditional assays. In order to test the possibility of complementary strand synthesis AAV virions were produced in which the DNA was radio-labelled with ^{32}P and density-labelled by substitution of bromodeoxyuridine for thymidine (R. Bohenzky, J. M. Ostrove & K. I. Berns, unpublished data). HeLa cells were infected with the heavy virus and the fate of the radioactive high-density parental DNA followed. At various times after infection total DNA was extracted and banded in isopycnic CsCl gradients. A shift of the parental radio-label to low density could imply either integration of intact viral genomes or breakdown of the input viral DNA with subsequent utilization of the breakdown products for cellular DNA synthesis. Viral complementary strand DNA synthesis would be indicated by banding of the viral radio-label at an intermediate- or hybrid-density position in the gradient. At 24 h post-infection the radio-label was found only at the original high-density position and at the light position corresponding to cellular DNA. By 36 h post-infection, however, there was a band of parental radio-label at the hybrid-density position in the gradients, suggesting that complementary strand synthesis had occurred. To confirm the impression that material banding at the high- and hybrid-density positions was indeed extra-chromosomal, the high molecular weight chromosomal DNA was pelleted by sedimentation through a linear sucrose gradient. The material which had not pelleted was only of high or hybrid density. Because of the use of the 3' terminal hairpin structure in AAV DNA as a primer for replication, replicative intermediates have been identified in which the parental and progeny strands are covalently cross-linked, presumably by the terminal hairpin (Straus, Sebring & Rose, 1976; Hauswirth & Berns, 1979). The cross-linking is assayed by the ability of such molecules to 'snap back' to a duplex structure after denaturation. When tested in this way the hybrid-density material containing parental radio-label demonstrated the level of 'snap back' characteristic of authentic replicative intermediates. The extent of 'snap back' observed was much greater than seen in the high-density material which presumably represented unreplicated parental DNA. Thus, it does appear that input AAV DNA can undergo one round of complementary strand DNA synthesis in the absence of helper virus co-infection. This result would support the notion that integration of AAV DNA into cellular DNA involves duplex viral DNA.

The mechanism by which the AAV genome is rescued from the latent state upon superinfection by a helper virus is equally unclear. Both adenovirus and herpesviruses are able to rescue AAV (Hoggan et al., 1972). Whether rescue involves the excision of a copy of the AAV genome followed by replication or whether a free copy is replicated using an integrated genome as template (Botchan, Topp & Sambrook, 1979) is unknown. Because it is the natural helper virus and is smaller and, thus, better characterized, emphasis has been given to the study of adenovirus mutants in an effort to determine the nature of the helper function(s). From studies involving micro-injection of adenovirus mRNAs into Vero cells, it appears that all of the helper function(s) are coded for by adenovirus early genes (Richardson, Carter & Westphal, 1980). No adenovirus temperature sensitive mutant, even if DNA minus, has been found to be defective in supporting AAV DNA replication, either in co-infection or in rescue of AAV from the latent state (Straus, Ginsberg & Rose, 1976; Mayor & Young, 1978; Ostrove & Berns, 1980). Recently, adenovirus host range mutants have been described (Harrison, Graham & Williams, 1977) which can replicate in adenovirus-transformed human cells (293 cells) but not in HeLa cells. These host range mutants are defective either in early region Ia or early region Ib. The required functions are provided by expression of those regions by the portion of the adenovirus genome integrated into the DNA of the permissive 293 cells. At a low m.o.i. the host range mutants in early region Ia can neither rescue AAV from the latent state nor support AAV DNA replication except in an exogenous co-infection of 293 cells (the 293 cells do not help AAV DNA replication in the absence of exogenous adenovirus co-infection, Ostrove & Berns, 1980). The host range mutants in early region Ib are DNA positive, but have decreased late protein synthesis and greatly reduced virion production. They are also deficient in synthesis of the 58K adenovirus t-antigen which has been associated with cellular transformation. These mutants can neither abortively nor stably transform cells in culture even at a high m.o.i. (>5) (Graham, Harrison & Williams, 1978). In exogenous co-infections the adenovirus host range mutants in early region Ib do support AAV DNA replication. However, these mutants are unable to rescue AAV from the latent state in the sense that superinfection of AAV latently infected Detroit 6 cells by these viruses does not lead to replication of AAV DNA in the free state. If there is co-infection of the latently infected Detroit 6 cells by both the early region Ib host range mutant and AAV, then AAV DNA synthesis is observed. Thus, it appears that there is a specific adenovirus function(s) required for rescue of AAV from the latent state and this function(s) is distinguishable from those adenovirus functions required to help AAV DNA replication. There are at least two gene products coded for by early region Ib, but the predominant product would appear to be the 58K t-antigen. Therefore, it appears quite possible that this protein supplies the helper function required for AAV rescue. It would be interesting if a gene product required for the establishment and possibly also the maintenance of adenovirus transformation were also to be involved in rescue of AAV from the latent state. This type of correlation may prove useful in finally elucidating the function of the 58K t-antigen. These data are reminiscent of the observation that an intact A gene (an early gene, required for transformation) is also required for rescue of integrated polyoma virus DNA (Basilico, Gattoni, Zouzias & Valle, 1979).

The possible association between AAV latent infection and an adenovirus tumour antigen may be related to the ability of AAV to inhibit the oncogenicity of both adenovirus and HSV in animals (Casto & Goodheart, 1972; Kirschstein, Smith & Peters, 1968; Blacklow, Cukor, Kibrick & Quinman, 1978; Mayor, Houlditch & Mumford, 1973). It has proven possible to establish a model system in which the molecular mechanisms involved in this inhibition can be studied (J. Ostrove, D. Duckworth, K. Berns, 1981). Hamster embryo cells transformed by Ad5 (H14b cells) induce tumour formation when inoculated into newborn Syrian Hamsters. Infection of these cells in culture by AAV greatly reduces the ability of the cells to induce tumours. The incidence of tumours is reduced, the induction time is increased, and the volume of those tumours induced by the AAV-infected cells is decreased by 1000 fold. *In vitro* the AAV-infected H14b cells have a decreased saturation density and increased anchorage dependence. A comparison of Southern (1975) blots of DNA after restriction enzyme digestion from control H14b cells or those infected with AAV showed no difference in the patterns of integrated adenovirus DNA sequences (H14b cells contain 5.5 copies of the left 40% of the adenovirus genome). Analysis by northern blots (Alwine, Kemp & Stark, 1977) of adenovirus-specific mRNAs from the two types of cells showed no difference in the major poly-A-containing RNA species, but there was a consistent alteration in some very faint high molecular weight species. The major change observed was in the amount of adenovirus-specific proteins after immunoprecipitation. The amount of the 58K t-antigen was reduced by 80% in extracts from AAV-infected H14b cells as compared to control cells. Thus, it would seem that the AAV inhibition of adenovirus-induced oncogenicity is closely connected to the decrease of the 58K t-antigen. Whether the decrease in the 58K t-antigen was a consequence of an inhibition of RNA processing, a decrease in RNA stability, an effect upon translation or represents a direct interaction between AAV DNA and the 58K t-antigen is unclear. What does seem likely is that there is a close association between AAV and the 58K t-antigen in the cases of at least two biological phenomena: rescue of AAV from the latent state and inhibition by AAV of adenovirus oncogenicity.

From the point of view of the virus the biological significance of AAV latent infection seems quite straightforward. By means of the latent infection the AAV genome can maintain its biological continuity as an integral part of the host genome. Upon superinfection by a competent helper virus the AAV genome can be actively replicated, encapsidated, and then transmitted to other cells and intact hosts. On the other hand the effects on, or consequences for, the host of both productive and latent infections by AAV are very uncertain. No known disease is associated with AAV infection of either humans or lower animals. AAV infection is very common in people. In the United States more than 80% of the population is seropositive for AAV antibodies but infectious AAV has only been isolated from individuals in whom there is evidence for a concomitant adenovirus infection (Blacklow, Hoggan & Rowe, 1968). It seems likely that a latent infection with AAV might result from many of the initial active infections in much the same way that this occurs with adenovirus infections as well as with infections by most of the herpesviruses. No serious epidemiological survey to answer this question has ever been undertaken but the data derived by Hoggan *et al.* (1972) in the vaccine screening trials in which they discovered AAV latent infection would tend to support this possibility. As cited above, these authors found that 20% of the lots of primary African green monkey kidney cells which were tested readily yielded infectious AAV upon challenge with adenovirus. The fact that only 1–2% of the human embryonic kidney cell lots tested were positive may simply reflect the possibility that latent infection of the foetus with AAV is the result of a primary maternal infection during gestation.

With the likelihood that AAV latent infection may be quite common *in vivo*, the potential consequences of such a latent infection become even more intriguing. One interesting possibility is that a latent AAV infection may actually have some sort of protective effect upon the host. The ability to demonstrate in the laboratory that AAV can inhibit the oncogenicity of both adenovirus and HSV leads to the question of whether this phenomenon occurs in nature. In the case of adenovirus this question may be moot for humans since no causal relationship has ever been demonstrated between adenovirus and human cancer. However, several herpesviruses have been epidemiologically associated with human cancers; in particular HSV Type 2 has been associated with cancers of the cervix and prostate. Therefore, Mayor, Drake, Stahmann & Mumford (1976) conducted a seriological survey of patients with these two diseases at the M. D. Anderson Hospital in Houston, Texas. Levels of antibodies to both HSV and AAV were measured in these patients and a matched control group. Both groups were highly seropositive (>85%) for HSV but, whereas >90% of the control group were seropositive for AAV, only 14% of the cancer patients had antibodies against AAV. Similar results have been obtained by other groups (Sprecher-Goldberger *et al.*, 1971). These results are consistent with the hypothesis that AAV latent infection may be protective against cancers induced by some viruses. Data in cell culture indicate that in order for lymphocytes latently infected with EBV to become transformed the latent viral genome must be activated so that infectious extracellular virions are produced. These virions then cause transformation by reinfection.

Hypothetically, herpesvirus infection of a cell latently infected with AAV would rescue the AAV. In turn the infectious AAV produced could possibly inhibit the oncogenicity of the herpesvirus.

In summary, AAV latent infections in nature seem to be common. It has been possible to establish a model system in cell culture which makes it possible to probe the molecular mechanisms of the latent infection. In some ways the model system developed for AAV latent infection may more closely mimic the natural situation than is currently possible with any of the other DNA viruses. This is a consequence of the defectiveness of AAV. All of the animal DNA viruses which replicate in the nucleus of the cell can cause latent infections in the intact animal. But it is very difficult to establish model systems in cell cultures for most of these viruses because in culture the cells from the natural host are permissive for lytic infection. Thus non-permissive cells from other species must be used and the parallels that may be drawn with the situation *in vivo* are open to question. In the case of AAV the cells of the natural host are nonpermissive for lytic infection in the absence of helper virus so that a latent infection may be readily established in cell culture under conditions which may well mimic those *in vivo*. The efficiency with which the latent AAV in these cells is rescued by superinfection with adenovirus supports this idea. By use of this model system it has been possible to demonstrate that in latent infection the AAV genome is integrated as a head-to-tail tandem repeat into cellular DNA and that the joint between viral and cellular DNAs involves the termini of the AAV genome. Further it has been possible to demonstrate that an early region Ib gene product is required for adenovirus to rescue AAV and the adenovirus 58K t-antigen has been implicated in this process.

We wish to thank Drs W. W. Hauswirth and G. E. Gifford for critical reading of the manuscript.

Some of the work cited was supported by National Science Foundation research grant PCM 79–09354 and National Institutes of Health research grant 1 RO1 AI 16326.

J. M. Ostrove was supported by USPHS Training Grant 1 T32 AI0 7110. M. Lewis was supported by a National Institute of Health Biomedical Research Grant #2S07–RR05362.

REFERENCES

ALWINE, J. C., KEMP, D. J. & STARK, G. R. (1977). Method for detection of specific RNAs in agarose gels by transfer to diazobenzyloxy-methyl-paper and hybridization with DNA probes. *Proceedings of the National Academy of Science, USA*, **74**, 5350–4.

ATCHISON, R. W. (1970). The role of herpesviruses in adenovirus-associated virus replication *in vitro*. *Virology*, **42**, 155–62.

ATCHISON, R. W., CASTO, B. C. & HAMMON, W.McD. (1965). Adenovirus-associated defective virus particles. *Science*, **194**, 754–6.

BASILICO, C., GATTONI, J., ZOUZIAS, D. & VALLE, G. D. (1979). Loss of integrated viral DNA sequences in polyoma-transformed cells is associated with an active viral A function. *Cell*, **17**, 645–59.

BERNS, K. I. (1974). Molecular biology of adeno-associated viruses. *Current Topics in Microbiology and Immunology*, **65**, 1–21.

BERNS, K. I. & ADLER, S. (1972). Separation of two types of adeno-associated virus particles containing complementary polynucleotide chains. *Journal of Virology*, **9**, 394–6.

BERNS, K. I. & HAUSWIRTH, W. W. (1979). Adeno-associated viruses. *Advances in Virus Research*, **25**, 407–50.

BERNS, K. I., PINKERTON, T. C., THOMAS, G. F. & HOGGAN, M. D. (1975). Detection of adeno-associated virus (AAV)–specific nucleotide sequences in DNA isolated from latently infected Detroit 6 cells. *Virology*, **68**, 556–60.

BERNS, K. I. & ROSE, J. A. (1970). Evidence for a single-stranded adenovirus-associated virus genome: isolation and separation of complementary single strands. *Journal of Virology*, **5**, 693–9.

BLACKLOW, N. R., CUKOR, G. KIBRICK, S. & QUINMAN, G. (1978). Interactions of adeno-associated viruses with cells transformed by herpes simplex virus. In *Replication of Mammalian Parvoviruses*, ed. D. C. Ward & P. Tattersall, pp. 87–98. Cold Spring Harbor Laboratory.

BLACKLOW, N. R., DOLIN, R. & HOGGAN, M. D. (1971). Studies of the enhancement of an adenovirus-associated virus by herpes simplex virus. *Journal of General Virology*, **10**, 29–36.

BLACKLOW, N. R., HOGGAN, M. D. & McCLANAHAN, M. S. (1970). Adenovirus-associated viruses: enhancement by human herpes viruses. *Proceedings of the Society of Experimental Biology and Medicine (NY)*, **134**, 952–4.

BLACKLOW, N. R., HOGGAN, M. D. & ROWE, W. P. (1968). Serologic evidence for human infection with adenovirus-associated viruses. *Journal of the National Cancer Institute*, **40**, 319–27.

BOTCHAN, M., TOPP, W. & SAMBROOK, J. (1976). The arrangement of simian virus 40 sequences in transformed cell DNA. *Cell*, **9**, 269–87.

BOTCHAN, M., TOPP, W. & SAMBROOK, J. (1979). Studies on SV40 excision from cellular chromosomes. In *Cold Spring Harbor Symposia on Quantitative Biology*, vol. 43, pp. 709–19. Cold Spring Harbor Laboratory.

CASTO, B. C., ATCHISON, R. W. & HAMMON, W.McD. (1967). Studies on the relationship between adeno-associated virus type 1 (AAV-1) and adenovirus. I. Replication of AAV-1 in certain cell cultures and its effect on helper adenovirus. *Virology*, **32**, 52–9.

CASTO, B. C. & GOODHEART, C. R. (1972). Inhibition of adenovirus transformation *in vitro* by AAV-1. *Proceedings of the Society for Experimental Biology and Medicine (NY)*, **140**, 72–8.

CHEUNG, A. K.-M., HOGGAN, M. D., HAUSWIRTH, W. W. & BERNS, K. I. (1980). Integration of the adeno-associated virus genome into cellular DNA in latently infected human Detroit 6 cells. *Journal of Virology*, **33**, 739–48.

CRAWFORD, L. V., FOLLETT, E. A. C., BURDON, M. G. & MCGEOCH, D. J. (1969). The DNA of a minute virus of mice. *Journal of General Virology*, **4**, 37–46.

DOMOTO, K. & YANAGAWA, R. (1969). Properties of a small virus associated with infectious canine hepatitis virus. *Japanese Journal of Veterinary Research*, **17**, 32–41.

DUTTA, S. K. & POMROY, B. C. (1967). Electron microscope studies of quail bronchitis virus. *American Journal of Veterinary Research*, **28**, 296–9.

GELB, L. D., KOHNE, D. E. & MARTIN, M. A. (1971). Quantitation of simian virus 40 sequences in African green monkey, mouse, and virus-transformed genomes. *Journal of Molecular Biology*, **57**, 129–45.

GERRY, N. W., KELLY, T. J., JR & BERNS, K. I. (1973). Arrangement of nucleotide sequences in adeno-associated virus DNA. *Journal of Molecular Biology*, **79**, 207–25.

GRAHAM, F. L., HARRISON, T. & WILLIAMS, J. (1978). Defective transforming capacity of adenovirus type 5 host-range mutants. *Virology*, **86**, 10–21.

HANDA, H., SHIROKI, K. & SHIMOJO, H. (1977). Establishment and characterization of KB cell lines latently infected with adeno-associated virus type 1. *Virology*, **82**, 84–92.

HARRISON, T., GRAHAM, F. L. & WILLIAMS, J. (1977). Host range mutants of adenovirus type 5 defective for growth in HeLa cells. *Virology*, **77**, 319–29.

HAUSWIRTH, W. W. & BERNS, K. I. (1977). Origin and termination of adeno-associated virus DNA replication. *Virology*, **78**, 488–99.

HAUSWIRTH, W. W. & BERNS, K. I. (1979). Adeno-associated virus DNA replication: non unit-length molecules. *Virology*, **93**, 57–68.

HIRT, B. (1967). Selective extraction of polyoma DNA from mouse cell cultures. *Journal of Molecular Biology*, **26**, 365–9.

HOGGAN, M. D., BLACKLOW, N. R. & ROWE, W. P. (1966). Studies of small DNA viruses found in various adenovirus preparations: physical, biological and immunological characteristics. *Proceedings of the National Academy of Science, USA*, **55**, 1467–71.

HOGGAN, M. D., THOMAS, G. F., THOMAS, F. B. & JOHNSON, F. B. (1972). Continuous 'carriage' of adenovirus associated virus genome in cell cultures in the absence of helper adenoviruses. In *Proceedings of the Fourth Lepetit Colloquium, Cocoyac, Mexico*, pp. 243–9. North Holland Publishing Company, Amsterdam.

KETNER, G. & KELLY, T. J., JR (1976). Integrated simian virus 40 sequences in transformed cell DNA: analysis using restriction endonucleases. *Proceedings of the National Academy of Science, USA*, **73**, 1102–6.

KIRSCHSTEIN, R. L., SMITH, K. O. & PETERS, E. A. (1968). Inhibition of adenovirus 12 oncogenicity by adeno-associated virus. *Proceedings of the Society for Experimental Biology and Medicine (NY)*, **128**, 670–3.

LUCHSINGER, E., STROBBE, R., WELLEMANS, G., KEKEGEL, D. & GOLDBERGER, S. (1970). Hemagglutinating adeno-associated virus (AAV) in association with bovine adenovirus type 1. *Archive Gesamte Virusforschung*, **31**, 390–2.

LUSBY, E., FIFE, K. H. & BERNS, K. I. (1980). Nucleotide sequence of the inverted terminal repetition in adeno-associated virus DNA. *Journal of Virology*, **34**, 402–9.

MAYOR, H. D., DRAKE, S., STAHMANN, J. & MUMFORD, D. M. (1976). Antibodies to adeno-associated satellite virus and herpes simplex in sera from cancer patients and normal adults. *American Journal of Obstetrics and Gynecology*, **126**, 100–4.

MAYOR, H. D., HOULDITCH, G. S. & MUMFORD, D. M. (1973). Influence of adeno-associated satellite virus on adenovirus-induced tumours in hamsters. *Nature (London) New Biology*, **241**, 44–6.

MAYOR, H. D., JAMISON, R. M., JORDAN, L. E. & MELNICK, J. L. (1965). Structure and composition of a small particle prepared from a simian adenovirus. *Journal of Bacteriology*, **90**, 235–42.

MAYOR, M. D., TORIKAI, K., MELNICK, J. L. & MANDEL, M. (1969). Plus and minus single-stranded DNA separately encapsidated in adeno-associated satellite virions. *Science*, **166**, 1280–2.

MAYOR, H. D. & YOUNG, J. F. (1978). Complementation of adeno-associated virus by temperature-sensitive mutants of human adenovirus and herpesvirus. In *Replication of Mammalian Parvoviruses*, ed. D. C. Ward & P. Tattersall, pp. 109–18. Cold Spring Harbor, New York.

MELNICK, J. L., MAYOR, H. D., SMITH, K. I. & RAPP, R. (1965). Association of 20 millimicron particles with adenoviruses. *Journal of Bacteriology*, **90**, 271–4.

MEYERS, M. W. & CARTER, B. J. (1981). Adeno-associated virus replication. The effect of L-canavanine or a helper virus mutation on accumulation of viral capsids and progeny single-stranded DNA. *Journal of Biological Chemistry*, (in press).

OSTROVE, J. M. & BERNS, K. I. (1980). Adenovirus early region Ib gene function required for rescue of latent adeno-associated virus. *Virology*, **104**, 502–5.

OSTROVE, J. M., DUCKWORTH, D. H. & BERNS, K. I. (1981). Inhibition of adenovirus-transformed cell oncogenicity by adeno-associated virus. *Virology*, **113**, 521–33.

RICHARDSON, W. D., CARTER, B. J. & WESTPHAL, H. (1980). Vero cells injected with adenovirus type 2 mRNA produce authentic viral polypeptide patterns: early mRNA promotes growth of adenovirus-associated virus. *Proceedings of the National Academy of Science, USA*, **77**, 931–5.

ROSE, J. A. (1974). Parvovirus reproduction. *Comprehensive Virology*, **4**, 1–21.

ROSE, J. A., BERNS, K. I., HOGGAN, M. D. & KOCZOT, F. J. (1969). Evidence for a single-stranded adenovirus-associated virus genome: formation of a DNA density hybrid on release of viral DNA. *Proceedings of the National Academy of Science, USA*, **64**, 863–9.

ROSE, J. A. & KOCZOT, F. (1972). Adenovirus-associated virus multiplication. VII. Helper requirement for viral deoxyribonucleic acid and ribonucleic acid synthesis. *Journal of Virology*, **10**, 1–8.

SOUTHERN E. (1975). Detection of specific sequences among DNA fragments separated by gel electrophoresis. *Journal of Molecular Biology*, **98**, 503–18.

SPRECHER-GOLDBERGER, S., THIRY, L., LEFEBRE, N., DEKEGEL, D. & DE HALLEUX, F. (1971). Complement fixation antibodies to adenovirus-associated viruses, adenoviruses, cytomegaloviruses, and herpes simplex viruses in patients with tumours and in control individuals. *American Journal of Epidemiology*, **94**, 351–8.

STRAUS, S. E., GINSBERG, H. S. & ROSE, J. A. (1976). DNA-minus temperature-sensitive mutants of adenovirus type 5 help adenovirus-associated virus replication. *Journal of Virology*, **17**, 140–8.

STRAUS, S. E., SEBRING, E. D. & ROSE, J. A. (1976). Concatemers of alternating plus and minus strands are intermediates in adenovirus-associated virus DNA synthesis. *Proceedings of the National Academy of Science, USA*, **73**, 742–6.

THE PERSISTENCE OF RETROVIRUSES

ROBIN A. WEISS

Chester Beatty Research Institute, Fulham Road,
London SW3 6JB, UK

INTRODUCTION

Retroviruses are small RNA viruses that replicate via the formation of an intracellular DNA provirus. The retroviruses are classified into three subfamilies, the *Oncovirinae* or RNA tumour viruses, the *Lentivirinae* or slow viruses of which visna-maedi virus is the only well studied example, and *Spumavirinae* or foamy viruses. The numerous isolates of RNA tumour viruses are further subdivided by morphology into B-type, C-type and D-type particles, and are named either by host and disease (e.g. feline leukaemia virus) or by the discoverer (e.g. Rous sarcoma virus, Friend virus). Retrovirus infections are widespread among vertebrate hosts, especially the RNA tumour viruses which have been described in fish, reptiles, birds and mammals. Insects may also play host to retrovirus infection, and the transmission of equine infectious anaemia virus is probably arthropod-borne. There is a vast literature on retroviruses which will not be reviewed in detail here. However, a comprehensive treatise on retroviruses has recently been published (Weiss, Teich, Varmus & Coffin, 1981).

Persistent infections are a way of life for retroviruses. Their replication is not usually cytocidal so that infection is not necessarily pathogenic. Retroviruses have a diploid, single-stranded RNA genome which is used as a template by the viral enzyme reverse transcriptase to form a haploid, double-stranded DNA provirus in the infected cell. The provirus has repeated sequences at each end of the genome known as long terminal repeats whose structure resembles that of transposable genetic elements (Temin, 1980). Like transposable genetic elements, lysogenic bacteriophages and many DNA tumour viruses, retroviruses integrate into the chromosomal DNA of the host, and the site of provirus insertion is not specific with respect to the host. Thus, the viral genome may persist as a set of 'host' genes. The most extreme form of persistence is represented by the 'endogenous' C-type and B-type viruses which are transmitted through germ cells as heritable proviruses.

The persistent infection of cells by retroviruses is either productive or non-productive. Progeny virions are produced by budding from the plasma membrane. Non-productive infection may result from the incomplete expression of the virus or because the viral genome is defective and is unable to code for all the viral proteins and functions. The retrovirus genome comprises three genes in the order 5'-*gag-pol-env*-3' (Baltimore, 1975). The *gag* gene codes for a polypeptide precursor which is cleaved to form the virion core proteins; *pol* codes for reverse transcriptase; and *env* codes for a precursor to the sulphydryl-linked membrane or envelope proteins, one or both of which are glycosylated. Defective viruses have deletions in one or more of these three structural genes, often because host genes with oncogenic potential have been substituted in their place, like transducing bacteriophages. Such defective viruses can be propagated by non-defective 'helper' viruses which supply the missing proteins through phenotypic mixing. Defective viruses may be 'rescued' in infectious form by helper viruses after many cell generations of non-productive infection. Mixed infections of related retrovirus strains may also result in the generation of viral recombinants. Phenotypic mixing and recombination can lead to the production of antigenically changed viruses in double infections, including combinations of features coded by infecting and endogenous viral genes, as will be discussed below.

PERSISTENCE OF RETROVIRUSES IN HOST POPULATIONS

Horizontal and vertical transmission

The natural transmission of retroviruses occurs via three distinct routes, as exemplified in Fig. 1 for avian lymphoid leukaemia viruses. 1. Horizontal infection occurs through contact with hosts shedding virus, and in chickens usually causes a transient viraemia followed by an immune response, mainly the production of neutralizing antibodies reacting with the major envelope glycoprotein (Rubin, Fanshier, Cornelius & Hughes, 1962; Weyl & Dougherty, 1977). 2. Vertical transmission by congenital infection via the egg albumin is also commonly found in chicken flocks (Rubin *et al.*, 1962; Dougherty & Distefano, 1967). In this case virus infection occurs early in embryogenesis, apparently with no ill effect on development. The hatchlings remain viraemic throughout life, with viral replication in all tissues except neural tissue, and the birds

HORIZONTAL

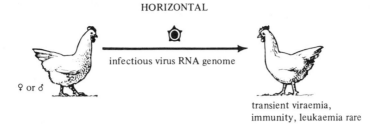

infectious virus RNA genome

♀ or ♂

transient viraemia,
immunity, leukaemia rare

VERTICAL

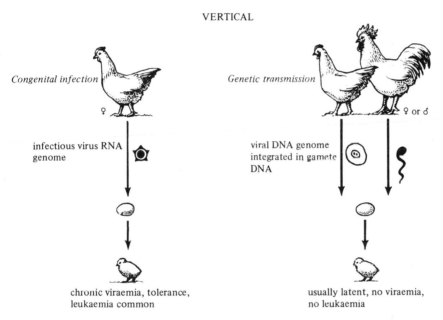

Congenital infection

♀

infectious virus RNA
genome

Genetic transmission

♀ or ♂

viral DNA genome
integrated in gamete
DNA

chronic viraemia, tolerance,
leukaemia common

usually latent, no viraemia,
no leukaemia

Fig. 1. Transmission of C-type retroviruses among chickens.

remain immunologically tolerant to viral antigens. Such birds have a
high incidence of lymphoid leukaemia. Although they lay fewer
eggs (Spencer, Gavora & Gowe, 1980), they do breed and thus pass
the virus congenitally from one generation to the next by maternal
transmission. Moreover, the viraemic, tolerant birds also act as
reservoirs for horizontal contact transmission to their flock-mates
because they chronically secrete virus into saliva, faeces and semen.
3. A distinct mode of vertical transmission is the inheritance of the
provirus through the germ line, when it is called an endogenous

virus. The leukaemogenic avian viruses are not transmitted genetically, but closely related sets of non-pathogenic viral genomes, with
distinct envelope antigenicity, are inherited in this way (Payne &
Chubb, 1968; Weiss & Payne, 1971; Astrin *et al.*, 1980).

The RNA tumour viruses of mammalian species also show these
three modes of transmission. The classical case of vertical transmission of infectious virus is the milk factor or mammary tumour virus
of C3H mice (Bittner, 1936). In GR strain mice, however, a very
similar mammary tumour virus is inherited as an integrated provirus
through the germ line (Bentvelzen & Daams, 1969). High
leukaemia strain laboratory mice, such as AKR, also transmit viral
genomes as germinal proviruses (Rowe, 1973) while a C-type virus
causing lymphoma and paralysis in wild mice (Gardner, Estes,
Casagrande & Rasheed, 1980) is transmitted by congenital infection. In cats, on the other hand, the virus causing feline leukaemia is
typically transmitted horizontally, and unlike chickens, infection of
adult cats occasionally leads to leukaemia or anaemia (Jarrett *et al.*,
1973; Essex, 1975; Jarrett, 1980). The feline leukaemia virus can
also be transmitted transplacentally, and is associated with abortion,
fetal resorption or runting, so that this is not a major route of
continuing transmission. Bovine leukosis virus is also transmitted
horizontally as well as through milk (Miller & Van der Maaten,
1979). In gibbons, the leukaemia virus is transmitted horizontally
and normally elicits a host immune response, but it is not uncommon for persistent viraemia to set in leading to chronic myelogenous
or lymphatic leukaemia (Kawakami & McDowell, 1980).

The transmission of lentiviruses and spumaviruses has not been
studied as extensively as that of oncoviruses. Visna virus spreads
horizontally through previously unexposed flocks of sheep, especially in the pneumonic (maedi) form of the disease, but the major
route of transmission in endemic flocks is from ewe to lamb by virus
secreted into the colostrum and milk (de Boer & Houwers, 1979).
Little is known about foamy virus transmission. Nine serotypes of
simian foamy or syncytial viruses have been described, as well as
bovine, feline and human isolates (Hooks & Gibbs, 1975). While
one school (Nemo, Brown, Gibbs & Gajdusek, 1978) has suggested
that human foamy viruses may be rare cases of infection by a simian
virus, recent seroepidemiological evidence indicates that infection is
widespread in East Africa (Muller *et al.*, 1980) and in Pacific islands
where there are no simians (Loh, Matsuura & Mizumoto, 1980).
Infection is thought to be horizontal, but vertical transmission has

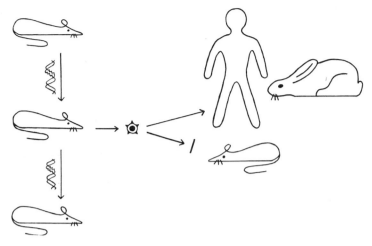

Fig. 2. Transmission of xenotropic C-type retrovirus in the mouse. The viral genome is perpetuated as a chromosomal element in the mouse, but it can be activated to synthesize virions. These virions cannot reinfect and replicate in mouse cells but can infect cells of 'foreign' species such as human or rabbit.

not been ruled out. There is no evidence of genetic transmission by lenti and foamy viruses.

Xenotropic endogenous viruses

Most genetically transmitted, endogenous retroviruses remain latent but can be activated to produce infectious progeny by a variety of inducing agents (Rowe *et al.*, 1971; Weiss, Friis, Katz & Vogt, 1971). The host range of many endogenous viruses is peculiar in that the virus is unable to re-infect the host in which the viral genome is transmitted as a genetic trait although it can infect cells of foreign species. Levy (1978) termed such viruses xenotropic. For example, xenotropic C-type viruses of mice cannot re-infect mouse cells but replicate to high titre in human, mink, rabbit or duck cells (Fig. 2). The infection of mouse cells by xenotropic viruses is restricted both at the receptor level and, if that is by-passed by phenotypic mixing, at a post-penetration stage of replication (Ishimoto, Hartley & Rowe, 1977; Levy, 1978).

Thus, xenotropic viruses are not able to spread horizontally from cell to cell within the host; with the exception of a few inbred strains that constitutively express high titres (Line 100 chickens, Crittenden, Smith, Weiss & Sarma, 1974; New Zealand Black mice, Levy,

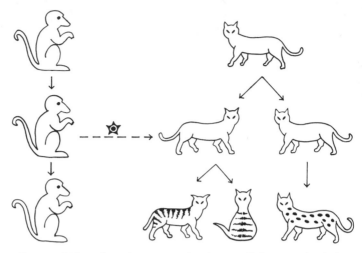

Fig. 3. Germ-line acquisition of a primate retrovirus by cats. Baboons and geladas carry a xenotropic endogenous viral genome which is readily activated to produce infectious virions. DNA sequence studies (Benveniste & Todaro, 1974) show that the germ-line of cats became infected with this virus after the divergence of Asian and Afro-European species.

1978), xenotropic viruses remain latent as the infectious cycle of replication is restricted. However, an opportunity for amplification through infection arises when cells of different species are brought into close proximity in laboratories, e.g. somatic cell hybridization or tumour xenograft procedures (Weiss, 1978). Human tumour xenografts frequently become infected with animal xenotropic retroviruses (Crawford *et al.*, 1979) which then replicate to high titre in the new host cells.

Interspecies transmission of retroviruses

Despite ready cross-infection in experimental situations, there is no evidence of natural transmission of xenotropic retroviruses from mouse to man although the two species share a common environment and diet. Yet some striking examples of natural trans-species infection by xenotropic viruses have been revealed. A C-type virus is transmitted genetically in cats which is quite distinct from exogenous feline leukaemia viruses (Livingston & Todaro, 1973). This virus was discovered as a contaminant of a human tumour xenograft passed through the brain of a fetal cat and it was at first thought to be a human retrovirus (McAllister *et al.*, 1972). It is closely related antigenically and by genome homology to an endogenous virus of baboons. Benveniste & Todaro (1974) have shown that North African and European species of *Felis*, such as the domestic cat and the European wild cat, carry this endogenous virus, as a Mendelian trait, whereas the related Asian leopard cat (*Felis bengalensis*) does not. Thus it appears that an endogenous virus of one species, when released in infectious form has become established in the germ line of another unrelated species during recent evolution (Fig. 3). Several other examples of trans-species infection of mammalian germ-lines and soma have been described (Todaro, 1980), and some are of ancient lineage in their hosts.

Inheritance, expression and possible functions of endogenous viral genes

Multiple viral elements in host DNA have been revealed by the analysis of endogenous genomes for B-type and C-type viruses by hybridization of viral probes to endonuclease restricted DNA in Southern blots (Cohen & Varmus, 1979; Frisby, MacCormick & Weiss, 1980; Steffen, Bird & Weinberg, 1980). One of the most detailed studies has been made on White Leghorn chickens by Astrin *et al.* (1980). Some sixteen separate endogenous virus (*ev*) elements have been defined which are stably inherited as Mendelian loci. Not all are complete genomes; some constitutively express viral proteins, while others remain entirely latent unless activated by inducing agents. For example, *ev-1* represents a complete, unexpressed, though inducible viral genome, *ev-3* is incomplete but expresses *gag* and *env* products, and *ev-6* is incomplete and expresses only *env* glycoproteins. Like the endogenous virus in the cat, the viral genes of chickens have been recently acquired in evolutionary terms, being represented as multiple *ev* loci in the red jungle fowl, *Gallus gallus*, from which the domestic chicken is derived, but being absent in the three other living species of *Gallus* (Frisby *et al.*, 1980). It would appear that the endogenous viral elements were acquired after speciation but before domestication in the genus *Gallus*. However, it is puzzling to find that related endogenous sequences are found in several other Galliform birds with no consistent phylogenetic or geographic distribution (Fig. 4).

As feral red jungle fowl express *gag* and *env* gene products in embryonic and adult tissues (Weiss & Biggs, 1972), it seems that natural selection is acting on the host species both for persistence and partial expression of endogenous retrovirus genomes. Yet it is

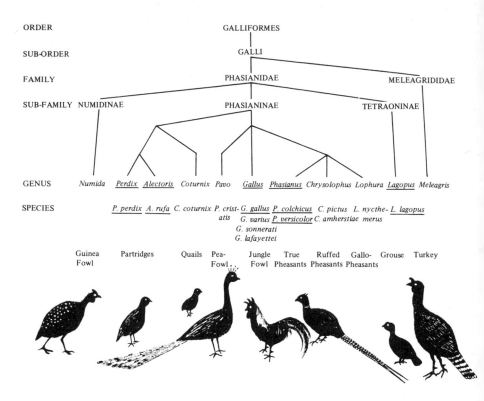

Fig. 4. Sporadic distribution among galliform birds of endogenous DNA sequences homologous to the genome of chicken leukaemia viruses (Frisby *et al.*, 1980). The species carrying viral DNA (underlined) do not fit any obvious phylogenetic or geographic distribution.

possible to breed viable domestic fowl which lack any *ev* loci (Astrin *et al.*, 1980). Infection with exogenous leukaemia viruses in fowl lacking *ev* loci, however, appears to be more pathogenic (L. B. Crittenden, personal communication); the immune response to *env* glycoprotein antigens after infection is broader than in chickens rendered partially tolerant by related endogenous *env* antigens. It is possible that the endogenous virus offers some protection against the virulence of related exogenous viruses inducing neoplasia or autoimmune disease.

All mice carry multiple *ev* loci for C-type viral genes, although there is much heterogeneity between inbred strains and in outbred populations (Steffen *et al.*, 1980). *Env* glycoproteins of different specifications are expressed on cell surfaces (Old & Stockert, 1977) and may have been entrained by the host to serve in recognition phenomena during cell differentiation, e.g. lymphocyte activation (Moroni & Schumann, 1977). *Env* glycoproteins are also secreted in substantial amounts into plasma, saliva and seminal fluid (Elder, Jensen, Bryant & Lerner, 1977). One may speculate that the viral glycoprotein in seminal fluid might function in the potentiation of sperm for recognizing ova; conceivably the heritable acquisition of new or recombinant *env* genes could play a role in the evolution of fertilization compatibility groups, leading to the divergence of host species.

Some wild mice lack endogenous DNA sequences of the B-type mammary tumour virus, while others have viral genetic elements at several chromosomal loci (Cohen & Varmus, 1979; Schlom *et al.*, 1980). With the murine mammary tumour virus we may be witnessing the process of germ-line acquisition occurring in different host populations. Retroviruses can be deliberately introduced into the germ line by experimental manipulation. This has been achieved by Jaenisch (1976, 1980) who has constructed several substrains of BALB/c mice with a complete Moloney murine leukaemia virus genome inserted at different chromosomal loci. Some of the genomes remain latent, while others are actively expressed causing secondary infection and leukaemia.

ANTIGEN SHIFT AND DRIFT

The life-long persistence of retroviruses in their hosts can lead to the emergence of antigenic variants that escape host immune mechanisms. The major envelope glycoprotein of retroviruses is typically the main target of neutralizing antibodies, whereas cellular immunity is directed not only to the envelope antigen but also to *gag*-protein-related antigens located in the membrane of infected cells, such as the Gross and Moloney cell surface antigens (GCSA and MCSA) of murine leukemia virus infections (Old & Stockert, 1977), and the feline oncovirus cell membrane antigen (FOCMA) of cats (Essex, 1975). The envelope glycoprotein is subject to rapid antigenic change when persistence includes successive infectious cycles. By contrast, the antigen specificity coded by inherited, endogenous *env* genes is remarkably well conserved. For example, the antigenic character of viral glycoprotein coded by endogenous *env* genes is not distinguishable between domestic Leghorn and feral red jungle fowl separated by thousands of generations (Weiss &

Biggs, 1972), whereas the *env* glycoproteins of infectiously transmitted avian leukosis viruses show major antigenic and host range groups (antigenic shift to the utilization of different host cell receptors) as well as variation of response to neutralizing antibodies within the major groups (antigenic drift).

There are three sources of envelope variation among retroviruses: mutation, recombination, and phenotypic mixing. Mutation and selection of envelope mutants is well exemplified in visna infection in individual sheep (Narayan, Griffin & Chase, 1977; Scott *et al.*, 1979). New serotypes arise which are not efficiently neutralized but the infected animal subsequently develops new antibodies which specifically neutralize the variant (Narayan, Griffin & Clements, 1978). Oligonucleotide mapping (Clements, Pedersen, Narayan & Haseltine, 1980) indicates that each variant is a new but stable mutant in the *env* gene. In this way, a series of new antigenic serotypes are generated during the long course of infection in a single host.

Genetic recombination between related C-type retroviruses occurs at high frequency, including recombination between exogenous viruses and endogenous *env* genes (Weiss, Mason & Vogt, 1973). The substitution of a new *env* gene can radically alter not only the serotype but the host range of the virus as the *env*-coded glycoproteins interact with highly specific cell surface receptors (Weiss, 1981). There is also evidence for intragenic *env* recombination yielding viruses with extended host range (Hartley, Wolford, Old & Rowe, 1977; Tsichlis, Conklin & Coffin, 1980). In AKR mice, two endogenous viral genomes, one ecotropic and one xenotropic recombine to yield 'dualtropic' viruses which are more lymphomagenic than either parent (Hartley *et al.*, 1977; Cloyd, Hartley & Rowe, 1980). Recombination requires virus production and probably occurs through the formation of heterozygote intermediates (Weiss *et al.*, 1973). Thus endogenous viral genomes which are stably inherited at chromosomal loci in unaltered form for innumerable host generations may rapidly recombine upon activation or superinfection with exogenous virus to generate new forms with altered serotype, host range and pathogenicity.

Phenotypic mixing between retroviruses leads to non-genetic, transient changes in serotype and host range (Vogt, 1967; Ishimoto *et al.*, 1977; Boettiger, 1979; Schnitzer, 1979; Weiss, 1981). Many acutely transforming C-type viruses are defective for *env* and other virion structural genes, and depend on non-defective helper viruses

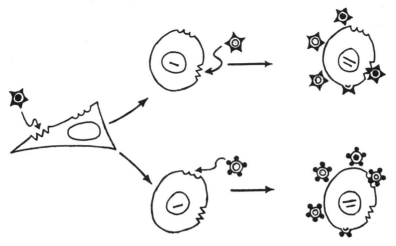

Fig. 5. Wolves in sheep's clothing: antigenic shift of retrovirus pseudotypes. A replication-defective sarcoma virus infects a cell via a receptor specific to the envelope glycoprotein of its pseudotype, and the provirus of the defective sarcoma virus integrates into the host chromosomal DNA. The mitotic descendants of the infected cell become transformed but do not produce infectious viral progeny without co-infection by a non-defective, 'helper' leukaemia virus, in which case both helper virus progeny and sarcoma virus pseudotypes are released. The pseudotypes carry the antigenicity and receptor specificity of the helper virus. If the rescuing helper virus is the same as that determining the original pseudotype (top row), no change in viral phenotype occurs. But rescue by a different helper virus (bottom row) can cause a major shift of serotype and host range.

for infectious transmission (Hanafusa, 1964). Under conditions of persistent infection, such defective viruses may be rescued by serologically distinct helper viruses and cause antigenic shifts of phenotype faster than genetic mutation or recombination would allow (Fig. 5). The infectious virions bearing antigens coded by the helper virus are called pseudotypes (Rubin, 1965), because the phenotype is changed without alteration of genotype.

Phenotypic mixing of membrane antigens also occurs between quite unrelated enveloped viruses (Boettiger, 1979). For example, pseudotypes of defective Rous sarcoma virus can be made with the envelope antigens of mammalian leukaemia viruses (Weiss & Wong, 1977; Levy, 1978) rendering the avian virus infectious for mammalian cells. Even more striking is phenotypic mixing across major virus groups. Pseudotypes of rhabdovirus with retrovirus envelopes (Zavada, 1972) and of retrovirus with rhabdovirus envelopes (Weiss, Boettiger & Love, 1975) are readily formed. Herpes viruses (Schnitzer & Gonczol, 1979) and togaviruses (Zavadova, Zavada & Weiss, 1977) also donate envelope glycoproteins to retroviruses but do not appear reciprocally to form functional

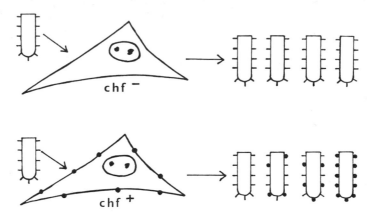

Fig. 6. Effect of endogenous retrovirus antigens on other enveloped viruses. Vesicular stomatitis virus (VSV) is propagated in chick embryo cells not expressing (chf⁻) or expressing (chf⁺) endogenous viral glycoproteins at the cell surface. Phenotypically mixed virions are released from the chf⁺ cells. In addition to particles with mosaic envelopes, a small proportion of VSV progeny have envelopes completely substituted by retrovirus glycoprotein (Love & Weiss, 1974). These VSV pseudotypes are neutralized by antiretrovirus antiserum but not by anti-VSV antiserum, and their host range is restricted to that of the retrovirus.

pseudotypes themselves. Because endogenous retroviruses form persistent infections in which envelope glycoproteins are commonly synthesized, phenotypic mixing upon acute or persistent infection of the same tissues by other enveloped viruses may be a common natural occurrence causing biologically significant host range and antigenic shifts (Fig. 6; Love & Weiss, 1974; Schnitzer & Gonczol, 1979).

PATHOGENESIS

The persistence of retroviruses in their hosts is manifest in a variety of pathological effects. Neoplasia is the most notorious but by no means the only disease associated with retrovirus infection, and should itself be divided into acute and 'slow' forms.

Apathogenic infections

Among the exogenous retroviruses, only the foamy viruses have not been associated with disease. These viruses frequently occur in neural tissue (Hooks & Gibbs, 1975) and while infected cultures form vacuolated (foamy) syncytia, infection in vivo is not known to be a cause of pathological symptoms.

Among the endogenous retroviruses, only those resident in

inbred strains of mice specially selected for high disease incidence have been found to have oncogenic potential, e.g. the leukaemia virus genome, Akv-1, in AKR mice, and the mammary tumour virus genome of GR mice. Under natural selection, the intimate persistence of endogenous retroviruses over millions of years has probably ensured a non-pathogenic, commensal relationship between virus and host. Endogenous viruses may effect some protection against infection or pathogenesis of related exogenous retroviruses; on the other hand, pathogenic variants and recombinants may arise from endogenous retrovirus genomes.

Acute neoplasms

Some C-type retroviruses cause the appearance of malignant tumours within a few days of infection. For instance Rous sarcomas in the wing-web of a chicken may reach 1 cm diameter within a week of virus inoculation, and chickens die of acute erythroblastic leukaemia within two to three weeks of infection by the causative virus. Acutely neoplastic viruses carry special transforming genes called oncogenes or *onc* (Baltimore, 1975). The molecular biology and pathogenesis of viruses carrying *onc* genes are reviewed in detail elsewhere (Weiss *et al.*, 1981).

About twenty *onc* genes have been identified in different acutely transforming oncoviruses. The majority of viral (*v-onc*) genes originate from cellular genes (*c-onc*), and exert their oncogenic effect when recombined into a viral genome governed by a strong promoter in the long terminal repeat sequence of the virus. Several *onc* genes code for protein kinases which phosphorylate tyrosine residues. *Onc* kinase activity in cells transformed by sarcoma viruses is also interconnected with the activity of mitogenic polypeptide hormones. For example, the receptor for epidermal growth factor becomes phosphorylated and itself has phosphokinase activity; many retrovirus-transformed cells produce a growth factor that interacts with this receptor. Some *c-onc* genes have been separately transduced by different acutely transforming retroviruses – an avian and a feline sarcoma virus have acquired homologous *c-onc* genes from their respective hosts independently. Some *onc* genes of acute leukaemia viruses are not related to the cell membrane tyrosine phosphokinases. These *v-onc* genes appear to be related to cellular genes active in specific pathways of haemopoietic differentiation, and the transforming virus blocks cell maturation (Graf, Beug &

Hayman, 1980). Some *v-onc* genes, (e.g. that of Friend erythroblastosis virus) may be derived largely from viral *env* genes rather than from cellular genes.

In productive infection the tumours induced by acutely tranforming viruses are usually not clonal (Ponten, 1964) and the tumour grows largely by recruitment, i.e. infecting and transforming new cells. Although acutely oncogenic viruses have been studied more intensively than other retroviruses, they are found only rarely in nature and are probably not transmitted readily from one host to another. None are endogenous even though their transforming genes are of host origin. They are so rapidly pathogenic that there must be strong selection against natural persistence.

Slow neoplasms

These are tumours induced by retroviruses which do not carry oncogenes as part of their genome. As the tumour may be fast growing and highly malignant, the 'slowness' indicates a long latent period between infection and manifestation of the disease. Tumours induced by slow retroviruses usually appear in animals that are chronically viraemic, whether chickens (Rubin *et al.*, 1962), mice (Rowe, 1973), cats (Essex, 1975) or gibbons (Kawakami & McDowell, 1980). However, restriction enzyme analysis of host–viral integration sites, for instance in mouse mammary cancer (Cohen & Varmus, 1980) and in avian lymphoma (Neiman, Jordan, Weiss & Payne, 1980), indicates that the tumours are clonal in origin.

It would appear that large numbers of potential target cells for malignancy become infected but only one or a few cells develop into fully malignant clones. This is clearly seen in avian lymphomagenesis (Neiman *et al.*, 1980). Within one month after infection of newly-hatched chicks, several hundred follicles in the bursa of Fabricius become filled with immature blast cells. Several months later one or two of these blastogenic follicles progress to large, neoplastic nodules and finally metastasize to disseminated lymphoma or leukaemia. The virus may exert its oncogenic effect by switching on the expression of endogenous *c-onc* genes of acutely transforming viruses. Hayward, Neel & Astrin (1981) have shown that avian lymphoma viruses integrate, as a rare event, adjacent to a cellular homologue, *c-myc*, of the viral oncogene for acute myelocytomatosis. The long terminal repeat (LTR) sequences contain a viral promoter that initiates transcription. The promoter in the LTR of 'right-hand end' of the viral genome initiates 'downstream' transcription into the host *c-myc* gene. Thus slow transforming viruses may act through promoter insertion triggering ectopic cellular gene expression.

In murine lymphomagenesis recombinants of xenotropic and ecotropic viruses arise in chronically viraemic animals and these recombinants may also have strong right-hand promoter sequences. There is evidence, however, that recombination within the *env* gene is important for the generation of lymphomagenic murine viruses (Hartley *et al.*, 1977; Cloyd *et al.*, 1980). It has been suggested by McGrath & Weissman (1979) that viral glycoproteins encoded by recombinant *env* genes recognize unique receptors on the surface of individual T-lymphocytes. These authors postulate that the interaction with the receptor not only allows entry of the virus but is also mitogenic. The infected lymphocyte thus synthesizes more *env* glycoprotein which acts as an auto-stimulatory mitogen leading to the clonal proliferation of a tumour. In this model the recombinant *env* gene is effectively an oncogene.

Auto-immune disease and anaemia

Retroviruses are also associated with slow, progressive auto-immune disease. New Zealand Black (NZB) mice are chronically viraemic with endogenous xenotropic C-type virus (Levy, 1978). These mice develop an auto-immune syndrome resembling human systemic lupus erythematosus (Talal & Steinberg, 1974), involving haemolytic anaemia and glomerulo-nephritis. However, in crosses between NZB mice and other strains, the viraemia and auto-immune disease are not closely linked so it is not clear whether the virus causes the auto-immune disease, although viraemia does appear to exacerbate it.

Feline leukaemia virus, especially the subgroup C serotype, can cause aplastic or haemolytic anaemia as frequently as leukaemia (Jarrett, 1980). Equine infectious anaemia, a febrile illness of worldwide distribution, is caused by a retrovirus (Charman, Bladen, Gilden & Coggins, 1976) transmitted horizontally by biting insects or vertically from mare to foal. The chronic phase of the disease involves lymphoproliferative disorders, glomerulo-nephritis and haemolytic anaemia.

Visna ('wasting') and maedi (shortness of breath) are neurological and pulmonary manifestations of persistent infection of the same retrovirus (Haase, 1975; Petursson *et al.*, 1979). Both forms of the disease are inflammatory, involving infiltration of lymphocytes and macrophages in the central nervous system, or in the septa dividing the alveoli of the lung. Caprine encephalitis virus causes an acute leucoencephalitis that resembles the disease produced in goats inoculated with visna virus. Persistent infection of adult goats causes a chronic arthritis in which the joints become infiltrated with lymphocytes, the synovial cells proliferate and cartilage is broken down (Crawford, Adams, Cheevers and Cork, 1980).

It is apparent, then, that retroviruses are associated with a wide variety of auto-immune disorders. The mechanisms of pathogenesis are not well understood, although antibody complexed with viral antigens is implicated in haemolytic anemia and glomerulo-nephritis.

Neurological disease

The inflammatory degeneration of the central nervous system during the long course of visna infection (Haase 1975; Petursson *et al.*, 1979) has already been alluded to as an 'auto-immune' disease reflecting the host response to chronic viral infection. The antigenic changes in virus envelope glycoproteins and subsequent production of matching neutralizing antibodies has been discussed and indicates a continuing immunological response to chronic infection by visna, to the detriment of the nervous system.

A different kind of neurological disorder is found in the infection of feral Californian mice with a congenitally transmitted C-type virus (Andrews & Gardner, 1974; Gardner *et al.*, 1980). This virus causes lymphoma and hind-limb paralysis; the two diseases are not linked, and they both occur relatively late in life. The paralysis is caused by a non-inflammatory destruction of motor neurones in the spinal cord, resembling poliomyelitis. Persistent viraemia is essential for the development of the disease, and passive immunization of new-born mice is highly protective (Gardner *et al.*, 1980). Numerous viral particles are seen in affected neural tissue, including bizarre rod-shaped virions atypical of C-type particles (Andrews & Gardner, 1974).

Retroviruses and human disease

Many claims have been made for the association of retroviruses with human malignancies and auto-immune diseases (Weiss *et al.*, 1981) but most remain enigmatic or have been discredited. Recently, a C-type virus has been isolated from patients with cutaneous T-cell lymphoma (Sezary syndrome) which may prove to be a horizontally transmitted, causative agent (Reitz, Poiesz, Ruscetti & Gallo, 1980). C-type viruses are rarely but consistently observed in normal human placentae (Kalter *et al.*, 1973; Nelson, Leong & Levy, 1978). Foamy viruses occur naturally in humans (Muller *et al.*, 1980; Loh *et al.*, 1980), apparently without ill effect.

CONCLUDING REMARKS

From this brief review it can be seen that the biology of retroviruses provides everything that the persistent virologist might desire, including, through their oncogenic properties, grant funding. Retroviruses are unique in their replicative cycle and perhaps in the most intimate form of viral persistence as host Mendelian traits. Not only have retrovirus genes become adopted as host genetic elements, but host genes have been transduced by retroviruses to generate a variety of highly oncogenic viruses. These viral oncogenes are giving us insight into normal cell regulation and differentiation, and its perversion as cancer. It is likely that other kinds of host gene can be transduced by retroviruses too. With the ability of retroviruses to move between germ lines of different host species, they may conceivably play a parasexual evolutionary role in the dispersion of genetic information among vertebrates.

REFERENCES

ANDREWS, J. M. & GARDNER, M. B. (1974). Lower motor neuron degeneration associated with type C RNA virus infection in mice: neuropathological features. *Journal of Neuropathology and Experimental Neurology*, **33**, 285–307.

ASTRIN, S. M., ROBINSON, H. L., CRITTENDEN, L. B., BUSS, E. G., WYBAN, J. & HAYWARD, J. (1980). Ten genetic loci in the chicken that contain structural genes for endogenous avian leukosis viruses. *Cold Spring Harbor Symposium of Quantitative Biology*, **44**, 1105–10.

BALTIMORE, D. (1975). Tumor Viruses: 1974. *Cold Spring Harbor Symposium of Quantitative Biology*, **39**, 1187–200.

BENTVELZEN, P. & DAAMS, J. H. (1969). Hereditary infections with mammary tumour viruses in mice. *Journal of the National Cancer Institute*, **43**, 1025–35.

BENVENISTE, R. E. & TODARO, G. J. (1974). Evolution of C-type viral genes: inheritance of exogenously acquired viral genes. *Nature* **252**, 456–9.

BITTNER, J. J. (1936). Some possible effects of nursing on the mammary tumor incidence in mice. *Science*, **84**, 162–4.

BOETTIGER, D. (1979). Animal virus pseudotypes. *Progress in Medical Virology*, **25**, 37–68.

CHARMAN, H. P., BLADEN, S., GILDEN, R. V. & COGGINS, L. (1976). Equine infectious anemia virus: evidence favoring classification as a retrovirus. *Journal of Virology*, **19**, 1973–79.

CLEMENTS, J. E., PEDERSEN, F. S., NARAYAN, O. & HASELTINE, W. A. (1980). Genomic changes associated with antigenic variation of visna virus during persistent infection. *Proceedings of the National Academy of Sciences of the United States of America*, **77**, 4454–8.

CLOYD, M. W., HARTLEY, J. W. & ROWE, W. P. (1980). Lymphomagenicity of recombinant mink cell focus-forming murine leukemia viruses. *Journal of Experimental Medicine*, **151**, 542–52.

COHEN, J. C. & VARMUS, H. E. (1979). Endogenous mammary tumour virus DNA varies among wild mice and segregates during inbreeding. *Nature*, **278**, 418–20.

COHEN, J. C. & VARMUS, H. E. (1980). Proviruses of mouse mammary tumor virus in normal and neoplastic tissues from GR and C3Hf mouse strains. *Journal of Virology*, **35**, 298–305.

CRAWFORD, D. H., ACHONG, B. G., TEICH, N. H., FINNERTY, S., THOMPSON, J. L., EPSTEIN, M. A. & GIOVANELLI, B. C. (1979). Identification of murine endogenous xenotropic retrovirus in cultured multicellar tumour spheroids from nude-mouse-passaged nasopharyngeal carcinoma. *International Journal of Cancer*, **23**, 1–7.

CRAWFORD, T. B., ADAMS, D. S., CHEEVERS, W. P. & CORK, L. C. (1980). Chronic arthritis in goats caused by a retrovirus. *Science*, **207**, 997–9.

CRITTENDEN, L. B., SMITH, E. J., WEISS, R. A. & SARMA, P. S. (1974). Host gene control of endogenous avian leukosis virus production. *Virology* **57**, 128–38.

DEBOER, G. F. & HOUWERS, D. J. (1979). Epizootiology of maedi/visna in sheep. In *Aspects of slow and persistent virus infections*, ed. D. A. J. Tyrrell, pp. 198–220. The Hague: M. Nijhoff.

DOUGHERTY, R. M. & DISTEFANO, H. S. (1967). Sites of avian leukosis virus multiplication in congenitally infected chickens. *Cancer Research*, **27**, 322–32.

ELDER, J. H., JENSEN, F. C., BRYANT, M. L. & LERNER, R. A. (1977). Polymorphism of the major envelope glycoprotein (gp70) of murine C-type viruses: virion associated and differentiation antigens encoded by a multigene family. *Nature*, **267**, 23–8.

ESSEX, M. (1975). Horizontally and vertically transmitted oncornaviruses of cats. *Advances in Cancer Research*, **21**, 175–248.

FRISBY, D., MACCORMICK, R. & WEISS, R. A. (1980). Origin of RAV-O, the endogenous retrovirus of chickens. In *Viruses in naturally occurring cancers*, ed. M. Essex, G. J. Todaro & H. zur Hausen, pp. 509–17. Cold Spring Harbor Laboratory.

GARDNER, M. B., ESTES, J. D., CASAGRANDE, J. & RASHEED, S. (1980). Prevention of paralysis and suppression of lymphoma in wild mice by passive immunization to congenitally transmitted murine leukemia virus. *Journal of the National Cancer Institute*, **65**, 359–64.

GRAF, T., BEUG, H. & HAYMAN, M. J. (1980). Target cell specificity of defective avian leukemia viruses. *Proceedings of the National Academy of Sciences of the United States of America*, **77**, 389–93.

HAASE, A. T. (1975). The slow infection caused by visna virus. *Current Topics in Microbiology and Immunology*, **72**, 101–56.

HANAFUSA, H. (1964). Nature of the defectiveness of Rous sarcoma virus. *National Cancer Institute Monograph*, **17**, 543–56.

HARTLEY, J. W., WOLFORD, N. K., OLD, L. J. & ROWE, W. P. (1977). A new class of murine leukemia virus associated with development of spontaneous lymphomas. *Proceedings of the National Academy of Sciences of the United States of America*, **74**, 789–92.

HAYWARD, W. S., NEEL, B. G. & ASTRIN, S. M. (1981). Activation of a cellular onc gene by promoter insertion in ALV-induced lymphoid leukosis. *Nature*, **290**, 475–80.

HOOKS, J. J. & GIBBS, C. J. (1975). The foamy viruses. *Bacteriological Reviews*, **39**, 169–85.

ISHIMOTO, A., HARTLEY, J. W. & ROWE, W. P. (1977). Detection and quantitation of phenotypically mixed viruses: Mixing of ecotropic and xenotropic murine leukemia viruses. *Virology*, **81**, 263–9.

JAENISCH, R. (1976). Germ line integration and Mendelian transmission of the exogenous Moloney leukemia virus. *Proceedings of the National Academy of Sciences of the United States of America*, **73**, 1260–4.

JAENISCH, R. (1980). Germ line integration and Mendelian transmission of exogenous type C viruses. In *Molecular Biology of RNA tumor viruses*, ed. J. Stephenson, pp. 131–41. New York: Academic Press.

JARRETT, O. (1980). Natural occurrence of subgroups of feline leukemia virus. In *Viruses in naturally occurring cancers*, ed. by M. Essex, G. J. Todaro & H. zur Hausen, pp. 603–11. Cold Spring Harbor Laboratory.

JARRETT, W. F. H., JARRETT, O., MACKEY, L., LAIRD, H. M., HARDY, W. D. & ESSEX, M. (1973). Horizontal transmission of leukemia virus and leukemia in cats. *Journal of the National Cancer Institute*, **51**, 833–41.

KALTER, S. S., HELINKE, R. J., HEBERLING, R. L., PANIGEL, M., FOWLER, A. K., STRICKLAND, J. E. & HELLMAN, L. (1973). C-type particles in normal human placentas. *Journal of the National Cancer Institute*, **50**, 1081–4.

KAWAKAMI, T. G. & McDOWELL, T. S. (1980). Factors regulating the onset of chronic myelogenous leukemia in gibbons. In *Viruses in naturally occurring cancers*, ed. M. Essex, G. J. Todaro & H. zur Hausen, pp. 719–27. Cold Spring Harbor Laboratory.

LEVY, J. A. (1978). Xenotropic type C viruses. *Current Topics in Microbiology and Immunology*, **79**, 111–213.

LIVINGSTON, D. M. & TODARO, G. J. (1973). Endogenous type C virus from a cat cell clone with properties distinct from previously described feline type C viruses. *Virology*, **53**, 142–51.

LOH, P. C., MATSUURA, F. & MIZUMOTO, C. (1980). Seroepidemiology of human syncytial virus: antibody prevalence in the Pacific. *Intervirology*, **13**, 87–90.

LOVE, D. N. & WEISS, R. A. (1974). Pseudotypes of vesicular stomatitis virus determined by exogenous and endogenous avian RNA tumour viruses. *Virology*, **57**, 271–8.

McALLISTER, R. M., NICHOLSON, M., GARDNER, M. B., RONGEY, R. W., RASHEED, S., SARMA, P. S., OROSZLAN, S., GILDEN, R. V., KABIGTUNG, D. & VERNON, L. (1972). C-type virus released from cultured human rhabdomyosarcoma cells. *Nature New Biology*, **235**, 3–6.

McGRATH, M. S. & WEISSMAN, I. L. (1979). AKR Leukemogenesis: identification and biological significance of thymic lymphoma receptors for AKR retroviruses. *Cell*, **17**, 65–75.

MILLER, J. M. & VAN DER MAATEN, M. J. (1979). Infectivity tests of secretions and excretions from cattle infected with bovine leukemia virus. *Journal of the National Cancer Institute*, **62**, 425–8.

MORONI, C. & SCHUMANN, G. (1977). Are endogenous C-type viruses involved in the immune system? *Nature*, **269**, 600–1.

MULLER, H. K., BALL, G., EPSTEIN, M. A., ACHONG, B. G., LENOIR, G. & LEVIN, A. (1980). The prevalence of naturally occurring antibodies to human syncytial virus in East African populations. *Journal of General Virology*, **47**, 399–406.

NARAYAN, O., GRIFFIN, D. E. & CHASE, J. (1977). Antigenic shift of visna virus in persistently infected sheep. *Science*, **197**, 376–8.

NARAYAN, O., GRIFFIN, D. E. & CLEMENTS, J. E. (1978). Virus mutation during 'slow infection': temporal development and characterization of mutants of visna virus recovered from sheep. *Journal of General Virology*, **41**, 343–52.

NEIMAN, P. E., JORDAN, L., WEISS, R. A. & PAYNE, L. N. (1980). Malignant lymphoma of the bursa of Fabricius: analysis of early transformation. In *Viruses in naturally occurring cancers*, ed. M. Essex, G. J. Todaro & H. zur Hausen, pp. 519–28. Cold Spring Harbor Laboratory.

NELSON, J., LEONG, J. A. & LEVY, J. A. (1978). Normal human placentas contain RNA-directed DNA polymerase activity like that in viruses. *Proceedings of the National Academy of Sciences of the United States of America*, **75**, 6263–7.

NEMO, G. J., BROWN, P. W., GIBBS, C. J. & GAJDUSEK, D. C. (1978). Antigen relationship of human foamy virus to the simian foamy viruses. *Infection and Immunity*, **20**, 69–72.

OLD, L. J. & STOCKERT, E. (1977). Immunogenetics of cell surface antigens of mouse leukemia. *Annual Review of Genetics*, **17**, 127–60.

PAYNE, L. N. & CHUBB, R. C. (1968). Studies on the nature and genetic control of an antigen in normal chick embryos which reacts in the COFAL test. *Journal of General Virology*, **3**, 379–91.

PETURSSON, G., MANTIN, J. R., GEORGSSON, G., NATHANSON, N. & PALSSON, P. A. (1979). Visna. The biology of the agent and the disease. In *Aspects of slow and persistent virus infections*, ed. D. A. J. Tyrrell, pp. 165–97. The Hague: M. Nijhoff.

PONTEN, J. (1964). The in vivo growth mechanism of avian Rous sarcoma. *National Cancer Institute Monograph*, **17**, 131–45.

REITZ, M. S., POIESZ, B. J., RUSCETTI, F. W. & GALLO, R. C. (1981). Characterization and distribution of nucleic acid sequences of a novel retrovirus isolated from neoplastic human T-lymphocytes. *Proceedings of the National Academy of Sciences of the United States of America*, **78**, 1887–91.

ROWE, W. P. (1973). Genetic factors in the natural history of murine leukemia virus infection. *Cancer Research*, **33**, 3061–8.

ROWE, W. P., HARTLEY, J. W., LANDER, M. R., PUGH, W. E. & TEICH, N. M. (1971). Non-infectious AKR mouse embryo cell lines in which each cell has the capacity to be activated to produce infectious murine leukemia virus. *Virology*, **46**, 866–76.

RUBIN, H. (1965). Genetic control of cellular susceptibility to pseudotypes of Rous sarcoma virus. *Virology*, **26**, 270–6.

RUBIN, H., FANSHIER, L., CORNELIUS, A. & HUGHES, W. F. (1962). Tolerance and immunity in chickens after congenital and contact infection with an avian leukosis virus. *Virology*, **17**, 143–56.

SCHLOM, J., DROHAN, W., TERAMOTO, Y. A., YOUNG, J. M. & HORAN HAND, P.

(1980). Diversity of mammary tumor virus genes and viral gene products in rodent species. In *Viruses in naturally occurring cancers*, ed. M. Essex, G. J. Todaro & H. zur Hausen, pp. 1115–32. Cold Spring Harbor Laboratory.

SCHNITZER, T. J. (1979). Phenotypic mixing between two primate oncoviruses. *Journal of General Virology*, **42**, 199–206.

SCHNITZER, T. J. & GONCZOL, E. (1979). Phenotypic mixing between murine oncoviruses and murine cytomegalovirus. *Journal of General Virology*, **43**, 691–5.

SCOTT, J. V., STOWRING, L., HAASE, A. T., NARAYAN, O. & VIGNE, R. (1979). Antigen variation in visna virus. *Cell*, **18**, 321–7.

SPENCER, J. L., GAVORA, J. S. & GOWE, R S. (1980). Lymphoid leukosis virus: natural transmission and non-neoplastic effects. In *Viruses in naturally occurring cancers*, M. Essex, G. J. Todaro & H. zur Hausen, pp. 553–64. Cold Spring Harbor Laboratory.

STEFFEN, D. L., BIRD, S. & WEINBERG, R. A. (1980). Evidence for the Asiatic origin of endogenous AKR-type murine leukemia proviruses. *Journal of Virology*, **35**, 824–32.

TALAL, N. & STEINBERG, A. D. (1974). The pathogenesis of autoimmunity in New Zealand Black mice. *Current Topics in Microbiology and Imunology*, **64**, 79–103.

TEMIN, H. M. (1980). Origin of retroviruses from cellular movable genetic elements. *Cell*, **21**, 599–600.

TODARO, G. J. (1980). Interspecies transmission of mammalian retroviruses. In *Viral oncology*, ed. G. Klein, p. 291-309. New York: Raven Press.

TSICHLIS, P. N., CONKLIN, K. F. & COFFIN, J. M. (1980). Mutant and recombinant avian retroviruses with extended host range. *Proceedings of the National Academy of Sciences of the United States of America*, **77**, 536–40.

VOGT, P. K. (1967). Phenotypic mixing in the avian tumor virus group. *Virology*, **32**, 708–18.

WEISS, R. A. (1978). Why cell biologists should be aware of genetically transmitted viruses. *National Cancer Institute Monograph*, **48**, 183–9.

WEISS, R. A. (1981). Retrovirus receptors and their genetics. In *Virus receptors*, Part 2, ed. K. Lonberg-Holm & L. Philipson pp. 187–202. London: Chapman & Hall.

WEISS, R. A. & BIGGS, P. M. (1972). Leukosis and Marek's disease viruses of feral red jungle fowl and domestic fowl in Malaya. *Journal of the National Cancer Institute*, **49**, 1713–25.

WEISS, R. A., BOETTIGER, D. & LOVE, D. N. (1975). Phenotypic mixing between vesicular stomatitis virus and avian RNA tumor virus. *Cold Spring Harbor Symposium of Quantitative Biology*, **39**, 913–18.

WEISS, R. A., FRIIS, R. R., KATZ, E. & VOGT, P. K. (1971). Induction of avian tumor viruses in normal cells by physical and chemical carcinogens. *Virology*, **46**, 920–38.

WEISS, R. A., MASON, W. S. & VOGT, P. K. (1973). Genetic recombinants and heterozygotes derived from endogenous and exogenous avian RNA tumor viruses. *Virology*, **52**, 535–52.

WEISS, R. A. & PAYNE, L. N. (1971). The heritable nature of the factor in chicken cells which acts as a helper virus for Rous sarcoma virus. *Virology*, **45**, 508–15.

WEISS, R. A., TEICH, N. M., VARMUS, H. E. & COFFIN, J. M. (eds.) (1981). *Molecular Biology of Tumor Viruses. 2nd Edn, Part 3. RNA Tumor Viruses*. Cold Spring Harbor Laboratory.

WEISS, R. A. & WONG, A. L. (1977). Phenotypic mixing between avian and mammalian RNA tumor viruses. I. Envelope pseudotypes of Rous sarcoma virus. *Virology*, **76**, 826–34.

WEYL, K. G. & DOUGHERTY, R. M. (1977). The contact transmission of avian leukosis virus. *Journal of the National Cancer Institute*, **58**, 1019–25.

ZAVADA, J. (1972). Pseudotypes of vesicular stomatitis virus with the coat of murine leukaemia and of avian myeloblastosis viruses. *Journal of General Virology*, **15**, 183–91.

ZAVADOVA, Z., ZAVADA, J. & WEISS, R. A. (1977). Unilateral phenotypic mixing of envelope antigens between togaviruses and vesicular stomatitis virus or avian RNA tumour virus. *Journal of General Virology*, **37**, 557–67.

INDEX

actinomycin D: prevents DNA-dependent RNA synthesis by host, but not RNA-dependent RNA synthesis by virus, 204

acyclovir, antiviral drug: prevents acute herpes infection and establishment of latency in mice, does not abolish existing latency, 146

adenoviruses, 8
latent in reticulo-endothelial system, 15
as natural helpers of AAV, 249, 258; mutants of, as helpers, 258–9; protein possibly involved in helper function, 259, 262
oncogenicity of cells infected by, inhibited by AAV, 259–60

adeno-associated viruses (AAV), defective parvoviruses requiring co-infection with helper virus, 249–50
adenoviruses as helpers for, *see under* adenoviruses
integration of DNA of, into host DNA, 253–4, 255–6; ssDNA converted to dsDNA before integration? 256–8
latent infection with, 251; in cell lines, 251–2; in humans, 260; in kidney cell cultures, 251, 261; protects against human cancers caused by adenoviruses? 261–2
structure of DNA of, 250–1

Adoxophyes orana, reactivation of viruses in larvae of, 80

adrenal gland, *in vitro* reactivation of HSV in, 142

Aedes taeniorhyncus, transmission of iridovirus in, 78

African swine fever virus, 8

agammaglobulinaemia
acute measles in, 110
(partial), SSPE in, 117

Aglais urticae, reactivation of virus in larvae of, 80

Agraulis vanillae, densonucleosis virus infecting, 63

Aleutian disease virus (mink), 3, 8

amino-acid sequences, of HA_1 and HA_2 polypeptide chains of influenza virus haemagglutinin, 218–19
differences in, between 1968 and 1976 viruses, 220, 221

anaemia, haemolytic: caused by feline leukaemia virus, 281, and by immune complexes, 282

animals, influenza A virus in, *see* horses, pigs

antibodies
antiviral, in multiple sclerosis, 119–21
to CDV, and course of disease in CNS, 119
circulating, may not reach epithelial sites, 7
with complement, unable to lyse virus-infected cells after removal of surface antigens, 194
to Epstein-Barr virus, 171
to hepatitis B virus, 40–1
to HSV: in establishment of latency, 148, 149, 151, 157; in mouse, 152–3
to influenza virus, react with HA_1 chain of haemagglutin, 219
to measles virus: after subsidence of acute measles, 197; in cases of SSPE, 100, 197; remove measles virus antigens from cell surfaces in both acute and persistent infections, 193–4; virus persists in presence of, 197–8
persistent virus infections in presence of excess of, 186
redistribution of antigens on cell surface in response to (capping), 193
stripping of viral antigens from cell surface by, 187, 199
to retroviruses, mainly to envelope protein, 275

antibody-dependent cell cytotoxicity, in immunity to HSV, 152, 156

antigen-antibody immune complexes, *see* immune complexes

antigenic drift (small changes), in influenza virus antigens, 17, 29, 215

antigenic shift (major changes), in virus antigens, 17, 276

antigens, viral
advantage to virus in minimal production of, 17
antibody-induced modulation of, as factor in persistence, 186, 188–9
of measles virus, removed from cell surface by antibodies in both acute and persistent infection, 193–4
persistence of, in absence of viral nucleic acid or viral replication, as cause of immune-complex diseases, 2, 8
redistribution of, on cell surface, in response to antibodies (capping), 193
removal of, from cell surface, associated with persistence, 187, 199

Apanteles melanoscelus, transmitter of NPV between larvae of *L. dispar*, during oviposition, 76

aphids, transmission of plant viruses by, 81